# Lecture Notes in Control and Information Sciences

Edited by M. Thoma and A. Wyner

Vol. 81: Stochastic Optimization
Proceedings of the International Conference, Kiew, 1984
Edited by I. Arkin, A. Shiraev, R. Wets
X, 754 pages, 1986.

Vol. 82: Analysis and Algorithms of Optimization Problems
Edited by K. Malanowski, K. Mizukami
VIII, 240 pages, 1986.

Vol. 83: Analysis and Optimization of Systems
Proceedings of the Seventh International Conference of Analysis and Optimization of Systems
Antiba, June 26-27, 1986
Edited by A. Bensoussan, J. L. Lions
XVI, 901 pages, 1986.

Vol. 84: System Modelling and Optimization
Proceedings of the 12th IFIP Conference
Budapest, Hungary, September 2–6, 1985
Edited by A. Prékopa, J. Szelezsán, B. Strazicky
XII, 1046 pages, 1986.

Vol. 85: Stochastic Processes in Underwater Acoustics
Edited by Charles R. Baker
V, 205 pages, 1986.

Vol. 86: Time Series and Linear Systems
Edited by Sergio Bittanti
XVII, 243 pages, 1986.

Vol. 87: Recent Advances in System Modelling and Optimization
Proceedings of the IFIP-WG 7/1 Working Conference
Santiago, Chile, August 27-31, 1984
Edited by L. Contesse, R. Correa, A. Weintraub
IV, 199 pages, 1987.

Vol. 88: Bruce A. Francis
A Course in $H_\infty$ Control Theory
XI, 156 pages, 1987.

Vol. 88: Bruce A. Francis
A Course in $H_\infty$ Control Theory
X, 150 pages, 1987.
Corrected - 1st printing 1987

Vol. 89: G. K. H. Pang/A. G. J. McFarlane
An Expert System Approach to Computer-Aided Design of Multivariable Systems
XII, 223 pages, 1987.

Vol. 90: Singular Perturbations and Asymptotic Analysis in Control Systems
Edited by P. Kokotovic, A. Bensoussan, G. Blankenship
VI, 419 pages, 1987.

Vol. 91 Stochastic Modelling and Filtering
Proceedings of the IFIP-WG 7/1 Working Conference
Rome, Italy, Decembre 10-14, 1984
Edited by A. Germani
IV, 209 pages, 1987.

Vol. 92: L. T. Grujić, A. A. Martynyuk, M. Ribbens-Pavella
Large-Scale Systems Stability Under Structural and Singular Perturbations
XV, 366 pages, 1987.

Vol. 93: K. Malanowski
Stability of Solutions to Convex Problems of Optimization
IX, 137 pages, 1987.

Vol. 94: H. Krishna
Computational Complexity of Bilinear Forms
Algebraic Coding Theory and Applications to Digital Communication Systems
XVIII, 166 pages, 1987.

Vol. 95: Optimal Control
Proceedings of the Conference on Optimal Control and Variational Calculus
Oberwolfach, West-Germany, June 15-21, 1986
Edited by R. Bulirsch, A. Miele, J. Stoer and K. H. Well
XII, 321 pages, 1987.

Vol. 96: H. J. Engelbert/W. Schmidt
Stochastic Differential Systems
Proceedings of the IFIP-WG 7/1 Working Conference
Eisenach, GDR, April 6-13, 1986
XII, 381 pages, 1987.

For information about Vols. 1–80 please contact your bookseller or Springer-Verlag.

# Lecture Notes in Control and Information Sciences

Edited by M. Thoma and A. Wyner

IFIP

149

K.-H. Hoffmann, W. Krabs (Eds.)

# Optimal Control of Partial Differential Equations

Proceedings of the IFIP WG 7.2 International Conference
Irsee, April 9-12, 1990

Springer-Verlag Berlin Heidelberg GmbH

**Editor of Conference Proceedings of the series:**
**Computational Techniques in Distributed Systems IFIP-WG 7.2**
Irena Lasiecka
Department of Applied Mathematics
Thornton Hall
University of Virginia
Charlottesville, VA 22903
USA

**Editors**
Prof. Karl-Heinz Hoffmann
Institut für Mathematik
der Universität Augsburg
Universitätsstraße 8
8900 Augsburg

Prof. Werner Krabs
Fachbereich Mathematik
der Technischen Hochschule Darmstadt
Schloßgartenstraße 7
6100 Darmstadt

ISBN 978-3-540-53591-1 ISBN 978-3-540-46883-7 (eBook)
DOI 10.1007/978-3-540-46883-7



Originally published by International Federation for Information Processing,
Geneva, Switzerland in 1991.



Offsetprinting: Mercedes-Druck, Berlin

61/3020-543210 Printed on acid-free paper

# Preface

These are the Proceedings of the IFIP WG7.2 Working Conference on Optimal Control of Partial Differential Equations held at the Monastery of Irsee in the Federal Republic of Germany from April 9 to 12, 1990. This conference was organized by K.-H. Hoffmann (University of Augsburg) and W. Krabs (Technical University of Darmstadt) under the auspices of the Technical Committee 7 of the International Federation for Information Processing (IFIP). It was financially supported by the Deutsche Forschungsgemeinschaft whose support is gratefully acknowledged.

Out of 31 participants who attended the conference 19 presented lectures 17 of which focussing on recent advances in control theory connected with partial differential equations. Within a broad spectrum of topics the problem of exact controllability played a central role and also shape control was given some special attention. Nonlinear problems were mainly treated under the aspect of optimality. Less represented were identification problems and also numerical aspects were regarded less than at other conferences of this type.

The organizers are grateful to all participants who made contributions to its success either by lecturing and publishing in the proceedings or by actively taking part in the discussions. The fruitful scientific and the agreeable social atmosphere that could be observed throughout the conference was certainly also due to the pleasing environment of the Monastery of Irsee and its excellent facilities for the organisation of conferences and the accomodation of the participants.

Last not least the organizers owe gratitude to all who assisted them in all the details that have to be considered in organizing such a conference.

(K.-H. Hoffmann)

(W. Krabs)

# Table of Contents

# RELAXED CONTROLS FOR STOCHASTIC BOUNDARY VALUE PROBLEMS IN INFINITE DIMENSION

N.U.Ahmed
University of Ottawa, Canada

*Dedicated to Professor L. Cesari*

## ABSTRACT

The paper deals with the question of boundary controls for a class of linear abstract stochastic initial boundary value problems. The objective is to control the mean state trajectory and the corresponding covariance operator in the presence of both boundary and distributed noises. Both existence of optimal relaxed controls and necessary conditions of optimality are presented. The question of practical realization of relaxed controls is briefly discussed as closing remarks.

## INTRODUCTION

Let $\Sigma$ be an open bounded connected subset of $R^n$ with smooth boundary, $X \equiv X(\Sigma)$ a Banach space of functions or generalized functions on $\Sigma$ and $E \equiv E(\partial\Sigma)$ a Banach space of functions or generalized functions on $\partial\Sigma$. Consider the initial boundary value problem governed by a system of partial differential equations with $L$ denoting the spatial differential operator and $\tau$ denoting the boundary differential operator

$$\text{IBVP} \begin{cases} \dfrac{\partial}{\partial t}\varphi = L\varphi + h, \quad t \geq 0, \\ \tau\varphi = g \\ \varphi(0) = \varphi_0. \end{cases} \tag{1}$$

We assume that $D(L) \subset D(\tau)$. The data $h$ and $g$ are $X$ and $E$ valued respectively and $\varphi_0 \in X$. Define

$$\begin{aligned} A &\equiv (L \,|_{\text{Ker } \tau}) : D(A) \to X \\ R &\equiv (\tau \,|_{\text{Ker } L})^{-1} : E \to X. \end{aligned} \tag{2}$$

Then the IBVP can be reformulated as an abstract Cauchy problem [1-5]:

$$\begin{cases} \dfrac{d\varphi}{dt} = A\varphi + \pi R g + h \\ \varphi(0) = \varphi_0 \end{cases} \tag{3}$$

where $\pi \equiv (\lambda I - A), \lambda \in \rho(A)(\neq \emptyset)$. More conveniently: setting $\varphi = \pi x$, we have,

$$\begin{cases} \dfrac{dx}{dt} = Ax + Rg + \Lambda h, \quad \Lambda = \pi^{-1} = R(\lambda, A) \\ x(0) = x_0 \qquad\qquad\qquad\qquad x_0 = \Lambda\varphi_0. \end{cases} \tag{4}$$

We consider

$$\begin{cases} \dfrac{dx}{dt} = Ax + Rg + \Lambda h \\ x(0) = x_0 \end{cases} \tag{5}$$

as the basic equation.

We are interested in the stochastic model controlled through the boundary. Suppose

$$\begin{aligned} h &= h_0 + \sigma_0 N_0 \quad \text{(on spatial domain) } \Sigma \\ g &= f(u) + \sigma(u) N \quad \text{(on boundary) } \partial\Sigma \end{aligned} \tag{6}$$

where $N_0$ and $N$ are considered as the spatial and boundary noises modelled as the distributional derivatives of certain Wiener processes $w_0$ and $w$ respectively.

The stochastic model is then given by:

$$\begin{cases} dx = Axdt + h_0 dt + Rfdt + \sigma_0 dw_0 + R\sigma dw \\ x(0) = x_0, \quad \varphi = \pi x. \end{cases} \tag{7}$$

We are interested in relaxed controls and hence the relaxed system,

$$\begin{cases} dx = Axdt + h_0 dt + R\nu_t(f)dt + \sigma_0 dw_0 + R\nu_t(\sigma)dw \\ x(0) = x_0, \quad \varphi = \pi x, \end{cases} \tag{8}$$

where,

$$\nu_t(\xi) \equiv \int_U \xi(u)\nu_t(du). \tag{9}$$

## BASIC NOTATIONS AND ASSUMPTIONS

$$\begin{aligned}
\text{State Space}: &\quad X \equiv \text{Hilbert Space}\\
\text{Boundary Space}: &\quad E \equiv \text{Hilbert Space}\\
\text{Control Space}: &\quad U \equiv \text{Compact Polish Space}\\
\text{State Space for Spatial Noise}: &\quad W_0 \equiv \text{Hilbert Space (Separable)}\\
\text{State Space for Boundary Noise}: &\quad W \equiv \text{Hilbert Space (Separable).}
\end{aligned}$$

$M(U) \equiv$ the space of probability measures on $B(U) \equiv$ Borel $U$.
$\mathcal{M} \equiv L_0([0,\infty), M(U)) \equiv$ the space of (weakly) Borel measurable functions from $[0,\infty)$ to $M(U)$ furnished with the Young topology $\tau_y$, given by, $\nu^n \xrightarrow{\tau_y} \nu^0$ as $n \to \infty$ if for every $\xi \in C_b(U,Y)$ $(C_b(U,\mathcal{L}_s(Z,Y)))$

$$\int_J \int_U \xi(u)\nu_t^n(du) \xrightarrow{s(\tau_s)} \int_J \int_U \xi(u)\nu_t^0(du) \text{ in } Y \tag{10}$$

for each $J \subset I$, where $Y$ is any Banach space ($\mathcal{L}_s(Z,Y) \equiv \mathcal{L}(Z,Y)$ furnished with strong operator topology). We assume

$$\left\{\begin{array}{l}
A \in G(X,M,\omega), M \geq 1, \omega \in R\\
Q_0 \equiv \text{Cov.}w_0 \in \mathcal{L}_n^+(W_0^*, W_0) = \mathcal{L}_n^+(W_0)\\
Q \equiv \text{Cov.}w \in \mathcal{L}_n^+(W^*, W) = \mathcal{L}_n^+(W)\\
f: U \to E, \quad \sigma: U \to \mathcal{L}(W,E).
\end{array}\right\} \tag{11}$$

## MOTIVATION

One of the physical problems that motivated us to this abstract stochastic boundary value problem is the boundary control of the Cantilever beam equation subject to random perturbations of the free end. This is described as follows:

$$\left\{\begin{array}{l}
\dfrac{\partial^2 y}{\partial t^2} + \Delta^2 y = u_1 + n_1 \equiv h, \quad x \in (0,l), t \geq 0\\
y|_{x=0} = 0, \quad Dy|_{x=0} = 0\\
D^2 y|_{x=l} = u_2 + h_2 \equiv g_1, \; D^3 y|_{x=l} = u_3 + n_3 \equiv g_2\\
y(0,x) = y_0(x), \quad y_t(0,x) = y_1(x), \quad x \in (0,l),
\end{array}\right. \tag{12}$$

where $u_1$ is the distributed control, $n_1$ the distributed noise; $u_2, u_3$ are the boundary controls and $n_2, n_3$ are the boundary noises which may be induced by turbulent flow of a fluid past the cantilever end.

Introducing $\varphi_1 = y$, $\varphi_2 = y_t$,

$$L \equiv \begin{pmatrix} 0 & 1 \\ -\Delta^2 & 0 \end{pmatrix}, \quad B = \begin{pmatrix} 0 \\ 1 \end{pmatrix}, \tag{13}$$

$$\tau\varphi \equiv \begin{pmatrix} D^2\varphi|_{x=l} \\ D^3\varphi|_{x=l} \end{pmatrix} \tag{14}$$

and the state space

$$X \equiv H_l^2 \times L_2(0,l) \tag{15}$$

where,

$$H_l^2 \equiv \{\psi \in H^2 : \psi|_{x=0} = 0, (\psi_x = D\psi)|_{x=0} = 0\}$$

and the boundary space

$$E = R^2, \tag{16}$$

we can rewrite the above equation in the semiabstract form as follows:

$$\begin{cases} \dfrac{d\varphi}{dt} = L\varphi + h \\ \tau\varphi = g \\ \varphi(0,\cdot) \equiv \varphi_0. \end{cases} \tag{17}$$

Clearly $D(L) \subset D(\tau)$. Define $A$ by $D(A) = \{\varphi \in X : L\varphi \in X \text{ and } \tau\varphi = 0\}$ and set $A\varphi = L\varphi$ for $\varphi \in D(A)$, that is, $A = L|_{\mathrm{Ker}(\tau)\cap X}$. Then define $R$ as $R \equiv (\tau|_{\mathrm{Ker}L})^{-1}$ obtained by solving the equation

$$\begin{cases} L\varphi = 0 \\ \tau\varphi = g \ \text{ in } X. \end{cases} \tag{18}$$

One can easily verify that this equation has a unique solution.

The operator $R$ is a matrix of multiplication operators given by $R \equiv \begin{pmatrix} R_{11} & R_{12} \\ R_{21} & R_{22} \end{pmatrix}$ where

$$R_{11} = x^2/2, \ R_{12} = (x^3/6 - Lx^2/2), R_{21} = R_{22} = 0. \tag{19}$$

The equation (12) or equivalently (17) can then be written as an abstract evolution equation

$$\begin{cases} d\varphi = A\varphi dt + Budt + Bdw + \pi Ru_b dt + \pi Rdw_b \\ \varphi(0) = \varphi_0 \end{cases} \tag{20}$$

where

$$u_b = \begin{pmatrix} u_2 \\ u_3 \end{pmatrix} \quad \text{and} \quad w_b = \begin{pmatrix} w_2 \\ w_3 \end{pmatrix}, \tag{21}$$

with $w_2, w_3$ being standard wiener processes corresponding to the white noises $n_2$ and $n_3$ respectively.

Note that, in this example, the boundary conditions are partly absorbed in the state space and partly by the operator $A$.

## RELAXED STOCHASTIC INITIAL BOUNDARY VALUE PROBLEM

Consider the boundary control system,

$$\begin{cases} dx = Axdt + R\nu_t(f)dt + R\nu_t(\sigma)dw \\ x(0) = x_0 \end{cases} \tag{22}$$

Objective 1

Find $\nu \in \mathcal{M}$ such that

$$J(\nu) \equiv \frac{1}{2}\int_0^\tau \{tr(PM_c) + \langle N_c(m - m^*, m - m^*_.\rangle\}dt$$

is minimum where $m(\cdot)$ is the mean of the process $\{x\}$ and $P(\cdot)$ is the covariance operator corresponding to $x$. $M_c$ and $N_c$ are $\mathcal{L}^+_n(X)$ and $\mathcal{L}^+(X)$ valued functions respectively.

Lemma 1

Let $A \in G(X, M, \omega)$, $f \in C_b(U, E)$, $\sigma \in C_b(U, \mathcal{L}(W, E))$ and $R \in \mathcal{L}(E, X)$. Suppose also $\mathcal{F}(x_0) \perp \mathcal{F}^w_t$ for all $t > 0$ and $x_0 \in L_2(\Omega, \mathcal{F}, P; X)$. Then for each $\nu \in \mathcal{M}$, the mean vector $m(\cdot)$ and the covariance operator $P(\cdot)$ satisfies the following differential equations

$$\begin{cases} \dfrac{dm}{dt} = Am + R\nu_t(f) \\ m(0) = m_0, \end{cases} \tag{23}$$

$$\begin{cases} \dfrac{d}{dt}(P(t)\xi, \eta) = (A^*\xi, P\eta) + (P\xi, A^*\eta) + (R\nu_t(\sigma)Q\nu_t(\sigma^*)R^*\xi, \eta) \\ P(0) = P_0, \quad \xi, \eta \in D(A^*), \quad t \in I = [0, \tau]. \end{cases} \tag{24}$$

Unfortunately the $\tau_y$-topology (Young topology) is too weak for the stochastic problem. We shall, instead, consider the topology of pointwise convergence $\tau_p$, in the sense that, $\nu^n \xrightarrow{\tau_p} \nu$ as $n \to \infty$ if

$$\nu^n_t(\xi) \equiv \int_U \xi(u)\nu^n_t(du) \xrightarrow{s} \int_U \xi(u)\nu^0_t(du) \equiv \nu^0_t(\xi)$$

in $Y$ as $n \to \infty$ for each $\xi \in C_b(U, Y)$ for almost all $t \geq 0$.

Lemma 2

Suppose the assumptions of Lemma 1 hold. Then, for each $\nu \in \mathcal{M}$, the equation (23) has a unique mild solution $m \in C(I,X)$ and the equation (24) has a weak solution $P \in C(I,\mathcal{L}_w^+(X))$. Further, $\nu \to m^\nu$ is continuous from $\mathcal{M}$ to $C(I,X)$ in $\tau_y$-topology and $\nu \to P^\nu$ is continuous from $\mathcal{M}$ to $C(I,\mathcal{L}_w^+(X))$ in the $\tau_p$-topology.

Theorem 3

Let $\mathcal{M}_p$ be a subset of $\mathcal{M}$, compact in the $\tau_p$-topology and suppose the assumptions of Lemma 1 hold and

$$M_c \in L_\infty(I,\mathcal{L}_n^+(X)), N_c \in L_\infty(I,\mathcal{L}^+(X)) \text{ and } m^* \in C(I,X).$$

Then there exists a $\nu^0 \in \mathcal{M}_p$ for which

$$J(\nu^0) \leq J(\nu) \text{ for all } \nu \in \mathcal{M}_p.$$

Proof

Follows from the facts that $\nu \to J(\nu)$ is continuous from $\mathcal{M}_p$ to $R$, $J(\nu) \geq 0$ and that $\mathcal{M}_p$ is compact in $\tau_p$-topology.

Objective 2 (Time Optimal Control)

Consider the uncontrolled system

$$d\xi = A\xi dt + \sigma_0 dw_0, \quad \xi(0) = x_0 \tag{25}$$

and the controlled system,

$$\begin{cases} dx = Axdt + \sigma_0 dw_0 + R\nu_t(f)dt + R\nu_t(\sigma)dw \\ x(0) = x_0. \end{cases} \tag{26}$$

Let $P_\xi$ and $P_x^\nu$ denote the covariance operators corresponding to the processes $\{\xi\}$ and $\{x\}$ respectively.

Let $P_\xi^\infty \equiv w \cdot \lim_{t\to\infty} P_\xi(t)$. The problem is to find a $\nu \in \mathcal{M}_p$ such that

$$P_x^\nu(t^0) = P_\xi^\infty \text{ in minimum time } t^0.$$

Lemma 4 [6]

Consider the uncontrolled system (25) and suppose $A \in G(X,M,-\delta)$ for some $\delta > 0$ and $\sigma_0 \in \mathcal{L}(W_0,X)$. Then there exists a $P_\xi^\infty \in \mathcal{L}_n^+(X)$ such that

$$P_\xi(t) \xrightarrow{\tau_{w_0}} P_\xi^\infty \text{ as } t \to \infty,$$

and further,

$$\mu_{\xi,t} \xrightarrow{w^*} \mu_\xi^\infty \text{ as } t \to \infty,$$

where $\mu_{\xi,t}(\Gamma) \equiv \text{Prob.}\{\xi(t) \in \Gamma\}, \Gamma \in B(X)$ and $\mu_\xi^\infty$ is a countably additive Gaussian measure with covariance operator $P_\xi^\infty$.

Note: We use the convention $\inf(\emptyset) = \infty$.

Theorem 5

Suppose the assumptions of Lemma 1 and Lemma 4 hold and further, there exists a $u^0 \in U$ such that $f(u^0) = 0, \sigma(u^0) = 0$. suppose there exists a $\nu \in \mathcal{M}_p$ such that

$$t(\nu) \equiv \inf\{t \geq 0 : P_x^\nu(t) = P_\xi^\infty\} < \infty.$$

Then there exists a $\nu^0 \in \mathcal{M}_p$ such that

$$t_0 \equiv t(\nu^0) \leq t(\nu) \ \ \forall \nu \in \mathcal{M}_p.$$

Proof

The proof is standard and follows from the expression,

$$\begin{aligned}
&(P^\nu(t)\xi, \eta) = (P_0 T^*(t)\xi, T^*(t)\eta) \\
&+ \int_0^t \langle \sigma_0 Q_0 \sigma_0^* T^*(t-\theta)\xi, T^*(t-\theta)\eta \rangle d\theta \\
&+ \int_0^t \langle \nu_\theta(\sigma) Q \nu_\theta(\sigma^*) R^* T^*(t-\theta)\xi, R^* T^*(t-\theta)\eta \rangle d\theta, \quad \xi, \eta \in D(A^*)
\end{aligned}$$

and the fact that $T^*$ is also a $C_0$-semigroup in $X$.

Objective 3

Find $\nu^0 \in \mathcal{M}$ such that $J(\nu^0) \leq J(\nu)$ for $\nu \in \mathcal{M}$ where

$$J(\nu) = \frac{1}{2} \int_0^\tau \{tr(P(t)M_c) + \langle N_c(m(t) - m^*(t)), m(t) - m^*(t) \rangle\} dt$$

subject to the dynamic constraints:

$$\begin{cases}
\dfrac{dm}{dt} = Am(t) + R\nu_t(f), \quad m(0) = m_0 \\
\dfrac{d}{dt}(P\xi, \eta) = (A^*\xi, P\eta) + (P\xi, A^*\eta) + \langle Q\nu_t(\sigma^*) R^*\xi, \nu_t(\sigma^*) R^*\eta \rangle, \quad \text{for } 0 \leq t \leq \tau < \infty \\
P(0) = P_0, \quad \xi, \eta \in D(A^*).
\end{cases}$$

Theorem 6

Suppose the assumptions of Lemma 1 hold and further,

$$m^* \in C(I,X),\ M_c \in L_\infty(I,\mathcal{L}_n^+(X)),\ \text{and } N_c \in L_\infty(I,\mathcal{L}^+(X)).$$

Then, in order that the triple $\{\nu^0, m^0, P^0\} \in \mathcal{M} \times C(I,X) \times C(I,\mathcal{L}_n^+(X))$ be optimal, it is necessary that the following equations and inequalities hold:

$$\begin{cases} \dfrac{dm^0}{dt} = Am^0 + R\nu_t^0(f), \quad m^0(0) = m_0 \\ \dfrac{d}{dt}(P^0\xi,\eta) = (A^*\xi, P^0\eta) + (P^0\xi, A^*\eta) + \langle Q\nu_t^0(\sigma^*)R^*\xi, \nu_t^0(\sigma^*)R^*\eta\rangle \\ P^0(0) = P_0, \quad \xi,\eta \in D(A^*), \quad \text{for } 0 \le t \le \tau < \infty \end{cases} \tag{27}$$

$$\begin{cases} -\dfrac{dp}{dt} = A^*p + N_c(t)(m^0(t) - m^*(t)), p(\tau) = 0 \\ \dfrac{d}{dt}(S(t)\xi,\eta) = (S(t)\xi, A\eta) + (A\xi, S(t)\eta) + (M_c(t)\xi,\eta) \\ S(\tau) = 0, \quad \xi,\eta \in D(A) \end{cases} \tag{28}$$

$$J^{(1)}(\nu^0, \nu - \nu^0) \equiv \int_0^\tau \{tr[(\nu_t(\sigma) - \nu_t^0(\sigma))Q\nu_t^0(\sigma^*)R^*SR] + \langle \nu_t(f) - \nu_t^0(f), R^*p\rangle\}dt \ge 0 \tag{29}$$

for all $\nu \in \mathcal{M}$.

Proof of theorem 6(outline)

Let $\nu^0 \in \mathcal{M}$ be optimal and $\nu \in \mathcal{M}$ arbitrary, and define

$$\nu^\varepsilon = \nu^0 + \varepsilon(\nu - \nu^0), \quad 0 \le \varepsilon \le 1.$$

Let $m^\varepsilon$ and $m^0$ correspond to $\nu^\varepsilon$ and $\nu^0$ respectively. Then one shows that

$$\frac{m^\varepsilon - m^0}{\varepsilon} \xrightarrow{\varepsilon \downarrow 0} \tilde{m} \text{ in } C(I,X), I = [0,T]$$

where $\tilde{m}$ is the mild solution of

$$(\tilde{m}) \qquad \frac{d\tilde{m}}{dt} = A\tilde{m} + R(\nu_t(f) - \nu_t^0(f)), \quad \tilde{m}(0) = 0.$$

Similarly one proves that

$$\frac{P^\varepsilon - P^0}{\varepsilon} \xrightarrow{\tau_{wo}} \tilde{P} \text{ as } \varepsilon \downarrow 0 \text{ in } C(I,\mathcal{L}_w^+(X))$$

where $\tilde{P}$ satisfies

$$(\tilde{P}) \qquad \begin{cases} \dfrac{d}{dt}(\tilde{P}\xi,\eta) = (A^*\xi, \tilde{P}\eta) + (\tilde{P}\xi, A^*\eta) + \langle Q(\nu - \nu^0)(\sigma^*)R^*\xi, \nu^0(\sigma^*)R^*\eta\rangle_{W,W^*} \\ \tilde{P}(0) = 0. \end{cases}$$

$$(\tilde{J}) \qquad J'(\nu^0, \nu - \nu^0) = 1/2 \int_0^T \{tr(\tilde{P}M_c) + \langle N_c(m^0 - m^*), \tilde{m}\rangle_{X^*,X}\}dt.$$

Using the variational equations $(\tilde{m}), (\tilde{P})$ and $(\tilde{J})$ and introducing the adjoint equations one obtains the result.

Remark 7

If one prefers to work with $\varphi = \pi x$, we may introduce

$$Y = [D(A^*)] \equiv (D(A^*), \text{ graph norm})$$

$$Y^* = [D(A^*)]^* \equiv \text{ dual of } Y.$$

Then

$$\mu_t^\varphi = \mu_t^x \pi^{-1} \quad \text{and}$$

$$J(\nu) = 1/2 \int_0^T \{tr(P_\varphi(t)M_c) + \langle N_c(m_\varphi - m^*), m_\varphi - m^*\rangle_{Y^*,Y}\}dt$$

where

$$M_c \in C(I, \mathcal{L}_n^+(Y^*, Y)),$$

$$N_c \in C(I, \mathcal{L}^+(Y^*, Y)),$$

$$\int_{Y^*} (\varphi, \xi)\mu_t^\varphi(d\varphi) = \int_X (x, \pi^*\xi)\mu_t^x(dx), \quad \xi \in Y$$
$$= (m_t^x, \pi^*\xi)_{X,X^*},$$

$$\int_{Y^*} (\varphi, \xi)^2\mu_t^\varphi(d\varphi) = \int_X (x, \pi^*\xi)^2\mu_t^x(dx)$$
$$= (P_x(t)\pi^*\xi, \pi^*\xi)_{X,X^*}$$
$$= (\pi P_x(t)\pi^*\xi, \xi)_{Y^*,Y}.$$

Remark 8

The results presented here also hold under the following (weaker) assumptions:

(i) $X$ is a reflexive Banach space, or $X$ is not necessarily reflexive, but $\overline{D(A^*)} = X^*$

and

(ii) $E$ is a Banach space, $W_0$ and $W$ are separable Banach spaces.

## PRACTICAL REALIZATION OF RELAXED CONTROLS

The controls $\{\nu_t\}$ can be approximated by sums of Dirac measures as

$$\nu_t(du) \cong \sum_{i=1}^N \alpha_i(t)\delta_{u_i}(du), \quad N < \infty$$

where $\{u_i\} \in U$.

In that case the inequality (29) takes the form

$$\int_0^T \{\sum_i (\alpha_i(t) - \alpha_i^0(t))\Big[\sum_{j=1}^n \alpha_j^0(t) tr[\sigma(u_i)Q\sigma^*(u_j)R^*SR] + \langle f(u_i), R^*p(t)\rangle_{E,E^*}\Big]\}dt \geq 0 \tag{30}$$

for $\alpha$'s satisfying

$$\sum \alpha_i(t) = 1, \quad 0 \leq \alpha_i \leq 1.$$

Let $\alpha^n$ denote the value of $\alpha$ at the $n$th stage and $\nu^n$ be the corresponding relaxed control. Using this $\nu^n$ in (27) and (28) one obtains $m^n, P^n, p^n$ and $S^n$. Substituting these in (30) and denoting the expression within the bracket $\Big[\,\Big]$ by $\beta_i^n$ one can obtain the update for $\alpha$ as

$$\alpha_i^{n+1} = \alpha_i^n - \varepsilon\beta_i^n, \ 1 \leq i \leq N, \quad \text{for } \varepsilon > 0,$$

giving

$$J(\alpha^{n+1}) = J(\alpha^n) - \varepsilon\|\beta^n\|^2.$$

This way one can obtain an approximate realization of optimal relaxed controls.

## REFERENCES


(1) H. Amann, *Parabolic evolution equations with nonlinear boundary conditions,* Nonlinear Functional Analysis and Applications, Proc. Symp. Pure Math. Vol 45, no 1, A.M.S., pp. 17-27, 1986.

(2) N.U. Ahmed, *Optimization and identification of systems governed by evolution equations on Banach space,* Pitman Research Notes in Mathematics, Vol 184, Longman Scientific and Technical, UK and John Wiley and Sons, New York, 1988.

(3) N.U. Ahmed, *Abstract stochastic evolution equations and related control and stability problems,* Lecture Notes in Control and Information Sciences, Springer-Verlag, Vol 97, pp. 107-120, 1987.

(4) N.U. Ahmed, *Stochastic initial boundary value problems for a class of second order evolution equations,* Proc. Int. Conf. on Theory and Applications of Differential Equations, (Ed. A.R. Aftabi Zadeh), Vol 1, Ohio Univ. Press, pp. 13-19, 1988.

(5) N.U. Ahmed, *Abstract stochastic evolution equations on Banach spaces,* Journal of Stochastic Analysis and Applications, 3, 4, pp. 397-432, 1985.

(6) N.U. Ahmed, *Stability of limit measures induced by stochastic differential equations on Hilbert space,* (invited paper, 29th CDC, Honolulu, Dec. 1990).

# STATE-CONSTRAINED CONTROL PROBLEMS OF QUASILINEAR ELLIPTIC EQUATIONS

Eduardo Casas and Luis A. Fernández
Departamento de Matemáticas, Estadística y Computación
Facultad de Ciencias - Universidad de Cantabria
39071-SANTANDER (SPAIN)

## 1 INTRODUCTION

In this work we continue the study of optimal control problems governed by quasilinear elliptic equations in divergence form. Up to now, we have only considered control constraints. Here we also consider state constraints (essentially of integral type). Our aim is to prove existence of solution and derive the optimality conditions. Moreover we utilize the optimality system to study the optimal control and optimal state regularity.

We consider the following differential operator:

$$Ay = -div(a(x, \nabla y)) + a_0(x, y)$$

where $a(x,\eta) = (a_1(x,\eta), \ldots, a_n(x,\eta))$ and the associated Dirichlet problem:

$$\left\{ \begin{array}{rl} Ay = u & \text{in } \Omega \\ y = 0 & \text{on } \Gamma \end{array} \right. \tag{1}$$

where $\Omega$ is a bounded and open subset of $R^n$ and $\Gamma = \partial\Omega$ is Lipschitz continuous.

In the sequel we will assume the conditions:

$$\left\{ \begin{array}{l} a_j(\cdot, \eta) \text{ is a measurable function in } \Omega \\ a_j(x, \cdot) \text{ belongs to } C^1(R^n) \quad j = 1, \ldots, n \end{array} \right. \tag{2}$$

$$\left\{ \begin{array}{l} a_0(\cdot, s) \text{ is a measurable function in } \Omega \\ a_0(x, \cdot) \text{ belongs to } C^1(R) \end{array} \right. \tag{3}$$

$$\sum_{i,j=1}^{n} \frac{\partial a_j}{\partial \eta_i}(x,\eta)\xi_i\xi_j \geq \Lambda_1(k + |\eta|)^{\alpha-2}|\xi|^2 \tag{4}$$

$$\sum_{i,j=1}^{n} \left| \frac{\partial a_j}{\partial \eta_i}(x,\eta) \right| \leq \Lambda_2(k + |\eta|)^{\alpha-2} \tag{5}$$

$$0 \leq \frac{\partial a_0}{\partial s}(x,s) \leq f(|s|) \tag{6}$$

$$a_0(x,0) = a_j(x,0) = 0 \quad j = 1,\dots,n \tag{7}$$

for some $k \in [0,1]$, some $\alpha \in (1,+\infty)$, some strictly positive constants $\Lambda_1, \Lambda_2$, some positive and non-decreasing function $f$, all $x \in \Omega$, all $s \in R$ and all $\eta, \xi \in R^n$.

We make the following additional assumption on $\alpha$:

$$\alpha > n/2 \tag{8}$$

Under hypotheses 2–8, for each $u \in L^2(\Omega)$ there exists a unique solution $y_u \in W_0^{1,\alpha}(\Omega) \cap L^\infty(\Omega)$ of Dirichlet problem 1. Moreover there exists an upper bound of $\|y_u\|_{L^\infty(\Omega)}$ depending only on $A$, $\Omega$ and $\|u\|_{L^2(\Omega)}$, see J.M. Rakotoson [9]. Let us note that $L^2(\Omega)$ is included in $W^{-1,\beta}(\Omega)$ thanks to 8, where $\frac{1}{\alpha} + \frac{1}{\beta} = 1$.

In this work we extend some results related to the control problems of linear and semilinear equations with state constraints (see for instance J.F. Bonnans and E. Casas [1]). These problems are included in the case $\alpha = 2$.

This paper is organized as follows: in Section 2 we present the control problems; in Section 3, we study the conditions under which the functional $v \longrightarrow y_v$ is differentiable; in Sections 4 and 5, we derive the optimality systems in the differentiable and non differentiable cases respectively. In this last case we introduce a family of approximating problems and we pass to the limit in the correspondent optimality conditions. In Section 6, we prove the qualification of the control problems in certain particular situations. Finally, in Section 7, we utilize the optimality conditions to derive some regularity results of the optimal control and state.

# 2 THE CONTROL PROBLEMS

Let $K$ be a non-empty, convex and closed subset of $L^2(\Omega)$ and $J : L^2(\Omega) \longrightarrow R$ the cost functional defined by

$$J(v) = \frac{1}{2}\int_\Omega |y_v - y_d|^2 dx + \frac{\nu}{2}\int_\Omega |v|^2 dx$$

with $y_d$ a fixed element of $L^2(\Omega)$ and $\nu$ a non-negative constant.

We are concerned with the following optimal control problems:

**PROBLEM 1 (Differentiable Constraints of Integral Type)**

$$(P_1)\begin{cases} \textit{Minimize } J(v) \\ \\ \textit{Subject to } v \in K \textit{ and } \int_\Omega g_j(x, y_v(x))dx \leq \delta_j \quad 1 \leq j \leq m. \end{cases}$$

**PROBLEM 2 (Non Differentiable Constraints of Integral Type)**

$$(P_2)\begin{cases} \textit{Minimize } J(v) \\ \\ \textit{Subject to } v \in K \textit{ and } \int_\Omega |y_v(x)|dx \leq \delta. \end{cases}$$

**PROBLEM 3 (Pointwise Constraints)**

$$(P_3)\begin{cases} \textit{Minimize } J(v) \\ \\ \textit{Subject to } v \in K \textit{ and } |y_v(x)| \leq \delta \ \ \forall x \in \Omega. \end{cases}$$

Existence of solutions can be established in a standard form using the continuity of functional $v \in L^2(\Omega) \longrightarrow y_v \in W_0^{1,\alpha}(\Omega)$, the dominated convergence theorem (for $(P_1)$ and $(P_2)$) and the Sobolev inclusions (for $(P_3)$).

**THEOREM 1** *Let us suppose the following conditions:*

1. *There exist an admissible point for* $(P_i)$, $1 \leq i \leq 3$.
2. *For* $i = 1$, $g_j : \Omega \times R \longrightarrow R$ *is a Caratheodory function and there exist* $h \in L^1(\Omega)$ *and* $\phi : R^+ \longrightarrow R^+$ *non decreasing such that*

$$|g_j(x,y)| \leq h(x)\phi(|y|) \ \ a.e.\ x \in \Omega \quad \forall y \in R \ \ 1 \leq j \leq m.$$

3. *Either* $K$ *is bounded in* $L^2(\Omega)$ *or* $\nu > 0$.

*Then there exists (at least) one solution of* $(P_i)$, $1 \leq i \leq 3$.

# 3 SENSITIVITY ANALYSIS

In order to derive the optimality system, the main question to investigate is the differentiability of functional $v \longrightarrow y_v$. For this study we will assume that $k \neq 0$. In fact, for $\alpha > 2$ and $k = 0$ this application is not necessarily differentiable. An example can be found in [3].

Now, given $y \in W_0^{1,\alpha}(\Omega)$, let us define the space $H_0^y(\Omega)$ as the completion of $D(\Omega)$ with respect to the norm:

$$\|z\| = \left(\int_\Omega (k + |\nabla y|)^{\alpha-2}|\nabla z|^2 dx\right)^{1/2}$$

It may be easily verified that $H_0^y(\Omega)$ is a Hilbert space with the inner product

$$(z_1, z_2) = \int_\Omega (k + |\nabla y|)^{\alpha-2}\nabla z_1 \nabla z_2 dx$$

Moreover we have

$$W_0^{1,\alpha}(\Omega) \subset H_0^y(\Omega) \subset H_0^1(\Omega) \ \text{ if } \ \alpha \geq 2$$
$$H_0^1(\Omega) \subset H_0^y(\Omega) \subset W_0^{1,\alpha}(\Omega) \ \text{ if } \ \alpha \leq 2$$

with continuous injections.

More general spaces of this type (weighted Sobolev spaces) have been studied by C.V. Coffman et al. [5], M.K.V. Murthy and G. Stampacchia [8] and N.S. Trudinger [13].

In the next theorem we prove that the relationship between the control and the state is differentiable in some cases.

**THEOREM 2** *Let us suppose $k \neq 0$ and one of the following assumptions:*

1. $\alpha \geq 2$.

2. $\alpha < 2$ *and* $n = 1$.

*Let $F : L^2(\Omega) \longrightarrow H_0^1(\Omega)$ (resp. $W_0^{1,\alpha}(\Omega)$ if $\alpha < 2$) be the functional defined by $F(u) = y_u$. Then $F$ is Gâteaux differentiable. Moreover, if $DF(u)v = z$, then $z$ belongs to $H_0^{y_u}(\Omega)$ and it is the unique solution in this space of problem*

$$\left\{ \begin{array}{ll} -div\left(\frac{\partial a}{\partial \eta}(x, \nabla y_u)\nabla z\right) + \frac{\partial a_0}{\partial s}(x, y_u)z = v & in\ \Omega \\ \\ z = 0 & on\ \Gamma \end{array} \right. \tag{9}$$

The proof of this theorem can be found in the paper of the authors [3]. Here we only mention some ideas. Given $u, v \in L^2(\Omega)$ and $0 < t < 1$, we consider the problems

$$\left\{ \begin{array}{ll} Ay_t = u + tv & \text{in } \Omega \\ y_t = 0 & \text{on } \Gamma \end{array} \right.$$

In case $\alpha \geq 2$, we prove that the sequence $\left\{\frac{y_t - y}{t}\right\}_{t>0}$ converges towards an element $z$ weakly in $H_0^{y_u}(\Omega)$ and strongly in $H_0^1(\Omega)$. This part requires a rather long development. For proving that $z \in H_0^{y_u}(\Omega)$ it is essential that

$$\frac{y_t - y}{t} \in W_0^{1,\alpha}(\Omega) \subset H_0^{y_u}(\Omega).$$

In case $\alpha < 2$, we can argue in the same form and we have also that $\frac{y_t - y}{t} \in W_0^{1,\alpha}(\Omega)$, but now $H_0^{y_u}(\Omega) \subset W_0^{1,\alpha}(\Omega)$ and then we can only prove that there exist subsequences converging to elements which are solutions of 9 in the distribution sense and belong to the space

$$V_0^{y_u}(\Omega) = \left\{ z \in W_0^{1,\alpha}(\Omega) : \int_\Omega (k + |\nabla y|)^{\alpha-2} |\nabla z|^2 dx < \infty \right\}$$

In fact, for $\alpha < 2$, the differentiability of functional $F$ is equivalent to the equality of all these limit points. This is true if $D(\Omega)$ is dense in $V_0^{y_u}(\Omega)$ or, which is equivalent, if $z = 0$ is the unique solution in $V_0^{y_u}(\Omega)$ of PDE

$$-div\left(\frac{\partial a}{\partial \eta}(x, \nabla y_u)\nabla z\right) + \frac{\partial a_0}{\partial s}(x, y_u)z = 0$$

If $n = 1$, it may be easily verified that $\frac{\partial a}{\partial \eta}(x, \nabla y_u)\nabla z \in W^{1,\infty}(\Omega)$ and then $z = 0$. If $n > 1$, we do not know any positive or negative result about this question. In this context, it is interesting to mention a Serrin's paper [11] where it is proved the existence of non null solutions in $W_0^{1,\alpha}(\Omega)$, with $\alpha < 2$, for the homogeneous Dirichlet problem associated to a linear elliptic operator with bounded measurable coefficients.

# 4 OPTIMALITY CONDITIONS I

In this section we will study the control problems corresponding to the situation described in theorem 2: the functional $v \longrightarrow y_v$ is differentiable. Thus we suppose $k \neq 0$ and also $\alpha \geq 2$ or $\alpha < 2$ and $n = 1$.

We are going to derive the optimality conditions for problems $(P_i)$. This is done by using the following result due to E. Casas:

**THEOREM 3** *Let $X, Y$ be Banach spaces, $\emptyset \neq K \subset X$, $C \subset Y$ convex sets with $\overset{o}{C} \neq \emptyset$. Let us consider the problem*

$$(P)\begin{cases} \textit{Minimize } J(x) \\ \textit{Subject to } x \in K \textit{ and } G(x) \in C \end{cases}$$

*where $G : X \longrightarrow Y$ and $J : X \longrightarrow (-\infty, +\infty]$ are two functions. Suppose that there exists a solution $\overline{x}$ of $(P)$ and that $G$ and $J$ are Gâteaux differentiable in $\overline{x}$. Then there exist a real number $\overline{\lambda} \geq 0$ and an element $\overline{\mu} \in Y'$ such that*

$$\overline{\lambda} + \|\overline{\mu}\| > 0 \tag{10}$$

$$< \overline{\mu}, y - G(\overline{x}) >_{Y',Y} \leq 0 \quad \forall y \in C \tag{11}$$

$$< \overline{\lambda} J'(\overline{x}) + [DG(\overline{x})]^* \overline{\mu}, x - \overline{x} >_{X',X} \geq 0 \quad \forall x \in K \tag{12}$$

*where $[DG(\overline{x})]^*$ denotes the adjoint functional of $DG(\overline{x})$. Moreover, if there exists $x_0 \in K$ such that*

$$G(\overline{x}) + DG(\overline{x}) \cdot (x_0 - \overline{x}) \in \overset{o}{C} \quad \textit{(Slater Condition)}$$

*we can take $\overline{\lambda} = 1$.*

The proof of this theorem is a simple application of the well known separation theorem of convex sets to

$$A = \{(y, \lambda) \in Y \times R : \exists x \in K \text{ such that } y = G(\overline{x}) + DG(\overline{x}) \cdot (x - \overline{x})$$

$$\text{and } \lambda = J'(\overline{x}) \cdot (x - \overline{x})\}$$

$$B = \overset{o}{C} \times (-\infty, 0)$$

which are non-empty disjoint convex sets.

With the aid of previous theorem we can easily obtain the first order optimality system in the differentiable cases.

**THEOREM 4** *Let us suppose that for each $j \in \{1, \ldots, m\}$, $g_j$ is a Caratheodory function of class $C^1$ with respect to the second variable and there exist functions $h \in L^2(\Omega)$ and $\phi : R^+ \longrightarrow R^+$ non decreasing such that*

$$\left| \frac{\partial g_j}{\partial y}(x, y) \right| \leq h(x)\phi(|y|) \quad \text{a.e. } x \in \Omega \quad \forall y \in R \quad 1 \leq j \leq m. \tag{13}$$

*Let $\overline{u} \in K$ be a solution of $(P_1)$, then there exist elements $\overline{\lambda} \geq 0$, $\overline{y} \in W_0^{1,\alpha}(\Omega)$, $\overline{p} \in H_0^{\overline{y}}(\Omega)$ and $\overline{\mu}_j \geq 0$, $1 \leq j \leq m$, such that*

$$\begin{cases} -div(a(x,\nabla\overline{y})) + a_0(x,\overline{y}) = \overline{u} & in\ \Omega \\ \overline{y} = 0 & on\ \Gamma \end{cases} \tag{14}$$

$$\overline{\lambda} + \sum_{j=1}^{m} \overline{\mu}_j > 0 \tag{15}$$

$$\begin{cases} -div\left(\left[\frac{\partial a}{\partial \eta}(x,\nabla\overline{y})\right]^T \nabla\overline{p}\right) + \frac{\partial a_0}{\partial s}(x,\overline{y})\overline{p} = \overline{\lambda}(\overline{y}-y_d) + \sum_{j=1}^{m}\overline{\mu}_j\frac{\partial g_j}{\partial y}(x,\overline{y}) \quad in\ \Omega \\ \overline{p} = 0 \quad on\ \Gamma \end{cases} \tag{16}$$

$$\overline{\mu}_j\left(\int_\Omega g_j(x,\overline{y}(x))dx - \delta_j\right) = 0 \quad 1 \leq j \leq m \tag{17}$$

$$\int_\Omega (\overline{p} + \overline{\lambda}\nu\overline{u}) \cdot (v-\overline{u})dx \geq 0 \quad \forall v \in K \tag{18}$$

*Proof.* We consider $(P_1)$ as a particular case of problem $(P)$ of theorem 3 with $X = L^2(\Omega)$, $Y = R^m$, $C = (-\infty,\delta_1] \times \cdots \times (-\infty,\delta_m]$, $G(v) = (G_1(v),\ldots,G_m(v))$ and

$$G_j(v) = \int_\Omega g_j(x,y_v(x))dx$$

From theorem 2 it follows that the functionals $G_j$ are differentiable and

$$DG_j(\overline{u}) \cdot v = \int_\Omega \frac{\partial g_j}{\partial y}(x,y_u)DF(\overline{u})vdx$$

Applying theorem 3, we deduce the existence of $\overline{\lambda} \geq 0$ and $\overline{\mu}_j \in R$, $1 \leq j \leq m$, such that

$$\overline{\lambda} + \sum_{j=1}^{m} |\overline{\mu}_j| > 0 \tag{19}$$

$$\sum_{j=1}^{m}\overline{\mu}_j\left(y_j - \int_\Omega g_j(x,\overline{y})dx\right) \leq 0 \quad \forall(y_1,\ldots,y_m) \in C \tag{20}$$

$$\overline{\lambda}J'(\overline{u})\cdot(v-\overline{u}) + \sum_{j=1}^{m}\overline{\mu}_j DG_j(\overline{u})\cdot(v-\overline{u}) \geq 0 \quad \forall v \in K \tag{21}$$

Denoting $DF(\overline{u})\cdot(v-\overline{u})$ by $z$, we have

$$J'(\overline{u})\cdot(v-\overline{u}) = \int_\Omega(\overline{y}-y_d)zdx + \nu\int_\Omega \overline{u}(v-\overline{u})dx.$$

Now let $\overline{p}$ be the unique solution of 16 in $H_0^{\overline{y}}(\Omega)$, then

$$\int_\Omega \left(\overline{\lambda}(\overline{y}-y_d)+\sum_{j=1}^m \overline{\mu}_j \frac{\partial g_j}{\partial y}(x,\overline{y})\right) z dx = \int_\Omega \nabla z^T \left(\frac{\partial a}{\partial \eta}(x,\nabla\overline{y})\right)^T \nabla\overline{p}dx + \int_\Omega z\frac{\partial a_0}{\partial s}(x,\overline{y})\overline{p}dx =$$

$$\int_\Omega \nabla\overline{p}^T \left(\frac{\partial a}{\partial \eta}(x,\nabla\overline{y})\right)\nabla z dx + \int_\Omega \overline{p}\frac{\partial a_0}{\partial s}(x,\overline{y})dx = \int_\Omega \overline{p}(v-\overline{u})dx$$

Last equality follows from theorem 2. So from 21 we obtain 18.

On the other hand, using that $G(\overline{y}) \in C$ and taking in 20

$$y_j < \int_\Omega g_j(x,\overline{y}(x))dx, \quad y_k = \int_\Omega g_k(x,\overline{y}(x))dx \quad \text{if } k \neq j$$

it follows that $\overline{\mu}_j \geq 0 \ \forall j \in \{1,\ldots,m\}$ and then 19 coincides with 15. Finally, if we take

$$y_j = \delta_j, \quad y_k = \int_\Omega g_k(x,\overline{y}(x))dx \quad \text{if } k \neq j$$

we derive 17. □

**THEOREM 5** *Let $\overline{u}$ be a solution of $(P_2)$. Then there exist a real number $\overline{\lambda} \geq 0$ and elements $\overline{y} \in W_0^{1,\alpha}(\Omega)$, $\overline{p} \in H_0^{\overline{y}}(\Omega)$ and $\overline{\mu} \in L^\infty(\Omega)$ verifying 14, 18 and*

$$\overline{\lambda} + \|\overline{\mu}\| > 0 \tag{22}$$

$$\begin{cases} -div\left(\left[\frac{\partial a}{\partial \eta}(x,\nabla\overline{y})\right]^T \nabla\overline{p}\right) + \frac{\partial a_0}{\partial s}(x,\overline{y})\overline{p} = \overline{\lambda}(\overline{y}-y_d)+\overline{\mu} \quad in \ \Omega \\ \overline{p} = 0 \quad on \ \Gamma \end{cases} \tag{23}$$

$$\int_\Omega \overline{\mu}(y-\overline{y})dx \leq 0 \quad if \ \|y\|_{L^1(\Omega)} \leq \delta \tag{24}$$

*Proof.* In this case it is enough to take $X = L^2(\Omega)$, $Y = L^1(\Omega)$, $C =$ the ball of $L^1(\Omega)$ with center at 0 and radio $\delta$ and $G(v) = y_v$. □

**REMARK 1** *It is possible to consider more general constraints of type*

$$\int_\Omega g(x,y_u(x))dx \leq \delta$$

*where $g : \Omega \times R \longrightarrow R$ is a measurable function with respect to the first variable and a convex function with respect to the second one, $g(\cdot,0) \in L^1(\Omega)$ and there exists $h \in L^\infty(\Omega)$ such that*

$$|g(x,y)-g(x,z)| \leq h(x)|y-z| \quad a.e.\ x \in \Omega \ \ \forall y,z \in R$$

*Nevertheless, it is necessary that the closed convex*

$$C = \{y \in L^1(\Omega) : \int_\Omega g(x,y(x))dx \leq \delta\}$$

*has a non empty interior.*

Before stating the optimality conditions for problem $(P_3)$, let us introduce some notations. Let $C_0(\Omega)$ be th space of real and continuous functions on $\overline{\Omega}$ that are null on $\Gamma$, endowed with the supremum-norm $\|\cdot\|_{L^\infty(\Omega)}$ and let $M(\Omega)$ be the space of real and regular Borel measures on $\Omega$ endowed with the norm

$$\|\mu\|_{M(\Omega)} = |\mu|(\Omega)$$

where $|\mu|$ is the total variation measure of $\mu$, see W. Rudin [10]. As consequence of Riesz representation theorem, it is known that $M(\Omega)$ is the dual space of $C_0(\Omega)$ with the duality product defined by

$$< \mu, y >= \int_\Omega y(x)d\mu(x) \quad \forall y \in C_0(\Omega)$$

**THEOREM 6** *Let us suppose $n = 1$ and let $\overline{u}$ be a solution of $(P_3)$. Then there exist $\overline{\lambda} \geq 0$, $\overline{y} \in W_0^{1,\alpha}(\Omega)$, $\overline{p} \in H_0^{\overline{y}}(\Omega)$ and $\overline{\mu} \in M(\Omega)$ verifying 14, 22, 23, 18 and*

$$\int_\Omega (y(x) - \overline{y}(x))d\overline{\mu}(x) \leq 0 \quad \forall y \in \overline{B}_\delta(0) \tag{25}$$

*where $\overline{B}_\delta(0)$ is the ball of $C_0(\Omega)$ with center at 0 and radio $\delta$.*

*Proof.* It follows from theorem 3 with $X = L^2(\Omega)$, $Y = C_0(\Omega)$, $C = \overline{B}_\delta(0)$ and $G(v) = y_v$.

**REMARKS 1** *1. From 25 it is possible to obtain certain properties of $\overline{\mu}$ as in E. Casas [2].*

*2. There exist two difficulties for considering the case $n > 1$: the first one is that, in general, the states are not continuous functions; the second one is the non existence of solution in $H_0^{\overline{y}}(\Omega)$ of problem 9 when the second term belongs to $M(\Omega)$. Let us note that*

$$H_0^{\overline{y}}(\Omega) \subset C_0(\Omega) \quad \textit{if and only if} \quad n = 1.$$

# 5 OPTIMALITY CONDITIONS II

In this section we will assume that one of the following conditions is satisfied:

1. $n > 1$, $\alpha < 2$ and $k \neq 0$.

2. $\alpha > 2$ and $k = 0$.

Therefore we do not know if the functional $v \longrightarrow y_v$ is differentiable (case 1.) or we know that it is not differentiable (case 2.). To deal with these situations we introduce a family of approximating problems $(P_\epsilon)$ that belongs to the differentiable case, we obtain the optimality system for each $(P_\epsilon)$ and we pass to the limit in these systems when $\epsilon \to 0$.

For each $\epsilon > 0$ let us consider the perturbed differential operator

$$A_\epsilon y = -\epsilon\Delta y + Ay$$

$A_\epsilon$ satisfies the hypotheses 2–7 (with exponent 2 if $\alpha < 2$). Hence, given $u \in L^2(\Omega)$ there exists a unique solution $y_\epsilon(u) \in W_0^{1,\alpha}(\Omega)$ (resp. $H_0^1(\Omega)$ if $\alpha < 2$) of problem

$$\begin{cases} A_\epsilon y = u & \text{in } \Omega \\ y = 0 & \text{on } \Gamma \end{cases}$$

Let $\overline{u}$ be a solution of $(P_i)$, $1 \leq i \leq 3$. Associated to this solution we define the cost functional

$$J_\epsilon(v) = \frac{1}{2}\int_\Omega |y_\epsilon(v) - y_d|^2 dx + \frac{\nu}{2}\int_\Omega |v|^2 dx + \frac{1}{2}\int_\Omega |v - \overline{u}|^2 dx$$

and the corresponding control problems

$$(P_1^\epsilon)\begin{cases} \text{Minimize} \quad J_\epsilon(v) \\ \text{Subject to} \quad v \in K \text{ and } \displaystyle\int_\Omega g_j(x, y_\epsilon(v))dx \leq \delta_j(\epsilon) \quad 1 \leq j \leq m \end{cases}$$

with $\delta_j(\epsilon) = \max\left\{\delta_j, \int_\Omega g_j(x, y_\epsilon(\overline{u}))dx\right\}$, $1 \leq j \leq m$.

$$(P_2^\epsilon)\begin{cases} \text{Minimize} \quad J_\epsilon(v) \\ \text{Subject to} \quad v \in K \text{ and } \displaystyle\int_\Omega |y_\epsilon(v)|dx \leq \delta(\epsilon) \end{cases}$$

with $\delta(\epsilon) = \max\left\{\delta, \int_\Omega |y_\epsilon(\overline{u})|dx\right\}$.

$$(P_3^\epsilon)\begin{cases} \text{Minimize} \quad J_\epsilon(v) \\ \text{Subject to} \quad v \in K \text{ and } |y_\epsilon(v)(x)| \leq \delta(\epsilon) \ \forall\, x \in \Omega \end{cases}$$

with $\delta(\epsilon) = \max\left\{\delta, \|y_\epsilon(\overline{u})\|_{L^\infty(\Omega)}\right\}$.

Let us observe that we have enlarged the set of state constraints in such a way that we can guarantee the existence of admissible points for each $(P_i^\epsilon)$: in fact $\overline{u}$ is an admissible point. Now we can deduce the existence of a solution $u_\epsilon$ of problem $(P_i^\epsilon)$ as in theorem 1. On the other hand, taking into account the Gâteaux differentiability of $J_\epsilon$, we can obtain the optimality conditions for $(P_i^\epsilon)$, $1 \leq i \leq 3$, as in theorems 4–6. For instance, if $i = 2$ this system is given by the following theorem.

**THEOREM 7** *For each $\epsilon > 0$ there exists (at least) one solution $u_\epsilon$ of $(P_2^\epsilon)$. Moreover there exist elements $\lambda_\epsilon \geq 0$, $y_\epsilon \in W_0^{1,\alpha}(\Omega)$ (resp. $H_0^1(\Omega)$ if $\alpha < 2$), $p_\epsilon \in H_0^{y_\epsilon}(\Omega)$ (resp. $H_0^1(\Omega)$) and $\mu_\epsilon \in L^\infty(\Omega)$ verifying*

$$\lambda_\epsilon + \|\mu_\epsilon\| > 0 \tag{26}$$

$$\begin{cases} -div[\epsilon\nabla y_\epsilon + a(x, \nabla y_\epsilon)] + a_0(x, y_\epsilon) = u_\epsilon & in\ \Omega \\ y_\epsilon = 0 & on\ \Gamma \end{cases} \tag{27}$$

$$\begin{cases} -div\left(\left[\epsilon I + \dfrac{\partial a}{\partial \eta}(x, \nabla y_\epsilon)\right]^T \nabla p_\epsilon\right) + \dfrac{\partial a_0}{\partial s}(x, y_\epsilon)p_\epsilon = \lambda_\epsilon(y_\epsilon - y_d) + \mu_\epsilon & in\ \Omega \\ p_\epsilon = 0 & on\ \Gamma \end{cases} \tag{28}$$

$$\int_\Omega \mu_\epsilon(y-y_\epsilon)dx \le 0 \quad if \quad \|y\|_{L^1(\Omega)} \le \delta(\epsilon) \tag{29}$$

$$\int_\Omega (p_\epsilon + \lambda_\epsilon[\nu u_\epsilon + u_\epsilon - \overline{u}])(v-u_\epsilon)dx \ge 0 \quad \forall v \in K \tag{30}$$

*where $I$ denotes the identity matrix $n \times n$.*

Similar theorems can be formulated for problems $(P_1^\epsilon)$ and $(P_3^\epsilon)$. In order to pass to the limit in these optimality systems we use the following results:

**THEOREM 8 (E. Casas and L.A. Fernández [3])** *For each $\epsilon > 0$ let $(y_\epsilon, v_\epsilon)$ belong to $W_0^{1,\alpha}(\Omega) \times L^2(\Omega)$ (resp. $H_0^1(\Omega) \times L^2(\Omega)$ if $\alpha < 2$) and satisfy*

$$\begin{cases} -div[\epsilon\nabla y_\epsilon(v_\epsilon) + a(x, \nabla y_\epsilon(v_\epsilon))] + a_0(x, y_\epsilon(v_\epsilon)) = v_\epsilon & in \ \Omega \\ y_\epsilon(v_\epsilon) = 0 & on \ \Gamma \end{cases}$$

*Assume that $v_\epsilon \to u$ weakly in $L^2(\Omega)$ as $\epsilon \to 0$, then $y_\epsilon(v_\epsilon) \to y_u$ in $W_0^{1,\alpha}(\Omega)$ as $\epsilon \to 0$ and there exists a constant $C > 0$ such that*

$$\|y_\epsilon\|_{L^\infty(\Omega)} \le C \quad \forall \epsilon > 0.$$

**THEOREM 9** *Let $u_\epsilon$ be a solution of $(P_i^\epsilon)$, $1 \le i \le 3$. Set $\overline{y} = y_{\overline{u}}$ and $y_\epsilon = y_\epsilon(u_\epsilon)$. Then we have*

$$u_\epsilon \to \overline{u} \quad in \ L^2(\Omega)$$

$$y_\epsilon \to \overline{y} \quad in \ W_0^{1,\alpha}(\Omega)$$

$$J_\epsilon(u_\epsilon) \to J(\overline{u})$$

*as $\epsilon \to 0$.*

The proof utilizes the same arguments that theorem 4.4 of [3]. Let us point out that if $y_\epsilon \to y_u$ in $W_0^{1,\alpha}(\Omega)$ and $\{y_\epsilon\}_{\epsilon>0}$ satisfies the state constraints of problems $(P_i^\epsilon)$, then $y_u$ satisfies the state constraints of $(P_i)$ thanks to the convergence of $y_\epsilon(\overline{u})$ towards $\overline{y}$ in $W_0^{1,\alpha}(\Omega)$.

## 5.1 PASSAGE TO THE LIMIT: $\alpha < 2$ and $k \neq 0$

Since we are assuming $\alpha < 2$ and 8, then obviously $n$ must be less than or equal to 3. However the case $n = 1$ has been studied in section 3, thus $n = 2$ or 3 in this section.

Next results can be proved using the same argumentation of [3], taking into account the following facts. For the family of problems $(P_2^\epsilon)$ we can suppose without loss of generality that $\lambda_\epsilon + \|\mu_\epsilon\|_{L^\infty(\Omega)} = 1$: in other case, it is enough to divide the expressions 28–30 by $\sigma_\epsilon = \lambda_\epsilon + \|\mu_\epsilon\|_{L^\infty(\Omega)}$ and to rename $p_\epsilon$, $\lambda_\epsilon$ and $\mu_\epsilon$ instead of $\sigma_\epsilon^{-1}p_\epsilon$, $\sigma_\epsilon^{-1}\lambda_\epsilon$ and $\sigma_\epsilon^{-1}\mu_\epsilon$. So there exist elements $\overline{\lambda} \ge 0$, $\overline{\mu} \in L^\infty(\Omega)$ and subsequences (denoted in the same for such that

$$\lambda_\epsilon \to \overline{\lambda} \quad \text{and} \quad \mu_\epsilon \to \overline{\mu} \quad \text{weakly}^* \text{ in } \ L^\infty(\Omega).$$

Hence, we can pass to the limit in 27–30. Now it remains to prove 22. Suppose that $\overline{\lambda}=0$, then $\|\mu_\epsilon\|_{L^\infty(\Omega)} \to 1$. We are going to see that $\overline{\mu}=0$ is not a weak* limit point of $\{\mu_\epsilon\}_{\epsilon>0}$. Let us consider $y_0 \in L^1(\Omega)$ such that

$$r = \delta - \|y_0\|_{L^1(\Omega)} > 0.$$

In virtue of 29 we have

$$\int_\Omega \mu_\epsilon y dx \leq \int_\Omega \mu_\epsilon(y_\epsilon - y_0)dx \quad \text{if} \quad \|y\|_{L^1(\Omega)} \leq r$$

Taking supremum in the first term, we get

$$r\|\mu_\epsilon\|_{L^\infty(\Omega)} \leq \int_\Omega \mu_\epsilon(y_\epsilon - y_0)dx$$

Letting $\epsilon$ tend to 0, we obtain

$$0 < r \leq \int_\Omega \overline{\mu}(\overline{y} - y_0)dx$$

therefore $\overline{\mu} \neq 0$.

For $(P_1^\epsilon)$ the arguments are similar. In this way we obtain the following theorems:

**THEOREM 10** *Let us suppose 13 and let $\overline{u}$ be a solution of $(P_1)$. Then there exist elements $\overline{\lambda} \geq 0$, $\overline{p} \in W_0^{1,\alpha}(\Omega)$ and $\overline{\mu}_j \geq 0$, $1 \leq j \leq m$ satisfying together with $\overline{u}$ and $\overline{y}$ the system 14–18. Moreover*

$$\int_\Omega \nabla\overline{p}^T \left(\frac{\partial a}{\partial \eta}(x, \nabla\overline{y})\right)^T \nabla\overline{p}dx + \int_\Omega \frac{\partial a_0}{\partial s}(x,\overline{y})\overline{p}^2 dx \leq \int_\Omega \left(\overline{\lambda}(\overline{y} - y_d) + \sum_{j=1}^m \overline{\mu}_j \frac{\partial g_j}{\partial y}(x,\overline{y})\right) \overline{p}dx$$

**THEOREM 11** *Let $\overline{u}$ be a solution of $(P_2)$. Then there exist $\overline{\lambda} \geq 0$, $\overline{p} \in W_0^{1,\alpha}(\Omega)$ and $\overline{\mu} \in L^\infty(\Omega)$ satisfying together with $\overline{u}$ and $\overline{y}$ the system 14, 22–24 and 18. Moreover*

$$\int_\Omega \nabla\overline{p}^T \left(\frac{\partial a}{\partial \eta}(x, \nabla\overline{y})\right)^T \nabla\overline{p}dx + \int_\Omega \frac{\partial a_0}{\partial s}(x,\overline{y})\overline{p}^2 dx \leq \int_\Omega [\overline{\lambda}(\overline{y} - y_d) + \overline{\mu}]\overline{p}dx$$

## 5.2 PASSAGE TO THE LIMIT: $\alpha > 2$ and $k = 0$

To treat case $k=0$, that corresponds to a degenerate equation, we need assume some additional hypotheses that guarantee $C^1$-regularity of states:

$$\left\{\begin{array}{l} a_j \in C^1(\Omega \times R^n) \;\; j = 1,\dots,n \\[2ex] \displaystyle\sum_{i,j=1}^N \left|\frac{\partial a_j}{\partial x_i}(x,\eta)\right| \leq \Lambda_2|\eta|^{\alpha-1} \\[2ex] \displaystyle 0 < \Lambda_3 \leq \frac{\partial a_0}{\partial s}(x,s) \leq f(|s|) \end{array}\right. \tag{31}$$

for all $x \in \Omega$, all $s \in R$ and all $\eta \in R^n$, where $f$ is a positive and non-decreasing function, and also the boundedness of the controls. So we will take in this section

$$K = \{v \in L^2(\Omega) : -\infty < m \leq v(x) \leq M < +\infty \quad a.e.\ x \in \Omega\}$$

Using a Tolksdorf's result [12], we deduce that $y_u \in C^{1,\sigma}(\Omega)$ for each $u \in K$ and some $\sigma \in (0,1)$. Now arguing as in the previous subsection and in the paper [3] we can prove the following theorems.

**THEOREM 12** *Let $\bar{u}$ be a solution of $(P_1)$ and let us assume 13. Then there exist elements $\bar{\lambda} \geq 0$, $\bar{\mu}_j \geq 0$, $1 \leq j \leq m$, and $\bar{p} \in L^2(\Omega) \cap H^1_{loc}(\Omega_0)$, where*

$$\Omega_0 = \{x \in \Omega : |\nabla \bar{y}(x)| > 0\}$$

*satisfying together with $\bar{u}$ and $\bar{y}$ the system 14–15, 17–18 and*

$$-\,div\left(\left[\frac{\partial a}{\partial \eta}(x, \nabla \bar{y})\right]^T \nabla \bar{p}\right) + \frac{\partial a_0}{\partial s}(x, \bar{y})\bar{p} = \bar{\lambda}(\bar{y} - y_d) + \sum_{j=1}^m \bar{\mu}_j \frac{\partial g_j}{\partial y}(x, \bar{y}) \quad in \ \Omega_0 \qquad (32)$$

**THEOREM 13** *Let $\bar{u}$ be a solution of $(P_2)$. Then there exist elements $\bar{\lambda} \geq 0$, $\bar{\mu} \in L^\infty(\Omega)$ and $\bar{p} \in L^2(\Omega) \cap H^1_{loc}(\Omega_0)$, where $\Omega_0$ is given as in previous theorem, satisfying together with $\bar{u}$ and $\bar{y}$ the system 14, 22, 24, 18 and*

$$-\,div\left(\left[\frac{\partial a}{\partial \eta}(x, \nabla \bar{y})\right]^T \nabla \bar{p}\right) + \frac{\partial a_0}{\partial s}(x, \bar{y})\bar{p} = \bar{\lambda}(\bar{y} - y_d) + \bar{\mu} \quad in \ \Omega_0 \qquad (33)$$

**THEOREM 14** *Let $\bar{u}$ be a solution of $(P_3)$ and let us assume $n = 1$. Then there exist elements $\bar{\lambda} \geq 0$, $\bar{\mu} \in M(\Omega)$ and $\bar{p} \in L^2(\Omega) \cap H^1_{loc}(\Omega_0)$, where $\Omega_0$ is given as in theorem 12, satisfying together with $\bar{u}$ and $\bar{y}$ the system 14, 22, 25, 18 and*

$$-\,div\left(\left[\frac{\partial a}{\partial \eta}(x, \nabla \bar{y})\right]^T \nabla \bar{p}\right) + \frac{\partial a_0}{\partial s}(x, \bar{y})\bar{p} = \bar{\lambda}(\bar{y} - y_d) + \bar{\mu} \quad in \ \Omega_0 \qquad (34)$$

# 6 CONSTRAINT QUALIFICATIONS

The optimality system can be viewed as being degenerate when $\bar{\lambda} = 0$ because the characteristic elements of functional $J$ to be minimized ($y_d$ and $\nu$) do not appear. Several supplementary conditions can be proposed under which it is possible to assert that $\bar{\lambda} \neq 0$ (in the terminology of F. H. Clarke [4] the problem is "normal"): for instance the Slater condition, see theorem 3. Other conditions are the following

**THEOREM 15** *1. If $K = L^2(\Omega)$ and $\left\{\frac{\partial g_j}{\partial y}(x,\overline{y})\right\}_{j=1}^m$ are linearly independent in $\Omega$ (resp. $\Omega_0$), then problem $(P_1)$ is normal.*

*2. If $K = L^2(\Omega)$ then problems $(P_2)$ and $(P_3)$ are normal.*

*3. If $K$ is bounded in $L^2(\Omega)$ or $\nu > 0$ and if there exists an admissible control for problem $(P_1^\delta)$, then problem $(P_1^\gamma)$ is normal for almost every $\gamma \in R^m$ such that $\gamma_j \geq \delta_j$, $1 \leq j \leq m$.*

*4. If $K$ is bounded in $L^2(\Omega)$ or $\nu > 0$ and if there exists an admissible control for problem $(P_i^\delta)$, with $i = 2$ or $3$, then problem $(P_i^\gamma)$ is normal for almost every $\gamma \in R$ such that $\gamma \geq \delta$.*

*Proof.* 1) If $\overline{\lambda} = 0$ it follows from 18 that $\overline{p} = 0$. Now, thanks to 16 or 32, we conclude that

$$\sum_{j=1}^m \overline{\mu}_j \frac{\partial g_j}{\partial y}(x,\overline{y}) = 0 \text{ in } \Omega \text{ ( or } \Omega_0)$$

then $\overline{\mu}_j = 0\ \forall j$ which contradicts 15.

3) It is a consequence of F.H. Clarke [4, theorem 3] applied to the function

$$\phi : R^m \longrightarrow (-\infty, +\infty]$$

$$\phi(\gamma) = \min\left\{J(v) : v \in K \text{ and } \int_\Omega g_j(x, y_v(x))dx \leq \gamma_j,\ 1 \leq j \leq m\right\}$$

Let us remark that $J$ and $G_j$ are locally Lipschitz functions from $L^2(\Omega)$ to $(-\infty, +\infty]$ because $v \longrightarrow y_v$ is a Lipschitz functional from $L^2(\Omega)$ to $H_0^1(\Omega)$ if $\alpha \geq 2$ and a locally Lipschitz functional from $L^2(\Omega)$ to $W_0^{1,\alpha}(\Omega)$ if $\alpha < 2$.

The proof of 2) and 4) can be carry out in the same way as 1) and 3) respectively.

# 7 REGULARITY OF THE OPTIMAL CONTROL AND STATE

In this section our aim is to derive some regularity results of optimal control and state from the previous optimality systems. Firstly we establish the boundedness of adjoint state $\overline{p}$.

**THEOREM 16 (E. Casas and L.A. Fernández [3])** *Let us suppose $k \neq 0$. If $\overline{\lambda} \neq 0$ assume also that $y_d \in L^\rho(\Omega) \cap L^2(\Omega)$, with $\rho > n/2$ and furthermore*

$$\rho > \frac{\alpha}{2\alpha - 2} \quad \textit{if} \quad \alpha < 2 \textit{ and } n = 2$$

$$\rho > \frac{3\alpha}{5\alpha - 6} \quad \textit{if} \quad \alpha < 2 \textit{ and } n = 3$$

*then $\overline{p} \in L^\infty(\Omega)$.*

**COROLLARY 1** *Let us suppose that $k$ and $\overline{\lambda}$ are non null, $y_d \in L^\rho(\Omega) \cap L^2(\Omega)$, with $\rho$ as in the previous theorem, $\nu$ is strictly positive and $K$ coincides with one of the following sets*

$$K_1 = \left\{ v \in L^2(\Omega) : \|v\|_{L^2(\Omega)} \le 1 \right\} \quad or$$

$$K_2 = \left\{ v \in L^2(\Omega) : -\infty \le m \le v(x) \le M \le +\infty \quad a.e.\ x \in \Omega \right\}$$

*then $\overline{u}$ belongs to $H^1(\Omega) \cap L^\infty(\Omega)$ if $\alpha \ge 2$ (resp. $W^{1,\alpha}(\Omega) \cap L^\infty(\Omega)$ if $\alpha < 2$).*

*Proof.* The inequality 18 characterizes $\overline{u}$ as the proyection of $-\dfrac{\overline{p}}{\nu\overline{\lambda}}$ in the convex $K$. If $K = K_1$ it follows easily that

$$\overline{u}(x) = -\frac{\overline{p}(x)}{\|\overline{p}\|_{L^2(\Omega)}} \quad a.e.\ x \in \Omega \ \text{ if } \ \|\overline{p}\|_{L^2(\Omega)} > \nu\overline{\lambda}$$

$$\overline{u}(x) = -\frac{\overline{p}(x)}{\nu\overline{\lambda}} \quad a.e.\ x \in \Omega \ \text{ if } \ \|\overline{p}\|_{L^2(\Omega)} \le \nu\overline{\lambda}$$

In the same way, if $K = K_2$ we deduce that

$$\overline{u}(x) = \max\left\{ m, \min\left\{ -\frac{\overline{p}(x)}{\nu\overline{\lambda}}, M \right\}\right\} \quad a.e.\ x \in \Omega.$$

The assertion follows from previous relations and theorem 16. □

When $K = K_2$ and $\overline{\lambda}$ or $\nu$ are null, it is easy to obtain from 18 that

$$\overline{u}(x) = \begin{cases} m & \text{if } \ \overline{p}(x) > 0 \\ M & \text{if } \ \overline{p}(x) < 0 \end{cases}$$

Thus $\overline{u}$ is essentially of Bang-Bang type. However, when $K = K_1$, we can still deduce some regularity properties of optimal control:

**COROLLARY 2** *Let us suppose $k \ne 0$, $\overline{\lambda}\nu = 0$, $\overline{p} \ne 0$ and $K = K_1$. Then $\overline{u}$ belongs to $H_0^1(\Omega) \cap L^\infty(\Omega)$ if $\alpha \ge 2$ (resp. $W_0^{1,\alpha}(\Omega) \cap L^\infty(\Omega)$ if $\alpha \le 2$).*

*Proof.* From 18 it can be deduced that $\overline{u}(x) = -\dfrac{\overline{p}(x)}{\|\overline{p}\|_{L^2(\Omega)}}$ a.e. $x \in \Omega$, which permits to conclude the assertion. □

Provided $a_j, a_0, y_d$ and $\Gamma$ are sufficiently smooth, we can combine the previous results with some regularity results of O. Ladyzhenskaya and N. Ural'tseva [7] and P. Tolksdorf [12] to obtain that

$$\overline{y} \in C^{0,\sigma}(\overline{\Omega}) \cap C^{1,\sigma}(\Omega) \ \text{ for some } \ \sigma \in (0,1) \ \text{ and } \ \overline{p}, \overline{u} \in C(\Omega)$$

For $\alpha = 2$, M. Giaquinta and E. Giusti [6] have proved that $\overline{y} \in C^{1,\sigma}(\overline{\Omega})$. In this case, we can deduce that $\overline{p} \in C^{0,\tau}(\overline{\Omega})$ for some $\tau \in (0,1)$ (see O. Ladyzhenskaya and N. Ural'tseva [7, theorem 14.1]) and we obtain higher regularity for $\overline{y}, \overline{p}$ and $\overline{u}$ (depending on $K$) with the aid of the usual bootstrap argumentation.

In case $k = 0$, we have supposed that $K$ is a bounded subset of $L^\infty(\Omega)$, thus we know that $\overline{y} \in C^{1,\sigma}(\Omega)$. As in the preceding case we can deduce that $\overline{p}$ and $\overline{u}$ are continuous in $\Omega_0$.

For $(P_3)$ we have only considered the one dimensional case. So the regularity of $\overline{u}$, $\overline{y}$ and $\overline{p}$ is a consequence of the Sobolev inclusions.

# References


[1] J.F. Bonnans and E. Casas. Contrôle de systèmes elliptiques semilinéaires comportant des contraintes sur l'état. In H. Brezis and J.L. Lions, editors, *Nonlinear Partial Differential Equations and Their Applications. Collège de France Seminar*, volume 8, pages 69–86. Longman Scientific & Technical, New York, 1988.

[2] E. Casas. Control of an elliptic problem with pointwise state constraints. *SIAM J. on Control & Optimiz.*, 24:1309–1318, 1986.

[3] E. Casas and L.A. Fernández. Distributed control of systems governed by a general class of quasilinear elliptic equations. To appear.

[4] F.H. Clarke. A new approach to Lagrange multipliers. *Math. Op. Res.*, 1(2):165–174, 1976.

[5] C.V. Coffman, V. Duffin, and V.J. Mizel. Positivity of weak solutions of non-uniformly elliptic equations. *Ann. Mat. Pura Appl.*, 104:209–238, 1975.

[6] M. Giaquinta and E. Giusti. Global $C^{1,\alpha}$-regularity for second order quasilinear elliptic equations in divergence form. *J. Reine Angew. Math.*, 351:55–65, 1984.

[7] O. Ladyzhenskaya and N. Ural'tseva. *Linear and Quasilinear Elliptic Equations.* Academic Press, New York and London, 1968.

[8] M.K.V. Murthy and G. Stampacchia. Boundary value problems for some degenerate elliptic operators. *Ann. Mat. Pura Appl.*, 80:1–122, 1968.

[9] J.M. Rakotoson. Réarrangement relatif dans les équations elliptiques quasi-linéaires avec un second membre distribution: Application à un théorème d'existence et de régularité. *J. Differential Equations*, 66(3):391–419, 1987.

[10] W. Rudin. *Real and Complex Analysis.* McGraw-Hill, London, 1970.

[11] J. Serrin. Pathological solutions of elliptic differential equations. *Ann. Scuola Norm. Sup. Pisa*, 18(3):385–387, 1964.

[12] P. Tolksdorf. Regularity for a more general class of quasi-linear elliptic equations. *J. Differential Equations*, 51:126–150, 1984.

[13] N.S. Trudinger. Linear elliptic equations with measurable coefficients. *Ann. Scuola Norm. Sup. Pisa*, 27:265–308, 1973.

# A NONLINEAR ABEL INTEGRAL EQUATION

Dang Dinh Ang
Department of Mathematics, HoChiMinh City University
Ho Chi Minh City, Vietnam

Rudolf Gorenflo
Fachbereich Mathematik, Freie Universität Berlin
Arnimallee 2-6, D-1000 Berlin 33, Germany

**Abstract**

For the general nonlinear Abel integral equation

$$\frac{1}{\Gamma(\alpha)}\int_0^x (x-t)^{\alpha-1}K(x,t,u(t))dt = f(x), \quad 0 \le x \le 1,\ 0 < \alpha < 1,$$

some theorems on existence and uniqueness of solutions in $L_p, 1 \le p \le \infty$, and in $C[0,1]$ are established. Furthermore, methods of regularization are described and stability estimates are given.

Keywords: Nonlinear Abel integral equation, existence and uniqueness of solutions, regularization, stability estimates. AMS(MOS) subject classifications: 45E10, 45G05, 45L05, 65R20.

## 1. Introduction

Consider the integral equation

$$\frac{1}{\Gamma(\alpha)}\int_0^x (x-t)^{\alpha-1}K(x,t,u(t))dt = f(x), \qquad 0 \le x \le 1, \tag{1}$$

where $0 < \alpha < 1$ and $K : T \times \mathbb{R} \longrightarrow \mathbb{R}$ and $f : [0,1] \longrightarrow \mathbb{R}$ with

$$T = \{(x,t) | (x,t) \in \mathbb{R}^2,\ 0 \le t \le x \le 1\}$$

are given functions. The following will be tacitly assumed throughout:

(A1) $K \in C(T \times \mathbb{R})$.

(A2) For $(t,w) \in [0,1] \times \mathbb{R}$, the function $x \longmapsto K(x,t,w)$ is differentiable on $[t,1]$ and $K_x \in C(T \times \mathbb{R})$.

(A3) There exists a constant $c > 0$ such that

$$(K(x,x,w_1) - K(x,x,w_2))(w_1 - w_2) \geq c(w_1 - w_2)^2$$

for all $x$ in $[0,1]$ and $w_1, w_2$ in $\mathbb{R}$.

(A4) $K_x$ is Lipschitzian with respect to the third variable, i.e. there exists a constant $M > 0$ such that

$$|K_x(x,t,w_1) - K_x(x,t,w_2)| \leq M|w_1 - w_2|.$$

As an example of a function $K$ satisfying (A1), (A2), (A3) and (A4), but not lying in $C^1(T \times \mathbb{R})$, take

$$K(x,t,w) = w + \frac{1}{4}\,\frac{1}{1+(x-t)^2+|w|}.$$

We shall use operators $J^\beta$ of fractional integration and $D^\beta$ of fractional differentation. For $0 < \beta < 1$ we define

$$J^\beta u(x) = \frac{1}{\Gamma(\beta)}\int_0^x (x-t)^{\beta-1}u(t)dt,\quad D^\beta f(x) = \frac{1}{\Gamma(1-\beta)}\frac{d}{dx}\int_0^x (x-t)^{-\beta}f(t)dt.$$

Note that $D^\beta = DJ^{1-\beta}$ with $D = \frac{d}{dx}$.

The purpose of this paper is to establish some existence and uniqueness results in $L_p(0,1)$, $1 \leq p \leq \infty$, and in $C[0,1]$ and to produce regularized solutions to equation (1). Stability estimates will be given.

We observe that equation (1) was studied by Branca [Br] for the case $\alpha = 1/2$, $K$ in $C^1(T\times\mathbb{R})$, with $K_w$ bounded away from zero and $K_x$ Lipschitzian with respect to the third variable. Branca's results were extended to the general case $0 < \alpha < 1$ by Brunner and Van der Houwen [BH] and, independently, by Gorenflo and Vessella [GV]. The basic tool used was Dini's implicit function theorem. Dini's theorem will not work here because of lack of differentiability. However, the kernel $K$ generates a monotone operator, and we can make use of a general theorem on monotone operators. Our strategy is as follows. We start with an existence theorem in the $L_2$-case , then move to the $L_\infty$-case and the continuous case, and end up with the $L_1$- and $L_2$-cases with stability estimates.

We first convert equation (1) into an equivalent equation, easier to handle. As in [GV], we apply the operator $J^{1-\alpha}$ to both sides of (1) to get

$$\int_0^x H(x,t,u(t))dt = J^{1-\alpha}f(x) \tag{2}$$

where

$$H(x,t,w) = \frac{\sin(\alpha\pi)}{\pi}\int_t^x (x-y)^{-\alpha}(y-t)^{\alpha-1}K(y,t,w)dy. \tag{3}$$

Note that $H$ satisfies

(i) $H \in C(T \times \mathbb{R})$.

(ii) $H_x$ is continuous on $T \times \mathbb{R}$ and is Lipschitz continuous with respect to the third variable with the same Lipschitz constant as for $K_x$.

(iii) $H(x,x,w) = K(x,x,w)$.

Differentiating (2) with respect to $x$, we get

$$K(x,x,u(x)) + \int_0^x H_x(x,t,u(t))dt = D^\alpha f(x). \tag{4}$$

Observe that if $J^{1-\alpha}f$ is absolutely continuous and if $u \in L_1 = L_1(0,1)$, then the foregoing differentiation process is valid. Furthermore, if $f(0) = 0$ and $f$ is absolutely continuous, then $J^{1-\alpha}f$ is absolutely continuous. However, $J^{1-\alpha}f$ need not be absolutely continuous if $f$ is only in $L_1$.

We have converted equation (2) into (4), which is an equation of second kind. This is possible if $J^{1-\alpha}f$ is absolutely continuous. In the absence of absolute continuity, one has to work with an equation of first kind. In the latter case, some kind of regularization is in order. This problem will be taken up in the final portion of the paper.

## 2. Existence and Uniqueness Results

We first give an existence theorem in the $L_2$-case.

**Theorem 1:** *Suppose $D^\alpha f$ is in $L_2 = L_2(0,1)$. Then equation (4) has a unique solution in $L_2$. If, furthermore, $J^{1-\alpha}f(0) = 0$, then this solution is the unique solution of (1).*

**Proof:** For $w = w(t)$ given in $L_2$, consider the following equation in $u = u(x)$ :

$$K(x,x,u(x)) = -\int_0^x H_x(x,t,w(t))dt + D^\alpha f(x). \tag{5}$$

Since $w$ is in $L_2$, the first term of the right hand side is continuous, and in particular in $L_2$. Thus the right hand side is in $L_2$. Let $A : L_2 \longrightarrow L_2$ be defined by

$$Au(x) = K(x, x, u(x)).$$

Then we can show that $A$ is monotone. In fact, we have

$$< Au - Av, u - v > \ \geq \ c \parallel u - v \parallel_2^2, \tag{6}$$

Here $< \cdot, \cdot >$ is the $L_2$ inner product, and $\parallel \cdot \parallel_2$ is the $L_2$ norm. By (6), $A$ is monotone and furthermore

$$< Au, u > / \parallel u \parallel_2 \longrightarrow \infty \text{ for } \parallel u \parallel_2 \longrightarrow \infty. \tag{7}$$

Since $H_x$ is Lipschitzian with respect to the third variable, $A$ takes bounded sets into bounded sets. Finally, $A$ is weakly continuous on lines. Hence by Theorem 2.1 of [L], p.171, there exists an element $u$ in $L_2$ such that

$$Au(x) = -\int_0^x H_x(x, t, w(t))dt + D^\alpha f(x). \tag{8}$$

The solution is clearly unique. Moreover, we have

$$c|u_1(x) - u_2(x)| \leq |Au_1(x) - Au_2(x)| = |v_1(x) - v_2(x)|. \tag{9}$$

Thus

$$|A^{-1}v_1(x) - A^{-1}v_2(x)| \leq |v_1(x) - v_2(x)|/c.$$

Consider the operator $A^{-1}B$, where $B$ is defined as

$$Bw(x) = -\int_0^x H_x(x, t, w(t))dt + D^\alpha f(x). \tag{10}$$

As shown earlier

$$u(x) = A^{-1}Bw(x). \tag{11}$$

We shall prove that $A^{-1}B$ has a unique fixed point $u = A^{-1}Bu$, and that $u$ can be computed by successive approximation. Put

$$\begin{aligned} u_0(x) &= 0, \\ &\vdots \\ u_n(x) &= A^{-1}Bu_{n-1}(x)\ , \end{aligned}$$

i.e.

$$u_n(x) = A^{-1}(-\int_0^x H_x(x,t,u_{n-1}(t))dt + D^\alpha f(x)). \tag{12}$$

Then, for $n \geq 1$, we have

$$|u_{n+1}(x) - u_n(x)| \leq (\frac{M}{c})^n \frac{1}{n!} \int_0^x |u_1(t)|dt \ .$$

Thus

$$\| u_{n+1} - u_n \|_2 \leq (M/c)^n \frac{1}{n!} \| u_1 \|_2 \ . \tag{13}$$

Hence, $(u_n)$ converges in $L_2$ to a function $u$, which, by the continuity of $A^{-1}B$, is a fixed point of $A^{-1}B$, i.e.,

$$u = A^{-1}Bu,$$

or equivalently,

$$K(x,x,u(x)) + \int_0^x H_x(x,t,u(t))dt = D^\alpha f(x). \tag{14}$$

The $L_2$-solution is unique since $(A^{-1}B)^n$ is a contraction for $n$ large. This completes the proof of Theorem 1.

We next consider the $L_\infty$-case and the continuous case.

**Theorem 2:** *Suppose $D^\alpha f$ is in $L_\infty = L_\infty(0,1)$. Then equation (4) has a unique solution in $L_\infty$. If $u_1, u_2$ are the solutions of (4) corresponding to $D^\alpha f_i$, $i = 1,2$, then the following holds:*

$$\| u_1 - u_2 \|_\infty \leq e^{-1} \exp(\frac{M}{c}) \| D^\alpha f_1 - D^\alpha f_2 \|_\infty \ . \tag{15}$$

**Proof:** Let $u$ be the $L_2$-solution of (4). It is sufficient to show that $u$ ist in $L_\infty$. Now, the second term of the left hand side of (4) is bounded since it is continuous. Since the right hand side is (essentially) bounded by hypothesis, is follows that $K(x,x,u(x))$ is essentially bounded. Then

$$c \| u \|_\infty \leq \| K(x,x,u(x)) - K(x,x,0) \|_\infty + \| K(x,x,0) \|_\infty \ .$$

Hence $u$ ist in $L_\infty$. Now, let $u_1, u_2$ be the $L_\infty$-solutions of (4) corresponding to $D^\alpha f_1, D^\alpha f_2$. Then,

$$|u_1(x) - u_2(x)| \leq \| D^\alpha f_1 - D^\alpha f_2 \|_\infty /c + (M/c) \int_0^x |u_1(t) - u_2(t)|dt \ .$$

By Gronwall's inequality,

$$\| u_1 - u_2 \|_\infty \leq \frac{e^{M/c}}{c} \| D^\alpha f_1 - D^\alpha f_2 \|_\infty .$$

QED.

**Theorem 3:** *Suppose $D^\alpha f$ is continuous on $[0,1]$. Then there exists a unique continuous solution of (4). If $u_i$ is the continuous solution of (4) corresponding to $D^\alpha f_i$, $i = 1,2$, then the following holds:*

$$|u_1(x)-u_2(x)| \leq \frac{1}{c}|D^\alpha f_1(x)-D^\alpha f_2(x)|+\frac{M}{c^2}\int_0^x \exp[\frac{M}{c}(x-s)]\cdot|D^\alpha f_1(s)-D^\alpha f_2(s)|ds. \quad (16)$$

**Proof:** Let $u$ be the $L_2$-solution of (4). Then, since the right hand side of (4) and the second term of the left side are continuous, it follows that $K(x,x,u(x))$ is continuous. Denoting it by $h(x)$, we have

$$|K(x',x',u(x)) - K(x',x',u(x')| = |K(x',x',u(x)) - K(x,x,u(x)) + h(x) - h(x')|.$$

Hence

$$|u(x) - u(x')| \leq c^{-1}|K(x',x',u(x)) - K(x,x,u(x))| + c^{-1}|h(x) - h(x')|.$$

Thus $u(x') \longrightarrow u(x)$ for $x' \longrightarrow x$. We have just proved that $u$ is continuous.

For a stability estimate, let $u_i$ be the continuous solution of (4) corresponding to $D^\alpha f_i$, $i = 1,2$. Then we have

$$|u_1(x) - u_2(x)| \leq \frac{1}{c}|D^\alpha f_1(x) - D^\alpha f_2(x)| + \frac{M}{c}\int_0^x |u_1(t) - u_2(t)|dt.$$

By Gronwall's generalized inequality [Hi]:

$$|u_1(x) - u_2(x)| \leq \frac{1}{c}|D^\alpha f_1(x) - D^\alpha f_2(x)| + \frac{M}{c^2}\int_0^x \exp(\frac{M}{c}(x-s))|D^\alpha f_1(s) - D^\alpha f_2(s)|ds.$$

This concludes the proof of Theorem 3.

We finally consider the $L_p$-case.

**Theorem 4:** *Suppose $D^\alpha f$ is in $L_1$. Then there exists a unique $L_1$-solution of (4). If $u_i$ is the $L_1$-solution of (4) corresponding to $D^\alpha f_i$, $i = 1,2$, then the following holds:*

$$\| u_1 - u_2 \|_1 \leq \frac{1}{c}\exp(\frac{M}{c}) \| D^\alpha f_1 - D^\alpha f_2 \|_1 . \quad (17)$$

**Proof:** Let $(g_n)$ be a sequence of continuous functions converging in $L_1$ to $D^\alpha f$. By Theorem 3, if $u_n$ is the continuous solution of (4) corresponding to $g_n$ in the right hand side, the following holds:

$$|u_n(x) - u_m(x)| \leq \frac{1}{c}|g_n(x) - g_m(x)| + \frac{M}{c^2}\int_0^x \exp(\frac{M}{c}(x-t))|g_n(t) - g_m(t)|dt.$$

Integrating over $x$ from 0 to 1 gives

$$\| u_n - u_m \|_1 \leq \frac{1}{c} \| g_n - g_m \|_1 + \frac{M}{c^2}\int_0^1 \int_0^x \exp(\frac{M}{c}(x-t))|g_n(t) - g_m(t)|dt\, dx \tag{18}$$

$$= \frac{1}{c} \| g_n - g_m \|_1 + \frac{M}{c^2}\int_0^1 \int_t^1 \exp(\frac{M}{c}(x-t))dx|g_n(t) - g_m(t)|dt \leq \frac{1}{c}e^{M/c} \| g_n - g_m \|_1 .$$

Thus $(u_n)$ is a Chauchy sequence in $L_1$, which converges to $u$, say. It is easily seen that $u$ is the $L_1$-solution of (4). The stability estimate (17) is derived by considering sequences of continuous functions $g_n^1, g_n^2$ converging in $L_1$ to $D^\alpha f_1, D^\alpha f_2$ respectively, and passing to the limits in [18], with $u_n, u_m$ replaced by $u_n^1, u_n^2$ respectively. QED.

**Theorem 5:** *Suppose $D^\alpha f$ is in $L_2$. Then the $L_2$ -solution of (4), which exists (and is unique) by Theorem 1, is stable with respect to variations in $D^\alpha f$. In fact, if $u_i$ is the $L_2$-solution of (4) corresponding to $D^\alpha f_i$, $i = 1, 2$ , then the followings holds:*

$$\| u_1 - u_2 \|_2 \leq \frac{1}{c}e^{M/c} \| D^\alpha f_1 - D^\alpha f_2 \|_2 . \tag{19}$$

**Proof:** We can (and shall) assume that $D^\alpha f_1$ and $D^\alpha f_2$ are continuous. The general case is obtained by passing to the limit. For $D^\alpha f_i \equiv g_i$ continuous, $i = 1, 2$, the corresponding (continuous) solutions $u_1, u_2$ of (4) satisfy, by Theorem 3,

$$|u_1(x) - u_2(x)| \leq \frac{1}{c}|g_1(x) - g_2(x)| + \frac{M}{c^2}\int_0^x \exp(\frac{M}{c}(x-t))|g_1(t) - g_2(t)|dt. \tag{20}$$

Consider the second term in the right hand side of (20) and denote it by $Q$, for brevity. Squaring and using Schwarz's inequality give

$$Q^2 \leq (\frac{M}{c^2})^2 e^{2Mx/c}\int_0^x \exp(-\frac{Mt}{c})dt \int_0^x \exp(-\frac{Mt}{c})|g_1(t) - g_2(t)|^2 dt$$

$$= \frac{M}{c^3} e^{2Mx/c}(1 - e^{-Mx/c}) \int_0^x \exp(-\frac{Mt}{c})|g_1(t) - g_2(t)|^2 dt.$$

Integrating the latter quantity from 0 to 1 gives, using Fubini's theorem and rearranging,

$$\frac{M}{c^3} \int_0^1 (e^{Mx/c} - 1) \int_0^x \exp(\frac{M}{c}(x-t))|g_1(t) - g_2(t)|^2 dt\, dx \tag{21}$$

$$\leq \frac{M}{c^3}(e^{M/c} - 1) \int_0^1 \int_t^1 \exp(\frac{M}{c}(x-t)) dx |g_1(t) - g_2(t)|^2 dt \leq \frac{1}{c^2}(e^{M/c} - 1)^2 \| g_1 - g_2 \|_2^2 .$$

Hence the $L_2$-norm of the right hand side of (20) is majorized by $\frac{1}{c} e^{M/c} \| g_1 - g_2 \|_2$. Thus

$$\| u_1 - u_2 \|_2 \leq \frac{1}{c} e^{M/c} \| g_1 - g_2 \|_2 . \tag{22}$$

QED.

**Remark 1:** From what precedes, it is clear that if $D^\alpha f \in L_p$, $1 < p < \infty$, then (4) admits a unique $L_p$-solution. Furthermore, the $L_p$-solution is stable with respect to variations in $D^\alpha f$, and stability estimates of the type (19) can be derived by using Hölder's inequality instead of Schwarz's inequality. Combining with the estimates (15) and (17) one then has

$$\| u_1 - u_2 \|_p \leq \frac{1}{c} e^{M/c} \| D^\alpha f_1 - D^\alpha F_2 \|_p \quad \text{for} \quad 1 \leq p \leq \infty.$$

We do not pursue this matter further.

**Remark 2:** It is observed that if $u \in L_1$, then the left hand side of (2) is an absolutely continuous function. As a first consequence, equation (2) is equivalent to equation (4). A second consequence is that (2) and (because of equivalence with (2)) (1) has a continuous (resp. $L_p$) solution only if $J^{1-\alpha} f(0) = 0$ and $J^{1-\alpha} f$ has a derivative that is continuous (resp. in $L_p$).

## 3. Regularization

Consider equation (2). We have seen in Section 2 that if $J^{1-\alpha} f$ is absolutely continuous, then (2), an equation of first kind, is equivalent to (4), an equation of second kind. The problem is then well posed in the sense that if $D^\alpha f$ belongs to $C[0,1]$ or $L_p(0,1)$, then a unique solution exists in the corresponding function space and depends continuously on $D^\alpha f$. We are now considering the case where $J^{1-\alpha} f$ is not supposed to be absolutely continuous but simply to be continuous or in $L_p$, and is known only approximately. In the case of the classical (linear) Abel equation, it is known that the problem is ill-posed in the usual (and most useful) function spaces. It can be shown that in the present non-linear case, the problem is also ill-posed in the usual function spaces. Hence some kind of regularization is required.

In the sequel, it will be assumed that $g$ is in $L_2$ (resp. $L_1$) and that $g_0$ is an $L_2$-function (resp. $L_1$-function) such that

$$\| g - g_0 \|_2 \leq \epsilon \quad (\text{resp.} \quad \| g - g_0 \|_1 \leq \epsilon). \tag{23}$$

It is assumed that $g_0$ is absolutely continuous with a derivative $g_0'$ in $L_2$ or $L_1$. Let $u$ be the solution of (4) corresponding to $g_0'$ in the right hand side ($g_0$ in place of $J^{1-\alpha}f$, $g_0'$ in place of $D^\alpha f$). It is our purpose to "construct" a function that depends continuously on $g$ and is $\delta$-close to $u$ where $\delta = \delta_{(\epsilon)} \longrightarrow 0$ for $\epsilon \longrightarrow 0$. Such a function will be called a *regularized solution* of (2).

It will be convenient to put

$$Jv(x) = \int_0^x v(t)dt \quad \text{for} \quad v \quad \text{in} \quad L_1(0,1).$$

Our regularization problem here consists in approximating the derivative of a function. We give two sample results.

**Theorem 6:** *Let $g$ and $g_0$ be in $L_2(0,1)$ such that*

$$\| g - g_0 \|_2 \leq \epsilon. \tag{24}$$

*Suppose*

$$g_0(x) = \int_0^x v(t)dt \tag{25}$$

*where $v \in H^1(0,1)$ with*

$$\| v \|_2 + \| v' \|_2 \leq E. \tag{26}$$

*For $\beta = \sqrt{\epsilon/E}$, let $v_\beta$ be given by*

$$v_\beta = (\beta I + J)^{-1} g \tag{27}$$

*(with $I$ as identity operator) and let $u_\beta$ be the solution of the equation*

$$K(x, x, u_\beta(x)) + \int_0^x H_x(x, t, u_\beta(t))dt = v_\beta(x). \tag{28}$$

*Suppose u is the solution of the equation*

$$\int_0^x H(x,t,u(t)dt = g_0(x). \tag{29}$$

*Then*

$$\| u - u_\beta \|_2 \leq (3/c)e^{M/c}\sqrt{E\epsilon}. \tag{30}$$

**Remark 3:** For application of Theorem 6 it is desirable to specify bounds on $g_0' = v$ in term of bounds on $u$. We propose to do this follows.

*Put* $h(x,w) = H(x,x,w)$ *(which is* $= K(x,x,w)$*). Suppose* $h : [0,1] \times I\!R \longrightarrow I\!R$ *is* $C^1$ *and* $K_{xx}(x,t,w)$ *is continuous on* $T \times I\!R$*. Assume that u, the solution of (29), is in* $H^1(0,1)$ *with* $\| u \|_2 + \| u' \|_2 \leq c$*. Put*

$$E = |h|_\infty + \| H_x \|_\infty + |h_x|_\infty + c|h_w|_\infty + |H_{xx}|_\infty + \| H_x \|_\infty$$

*where*

$$|\cdot| = \sup\{|\cdot| \;\; |0 \leq x \leq 1, |w| \leq c\},$$

$$\| \cdot \|_\infty = \sup\{|\cdot| \;\; |(x,t) \in T, \; |w| \leq c\}.$$

*Then* $\| v \|_2 + \| v' \|_2 \leq E$.

**Proof:** It can be shown (cf. [HA 1] and [Go] for methods of estimation) that $\| v - v_\beta \|_2 \leq 3\sqrt{E\epsilon}$. Combining this with (19) we have (30). QED.

**Theorem 7:** *Let* $g, g_o$ *satisfy* $\| g - g_0 \|_1 \leq \epsilon$*. Suppose* $g_0(x) = \int_0^x v(t)dt$ *where v is of bounded variation with* $var(v) \leq E$ *where* $var(v)$ *is the total variation of v on* $[0,1]$*. For* $0 < h < E/4$*, put*

$$g_h(x) = \frac{1}{h}(g(x+h) - g(x)) \;\; \text{if} \;\; 0 \leq x \leq 1-h, = \frac{1}{h}(g(x) - g(x-h)) \;\; \text{if} \;\; 1-h < x \leq 1.$$

*Let u be the solution of the equation*

$$\int_0^x H(x,t,u(t)dt = g_0(x) \tag{31}$$

*and let* $u^h$ *be the solution of*

$$K(x,x,u^h(x)) + \int_0^x H_x(x,t,u^h(t))dt = g_h(x). \tag{32}$$

*Then*

$$\| u^h - u \|_1 \leq (4/c)e^{M/c}\sqrt{E\epsilon}. \tag{33}$$

**Proof:** It is shown in [HA 2] that $\| v - g_h \|_1 \leq 4\sqrt{E\epsilon}$. Combining this with (17), we have (31).

**Remark 4:** *Suppose $K$ is in $C^1$ $(T \times I\!R)$ and $K_x(x,t,w)$ is Lipschitzian with respect to $x$, with Lipschitz constant $L$. Assume that $u$, the solution of (31), is of bounded variation, var $(u) \leq c$. Put*

$$E = cM_0 + 2 \| H_x \|_\infty + \| H_t \|_\infty + L,$$

*with $M_0$ as Lipschitz constant of $K(x,t,w)$ with respect to $w$ and*

$$\| \cdot \|_\infty = \sup\{ | \cdot | \quad |(x,t) \in T, \quad |w| \leq c \}.$$

*Then var $(v) \leq E$.*

**Acknowledgement:** This paper was written while the first author was visiting Freie Universität Berlin from January to April 1990 as a guest to the research group "Regularization", under a grant from Deutsche Forschungsgemeinschaft. The second named author is a member of this research group that is supported by the Research Commission of Freie Universität Berlin.

**References**

[Br ] H.W. BRANCA: The nonlinear Volterra equation of Abels's kind and its numerical treatment. Computing 20 (1978), 307-321.

[BH ] H. BRUNNER and P.J. VAN DER HOUWEN: The numerical solution of Volterra equations. North Holland, Amsterdam 1986.

[Go ] R. GORENFLO: On stabilizing the inversion of Abel integral operators. K. Nishimoto (ed.): Fractional calculus and its applications. College of Engineering, Nihon University, Tokyo 1970.

[GV ] R. GORENFLO and S. VESSELLA: Abel integral equations, analysis and applications. Submitted to Lecture Notes in Mathematics. Springer Verlag , Berlin, Heidelberg, New York, Tokyo.

[HA1 ] D.D. HAI and D.D.ANG: Regularization of Abel's integral equation. Proc. Roy. Soc. Edinburgh 107A (1987), 165-168.

[HA2 ] D.D. HAI and D.D. Ang: On nonsmooth solutions of Abel's integral equation. J. Diff. Integral Equations. 1988-1989.

[Hi ] E. HILLE: Lectures on ordinary differential equations. Addison-Wesley, Reading, Mass. 1969.

[L ] J. L. LIONS: Quellques méthodes de résolution des problèmes aux limites non linéaires. Dunod: Paris 1968.

[V ] S. VESSELLA: Stability results for Abel's equation. J. Integral Equations 9 (1985), 125-134.

# SHAPE DERIVATIVES FOR NONSMOOTH DOMAINS.*

M.C. DELFOUR
Centre de recherches mathématiques et
Département de mathématiques et de statistique,
Université de Montréal, C.P. 6128 Succ. A,
Montréal, Québec, Canada, H3C 3J7

J.P. ZOLÉSIO
Laboratoire de physique théorique,
Université des Sciences et Techniques du Languedoc,
Place Eugène Bataillon,
34060 - Montpellier, Cédex 02, France

ABSTRACT. The object of this paper is to study the Shape gradient and the Shape Hessian by the Velocity (Speed) Method for arbitrary domains with or without constraints. It makes the connection between methods using a family of transformations such as first or second order Perturbations of the Identity Operator. New definitions for Shape derivatives are given. They naturally extend existing theories for $C^k$ or Lipchitzian domains to arbitrary domains without any smoothness conditions on their geometric boundary. In this new framework extensions of the classical structure theorems are given for the Shape gradient and the Shape Hessian.

## 1. INTRODUCTION.

The object of this paper is to study the *Shape gradient* and the *Shape Hessian* by the *Velocity (Speed) Method* (cf. J. CÉA [1, 2] and J. P. ZOLÉSIO [1, 2]) for nonsmooth constrained and unconstrained domains and discuss their relationship to various methods based on perturbations of the identity operator. This extends basic results for $C^k$ and Lipschitzian domains to non-smooth domains.

In section 2 we extend the Velocity (Speed) Method to nonsmooth domains $\Omega$ which are constrained to lie within a fixed domain $D$. This is done by a double use of the *Viability Theory* and the introduction of Bouligand contingent and Clarke tangent cones. We obtain natural extension of *Hadamard's structure theorem* for both the Shape gradient and the Shape Hessian (cf. DELFOUR-ZOLÉSIO [2, 3, 4] for a description of the smooth case) and recover known results in the smooth case. The *canonical structures* of the gradient and the Hessian are given for time-varying velocity fields. We show that Methods of Perturbation of the Identity Operator (first and second order) are special cases corresponding to a time varying velocity fields and indicate how to construct the associated velocity.

For the Shape gradient, the different methods yield expression which may look different but are all equal. However this is no longer true for the Shape Hessian. In fact we shall show in section 4 that different perturbations of the identity yield final expressions which are not equal. It turns out that we can introduce an infinity of definitions based on perturbations of the identity. However we shall show that they always contain a *canonical bilinear term* plus the Shape gradient of the functional acting in the direction of an *acceleration* field which is characteristic of the chosen perturbation. The canonical

* The research of the first author has been supported in part by a Killam fellowship from Canada Council and by National Sciences and Engineering Research Council of Canada operating grant A-8730.

bilinear term exactly coincides with the second order Shape derivative obtained by the Velocity (Speed) Method for time-invariant velocity fields. Moreover each expression obtained by a method of perturbation of the identity can be strictly recovered by adding to the canonical term the Shape gradient acting in the direction of an appropriate acceleration field. In view of this we propose to refer to this canonical term as the *Shape Hessian*.

A few paper have dealt with the second variation of a Shape cost function for linear partial differential equations models. To our knowledge the first one by N. FUJII [**1**] used a second order perturbation of the identity along the normal to the boundary for second order linear elliptic problems. An extremely interesting paper by ARUMUGAN-PIRONNEAU [**1, 2**] used the Shape second variation to solve the *ribblet problem*. Finally J. SIMON [**1**] presented a computation of the second variation using a first order perturbation of the identity. The first general approach to the computation of Shape Hessians can be found in DELFOUR-ZOLÉSIO [**2, 3, 4**]. It uses the Velocity (Speed) Method and includes simple illustrative examples for the Neumann and Dirichlet problems.

In conclusion, we would like to reiterate that the Velocity method and methods using first and second order perturbations of the identity lead to three different second order Shape derivatives which are not equal. The Velocity method with constant velocity fields provides the canonical bilinear Shape Hessian and the expressions arising from the other method can be recovered by special choices of time-dependent velocity fields.

The proofs of the main theorems and lemmas are given in DELFOUR-ZOLÉSIO [**5**].

## 2. VELOCITY (SPEED) METHOD AND METHODS OF PERTURBATION OF THE IDENTITY.

In this section we review and extend the Velocity Method (cf. J.P. ZOLÉSIO [**1, 2**] and discuss its relationship to various methods based on perturbations of the identity. Under appropriate conditions we show how to construct a family of time-dependent transformations of $\mathbf{R}^N$ (or the closure of a subset $D$ of $\mathbf{R}^N$) from a family of time dependent velocity fields. Conversely we show how to construct the family of time-dependent velocity fields from a family of time-dependent transformations of $\mathbf{R}^N$ (or the closure of a subset $D$ of $\mathbf{R}^N$).

### 2.1. Unconstrained families of domains.

Let the real number $\tau > 0$ and the map $V : [0,\tau] \times \mathbf{R}^N \to \mathbf{R}^N$ be given. The map $V$ can be viewed as a family $\{V(t) : 0 \leq t \leq \tau\}$ of time-dependent velocity fields on $\mathbf{R}^N$ defined by

$$x \mapsto V(t)(x) = V(t,x) : \mathbf{R}^N \mapsto \mathbf{R}^N. \tag{1}$$

Assume that

(V1) $\quad \forall x \in \mathbf{R}^N,\ V(\cdot,x) \in C^0([0,\tau];\mathbf{R}^N)$,

where $V(\cdot,x)$ is the function $t \mapsto V(t,x)$, and that

(V2) $\quad \exists c > 0, \forall x,y \in \mathbf{R}^N, \forall t \in [0,\tau], \quad |V(t,y) - V(t,x)| \leq c|y-x|$.

Associate with $V$ the solution $x(t;V)$ of the ordinary differential equation

$$\frac{dx}{dt}(t) = V(t,x(t)), \quad t \in [0,\tau], \quad x(0) = X \in \mathbf{R}^N \tag{2}$$

and introduce the homeomorphism

$$X \mapsto T_t(V)(X) = x(t;V) : \mathbf{R}^N \to \mathbf{R}^N. \tag{3}$$

and the maps

$$(t,X) \mapsto T_V(t,X) \stackrel{\text{def}}{=} T_t(V)(X) : [0,\tau] \times \mathbf{R}^N \to \mathbf{R}^N, \tag{4}$$

$$(t,x) \mapsto T_V^{-1}(t,x) \stackrel{\text{def}}{=} T_t^{-1}(V)(x) : [0,\tau] \times \mathbf{R}^N \to \mathbf{R}^N. \tag{5}$$

NOTATION 2.1. *In the sequel we shall drop the $V$ in $T_V(t,X)$, $T_V^{-1}(t,x)$ and $T_t(V)$ whenever no confusion is possible.*

THEOREM 2.1.

(i) *Under hypotheses (V1) and (V2) the maps $T$ and $T^{-1}$ have the following properties*

$$(T1) \quad \begin{cases} \forall X \in \mathbf{R}^N, \quad T(\cdot,X) \in C^1([0,\tau];\mathbf{R}^N) \\ \exists c > 0, \forall X,Y \in \mathbf{R}^N, \quad \|T(\cdot,Y) - T(\cdot,X)\|_{C^1([0,\tau];\mathbf{R}^N)} \le c|Y - X|, \end{cases}$$

$$(T2) \quad \forall t \in [0,\tau], \quad X \mapsto T_t(X) = T(t,X) : \mathbf{R}^N \to \mathbf{R}^N \quad \text{is bijective,}$$

$$(T3) \quad \begin{cases} \forall x \in \mathbf{R}^N, \quad T^{-1}(\cdot,x) \in C^0([0,\tau];\mathbf{R}^N) \\ \exists c > 0, \forall x,y \in \mathbf{R}^N, \quad \|T^{-1}(\cdot,y) - T^{-1}(\cdot,x)\|_{C^0([0,\tau];\mathbf{R}^N)} \le c|y - x|. \end{cases}$$

(ii) *Given a real $\tau > 0$ and a map $T : [0,\tau] \times \mathbf{R}^N \to \mathbf{R}^N$ verifying hypotheses (T1), (T2) and (T3), then the map*

$$(t,x) \mapsto V(t,x) = \frac{\partial T}{\partial t}(t, T_t^{-1}(x)) : [0,\tau] \times \mathbf{R}^N \to \mathbf{R}^N, \tag{6}$$

*verifies hypotheses (V1) and (V2), where $T_t^{-1}$ is the inverse of $X \mapsto T_t(X) = T(t,X)$.* □

This first theorem is an equivalence result which says that we can either start from a family of velocity fields $\{V(t)\}$ on $\mathbf{R}^N$ or a family of transformations $\{T_t\}$ of $\mathbf{R}^N$ provided that the map $V, V(t,x) = V(t)(x)$, verifies (V1) and (V2) or the map $T, T(t,X) = T_t(X)$, verifies (T1) to (T3). When we start from $V$, we obtain the velocity method. Given an initial domain $\Omega$, the family of homeomorphisms $\{T_t(V)\}$ generates a family of transformed domains

$$\Omega_t = T_t(V)(\Omega) = \{T_t(V)(X) : X \in \Omega\}. \tag{7}$$

We shall see in sections 3 and 4 how this family of transformations of $\Omega$ can be used to define shape derivatives.

## 2.2. Perturbation of the identity operator.

In examples where we start from $T$, it is usually possible to verify hypotheses (T1) to (T3) and construct the corresponding velocity field $V$ defined in (6). For instance perturbations of the identity to the first or second order fall in that category:

$$T_t(X) = X + tU(X) + \frac{t^2}{2}A(X) \quad (A = 0 \text{ for the first order }),\ t \geq 0, X \in \mathbf{R}^N, \quad (8)$$

where $U$ and $A$ are transformations of $\mathbf{R}^N$. It turns out that for Lipschitzian transformations $U$ and $A$, hypotheses (T1) to (T3) are verified.

THEOREM 2.2. *Let $U$ and $A$ be two uniform Lipschitzian transformations of $\mathbf{R}^N$:*

$$\exists c > 0,\ \forall X,\ Y \in \mathbf{R}^N, \quad |U(Y) - U(X)| \leq c|Y - X|,\ |A(Y) - A(X)| \leq c|Y - X|.$$

*For $\tau = \min\{1, 1/4c\}$ and $T$ given by (8), the map $T$ verifies hypotheses (T1) to (T3) on $[0, \tau]$. Moreover the associated velocity $V$ is given by*

$$(t, x) \mapsto V(t, x) = U(T_t^{-1}(x)) + tA(T_t^{-1}(x)) : [0, \tau] \times \mathbf{R}^N \to \mathbf{R}^N, \quad (9)$$

*and it verifies hypotheses (V1) and (V2) on $[0, \tau]$.* □

REMARK 2.1. Observe that from (8) and (9)

$$V(0) = U, \quad \overset{\bullet}{V}(0)(x) = \frac{\partial V}{\partial t}(t, x)|_{t=0} = A - [DU]U. \quad (10)$$

where $DU$ is the Jacobian matrix of $U$. The term $\overset{\bullet}{V}(0)$ is an *acceleration* at $t = 0$ which will always be present even when $A = 0$. □

## 2.3. Constrained families of domains.

In many applications the family of admissible domains $\Omega$ is constrained to subsets of a fixed larger domain or *hold-all* $D$. To reflect that constraint we would like to consider transformations

$$T : [0, \tau] \times \bar{D} \to \mathbf{R}^N \quad (11)$$

with the following properties

$$(T1_D) \quad \begin{cases} \forall X \in \bar{D}, \quad T(\cdot, X) \in C^1([0, \tau]; \mathbf{R}^N) \\ \exists c > 0, \forall X, Y \in \bar{D}, \quad \|T(\cdot, Y) - T(\cdot, X)\|_{C^1([0,\tau];\mathbf{R}^N)} \leq c|Y - X|, \end{cases}$$

$$(T2_D) \quad \forall t \in [0, \tau], \quad X \mapsto T_t(X) = T(t, X) : \bar{D} \to \bar{D} \quad \text{is bijective},$$

$$(T3_D) \quad \begin{cases} \forall x \in \bar{D}, \quad T^{-1}(\cdot, x) \in C^0([0, \tau]; \mathbf{R}^N) \\ \exists c > 0, \forall x, y \in \bar{D}, \quad \|T^{-1}(\cdot, y) - T^{-1}(\cdot, x)\|_{C^0([0,\tau];\mathbf{R}^N)} \leq c|y - x|. \end{cases}$$

where under hypothesis $(T2_D)$ $T^{-1}$ is defined as

$$(t,x) \mapsto T^{-1}(t,x) = T_t^{-1}(x) : [0,\tau] \times \bar{D} \to \mathbf{R}^N. \tag{12}$$

Those three properties are the analogue for $\bar{D}$ of the same three properties obtained for $\mathbf{R}^N$. In fact Theorem 2.1 extends from $\mathbf{R}^N$ to $\bar{D}$ by adding one hypothesis to $(V1_D)$ and $(V2_D)$. Specifically we shall consider for $\tau > 0$ velocities

$$V : [0,\tau] \times \bar{D} \to \mathbf{R}^N \tag{13}$$

such that

$(V1_D)$ $\quad \forall x \in \bar{D}, \quad V(\cdot,x) \in C^0([0,\tau];\mathbf{R}^N)$

$(V2_D)$ $\quad \exists n > 0, \forall x, y \in \bar{D}, \quad \|V(\cdot,y) - V(\cdot,x)\|_{C^0([0,\tau];\mathbf{R}^N)} \le n|y-x|$

$(V3_D)$ $\quad \forall x \in \bar{D}, \forall t \in [0,\tau], \quad \pm V(t,x) \in T_D(x),$

where $T_D(x)$ is the Bouligand contingent cone to $\bar{D}$ at the point $x$ in $\bar{D}$ (cf. AUBIN–CELLINA [1, p. 176]).

THEOREM 2.3.

(i) *Let $\tau > 0$ and $V$ be a family of velocity fields verifying hypotheses $(V1_D)$ to $(V3_D)$ and consider the family of transformations*

$$(t,X) \mapsto T(t,X) = x(t;X) : [0,\tau] \times \bar{D} \to \mathbf{R}^N \tag{14}$$

*where $x(\cdot,X)$ is the solution of*

$$\frac{dx}{dt}(t) = V(t,x(t)), \quad 0 \le t \le \tau,\ x(0) = X. \tag{15}$$

*Then the family of transformations $T$ verifies conditions $(T1_D)$ to $(T3_D)$.*

(ii) *Conversely given a family of transformations $T$ verifying hypotheses $(T1_D)$ to $(T3_D)$, the family of velocity fields*

$$(t,x) \mapsto V(t,x) = \frac{\partial T}{\partial t}\left(t, T_t^{-1}(x)\right) : [0,\tau] \times \bar{D} \to \mathbf{R}^N \tag{16}$$

*verifies conditions $(V1_D)$ to $(V3_D)$ and the transformations constructed from this $V$ coincide with $T$.* □

REMARK 2.2. Under $(V1_D)$ to $(V3_D)$, $\{T_t : 0 \le t \le \tau\}$ is a family of homeomorphisms of $\bar{D}$ which map the interior $\overset{\circ}{D}$ (resp. the boundary $\partial D$) of $D$ onto $\overset{\circ}{D}$ (resp. $\partial D$) (cf. J. DUGUNJI [1, p. 87–88]). □

REMARK 2.3. Assumption $(V3_D)$ is a double viability condition. M. NAGUMO [1]'s usual viability condition

$$V(t,x) \in T_{\bar{D}}(x), \forall t \in [0,\tau], \forall x \in \bar{D} \tag{17}$$

is a necessary and sufficient condition for a *viable solution* to (15), that is

$$\forall t \in [0,\tau], \forall X \in \bar{D}, x(t;X) \in \bar{D} \text{ or } T_t(\bar{D}) \subset \bar{D} \tag{18}$$

(cf. AUBIN-CELLINA [1, p. 174 and p. 180]). Condition $(V3_D)$

$$\forall t \in [0,\tau], \forall x \in \bar{D}, \pm V(t,x) \in T_{\bar{D}}(x) \tag{19}$$

is a *strict viability condition* which not only says that $T_t$ maps $\bar{D}$ into $\bar{D}$ but also that

$$\forall t \in [0,\tau], \quad T_t : \bar{D} \to \bar{D} \quad \text{is a homeomorphism.} \tag{20}$$

In particular it keeps interior points in the interior and boundary points on the boundary. □

REMARK 2.4. Condition $(V3_D)$ is a generalization to arbitrary domains $D$ of the following condition used by J.P. ZOLÉSIO [1] in 1979: for all $x$ in $\partial D$

$$\begin{cases} V(t,x) \bullet n(x) = 0, & \text{if the outward normal } n(x) \text{ exists} \\ 0, & \text{otherwise .} \end{cases}$$ □

Theorem 2.2 is a generalization of Theorem 2.1 to arbitrary domains $D$. It shows that we can either start from a velocity $V$ or a transformation $T$.

## 2.4. Transformation of condition $(V3_D)$ into a linear constraint.

Condition $(V3_D)$ is equivalent to

$$\forall t \in [0,\tau], \forall x \in \bar{D}, \ V(t,x) \in \{-T_{\bar{D}}(x)\} \cap \{T_{\bar{D}}(x)\} \tag{21}$$

since $T_{\bar{D}}(x) = T_D(x)$. If $T_D(x)$ was convex, then the above intersection would be a closed linear subspace of $\mathbf{R}^N$. This is true when $D$ is convex. In that case $T_D(x) = C_D(x)$, where $C_D(x)$ is Clarke tangent cone and

$$L_D(x) = \{-C_D(x)\} \cap \{C_D(x)\} \tag{22}$$

is a closed linear subspace of $\mathbf{R}^N$. This means that $(V3_D)$ reduces to

$$\forall t \in [0,\tau], \forall x \in \bar{D}, \quad V(t,x) \in L_D(x). \tag{23}$$

It turns out that for continuous vector fields $V(t,\cdot)$ the equivalence of $(V3_D)$ and (23) extends to arbitrary domains $D$.

THEOREM 2.4. *Given a velocity field $V$ verifying $(V1_D)$ and $(V2_D)$, then condition $(V3_D)$ is equivalent to*

$(V3_C)$ $\qquad \forall t \in [0,\tau], \forall x \in \bar{D},\ V(t,x) \in L_D(x) = \{-C_D(x)\} \cap C_D(x),$

*where $C_D(x)$ is the (closed convex) Clarke tangent cone to $\bar{D}$ at $x$ which is defined by*

$$C_D(x) = \left\{ v \in \mathbf{R}^N : \lim_{\substack{h \to 0 \\ y \to_D x}} \quad d_D(y+hv)/h = 0 \right\}$$

*$d_D(y)$ is the minimum distance from $y$ to $D$, and $\to_D$ denotes the convergence in $\bar{D}$. Moreover $L_D(x)$ is a closed linear subspace of $\mathbf{R}^N$.* □

The equivalence of (V3) and $(V3_C)$ is a direct consequence of the following lemma.

LEMMA 2.1. *Given a vector field $W \in C^0(\bar{D};\mathbf{R}^N)$, the following two conditions are equivalent:*

$$\forall x \in \bar{D}, \quad W(x) \in T_D(x); \tag{25}$$

$$\forall x \in \bar{D}, \quad W(x) \in C_D(x). \quad \square \tag{26}$$

REMARK 2.5. Lemma 2.1 essentially says that for continuous vector fields we can relax the condition of M. NAGUMO [1]'s theorem from $(V3_D)$ involving Bouligand contingent cone to $(V3_C)$ involving the smaller Clarke convex tangent cone. In dimension $N = 3, L_D(x)$ is $\{0\}$ a line, a plane or the whole space. □

NOTATION 2.1. In the sequel it will be convenient to introduce the following spaces and subspaces

$$\mathcal{L} = \{V : [0,\tau] \times \mathbf{R}^N \to \mathbf{R}^N : V \text{ verifies (V1) and (V2) on } \mathbf{R}^N\} \tag{27}$$

and for an arbitrary domain $D$ in $\mathbf{R}^N$

$$\mathcal{L}_D = \{V : [0,\tau] \times \bar{D} \to \mathbf{R}^N : V \text{ verifies } (V1_D), (V2_D) \text{ and } (V3_C) \text{ on } \bar{D}\}. \tag{28}$$

For any integers $k \geq 0$ and $m \geq 0$ and any compact subset $K$ of $\mathbf{R}^N$ define the following subspaces of $\mathcal{L}$

$$\begin{cases} \mathcal{V}_K^{m,0} = C^m\left([0,\tau], \mathcal{D}^0(K,\mathbf{R}^N)\right) \cap \mathcal{L}, & \text{if } k = 0 \\ \mathcal{V}_K^{m,k} = C^m\left([0,\tau], \mathcal{D}^k(K,\mathbf{R}^N)\right), & \text{if } k \geq 1, \end{cases} \tag{29}$$

where $\mathcal{D}^k(K,\mathbf{R}^N)$ is the space of $k$–times continuously differentiable transformations of $\mathbf{R}^N$ with compact support in $K$. In all cases $\mathcal{V}_K^{m,k} \subset \mathcal{L}_K$. As usual $\mathcal{D}^\infty(K,\mathbf{R}^N)$ will de written $\mathcal{D}(K,\mathbf{R}^N)$. □

## 3. SHAPE GRADIENT.

Consider the set $\mathcal{P}(D)$ of subsets $\Omega$ of a fixed domain $D$ of $\mathbf{R}^N$ (possibly all of $\mathbf{R}^N$) which will play the role of a *hold-all*. Under the action of a velocity field $V$ in $\mathcal{L}_D$, the domain $\Omega$ in $\mathcal{P}(D)$ is transformed into a new domain

$$\Omega_t(V) = T_t(V)(\Omega) = \{T_t(V)(X) : X \in \Omega\}. \tag{1}$$

This will now provide our first notion of derivative for a shape functional, that is a map

$$\Omega \mapsto J(\Omega) : \mathcal{P}(D) \to \mathbf{R}. \tag{2}$$

DEFINITION 3.1. *Given a velocity field $V$ in $\mathcal{L}_D$, $J$ is said to have an* Eulerian semiderivative *at $\Omega$ in the direction $V$ if the following limit exists and is finite*

$$\lim_{t \searrow 0} [J(\Omega_t(V)) - J(\Omega)]/t. \tag{3}$$

*Whenever it exists, the limit will be denoted $dJ(\Omega; V)$.* □

This definition is quite general and may include situations where $dJ(\Omega;V)$ is not only a function of $V(0)$ but also of $V(t)$ in a neighbourhood of $t = 0$. This will not occur under some appropriate continuity hypothesis on the map $V \mapsto dJ(\Omega;V)$. This immediately raises the question of the choice of topology and eventually the choice of gradient when we specialize to time-invariant vector fields $V$. We choose to follow the classical philosophy of the Theory of Distributions (cf. L. SCHWARTZ [1]). Assume that $D$ is an open domain in $\mathbf{R}^N$. Domains $\Omega$ in $\mathcal{P}(D)$ will be perturbed by velocity fields $V(t)$ with values in $\mathcal{D}^k(K,\mathbf{R}^N)$ for some compact subset $K$ of $D$ and integer $k \geq 0$. More precisely we shall consider velocity fields in

$$\overrightarrow{\mathcal{V}}_D^{m,k} = \varinjlim_{K} \left\{ \mathcal{V}_K^{m,k} : \forall K \text{ compact in } D \right\} \tag{4}$$

where $\varinjlim$ denotes the inductive limit set with respect to $K$ endowed with its natural inductive limit topology. For time-invariant fields, the above construction reduce to

$$\mathcal{V}_D^k = \left\{ \begin{matrix} \mathcal{D}^0(D,\mathbf{R}^N) \cap \text{ Lip }(\mathbf{R}^N,\mathbf{R}^N), & k=0 \\ \mathcal{D}^k(D,\mathbf{R}^N), & 1 \leq k \leq \infty \end{matrix} \right\} \tag{5}$$

where $\mathrm{Lip}(\mathbf{R}^N,\mathbf{R}^N)$ denotes the space of uniformily Lipschitzian transformations of $\mathbf{R}^N$. In all cases hypotheses $(V1_D)$ to $(V3_D)$ are verified since for all $t \in [0,\tau], V(t,x) = 0$

for all $x$ in $\partial D$. When $D = \mathbf{R}^N$ we drop the index $D$ in the above definitions and simply write $\vec{\mathcal{V}}^{m,k}$ and $\mathcal{V}^k$.

THEOREM 3.1. *Let $\Omega$ be a domain in the fixed open hold-all $D$. Assume that there exist integers $m \geq 0$ and $k \geq 0$ such that*

$$\forall V \in \vec{\mathcal{V}}_D^{m,k}, \quad dJ(\Omega; V) \text{ exists,} \tag{6}$$

*and that the map*

$$V \mapsto dJ(\Omega; V) : \vec{\mathcal{V}}_D^{m,k} \to \mathbf{R} \tag{7}$$

*is continuous. Then*

$$\forall V \in \vec{\mathcal{V}}_D^{m,k}, \quad dJ(\Omega; V) = dJ(\Omega; V(0)), \tag{8}$$

*where $dJ(\Omega; V(0))$ is the Eulerian semiderivative for the time-independent vector field equal to $V(0)$.* □

By virtue of this theorem we can now specialize to time-invariant vector fields $V$ to further study the properties and the structure of $dJ(\Omega; V)$.

DEFINITION 3.2. *Let $\Omega$ be a domain in the open hold-all $D$ of $\mathbf{R}^N$.*

*(i) The functional $J$ is said to be shape differentiable at $\Omega$, if the Eulerian semiderivative $dJ(\Omega; V)$ exists for all $V$ in $\mathcal{D}(D, \mathbf{R}^N)$ and the map*

$$V \mapsto dJ(\Omega; V) : \mathcal{D}(D, \mathbf{R}^N) \to \mathbf{R} \tag{9}$$

*is linear and continuous.*

*(ii) The map (9) defines a vector distribution $G(\Omega)$ which will be referred to as the* shape gradient *of $J$ at $\Omega$.*

*(iii) When there exists some finite $k \geq 0$ such that $G(\Omega)$ is continuous for the $\mathcal{D}^k(D, \mathbf{R}^N)$-topology, we say that the shape gradient $G(\Omega)$ is of* order *$k$.* □

The next theorem gives additional properties of shape differentiable functionals.

NOTATION 3.1. Associate with a subset $A$ of $D$ and an integer $k \geq 0$ the set

$$L_A^k = \left\{ V \in \mathcal{D}^k(D, \mathbf{R}^N) : \forall x \in A, V(x) \in L_A(x) \right\}. \quad \square$$

THEOREM 3.2. *(Generalized Hadamard's structure theorem)*

*Let $\Omega$ be a domain with boundary $\Gamma$ in the open hold-all $D$ of $\mathbf{R}^N$ and assume that $J$ has a shape gradient $G(\Omega)$.*

(i) *The support of the shape gradient $G(\Omega)$ is contained in $\Gamma_D \stackrel{\text{def}}{=} \Gamma \cap D$.*

(ii) *If $\Omega$ is open or closed in $\mathbf{R}^N$ and the shape gradient is of order $k$ for some $k \geq 0$, then there exists $[G(\Omega)]$ in $(\mathcal{D}_D^k/L_\Omega^k)'$ such that for all $V$ in $\mathcal{D}_D^k \stackrel{\text{def}}{=} \mathcal{D}^k(D,\mathbf{R}^N)$*

$$dJ(\Omega;V) = \langle [G(\Omega)], q_L V\rangle_{\mathcal{D}_D^k/L_\Omega^k} \tag{11}$$

*where $q_L : \mathcal{D}_D^k \to \mathcal{D}_D^k/L_\Omega^k$ is the canonical quotient surjection. Moreover*

$$G(\Omega) = {}^*(q_L)[G(\Omega)] \tag{12}$$

*where ${}^*(q_L)$ denotes the transposed of the linear map $q_L$.* □

REMARK 3.1. When the boundary $\Gamma$ of $\Omega$ is compact and $J$ is shape differentiable at $\Omega$, the distribution $G(\Omega)$ is of finite order. Once this is known, the conclusions of Theorem 3.2(ii) apply with $k$ equal to the order of $G(\Omega)$. □

The quotient space is very much related to a trace on the boundary $\Gamma$ and when the boundary $\Gamma$ is sufficiently smooth we can indeed make that identification.

COROLLARY. *Assume that the hypotheses of Theorem 3.2 are verified for an open domain $\Omega$, that the order of $G(\Omega)$ is $k \geq 0$, and that the boundary $\Gamma$ of $\Omega$ is $C^{k+1}$. Then for all $x$ in $\Gamma$, $L_\Omega(x)$ is an $(N-1)$-dimensional hyperplane to $\Omega$ at $x$ and there exists a unique outward unit normal $n(x)$ which belongs to $C^k(\Gamma;\mathbf{R}^N)$. As a result the kernel of the map*

$$V \to \gamma_\Gamma(V) \bullet n : \mathcal{D}^k(D,\mathbf{R}^N) \to \mathcal{D}^k(\Gamma \cap D) \tag{13}$$

*coincides with $L_\Omega^k$ where $\gamma_\Gamma : \mathcal{D}^k(D,\mathbf{R}^N) \to \mathcal{D}^k(\Gamma \cap D,\mathbf{R}^N)$ is the trace of $V$ on $\Gamma \cap D$. Moreover the map $p_L(V)$*

$$q_L(V) \mapsto p_L(q_L(V)) = \gamma_\Gamma(V) \bullet n : \mathcal{D}_D^k/L_\Omega^k \to \mathcal{D}^k(\Gamma \cap D) \tag{14}$$

*is a well-defined isomorphism. In particular there exists a scalar distribution $g(\Gamma)$ in $\mathcal{D}^k(\Gamma \cap D)'$ such that for all $V$ in $\mathcal{D}^k(D,\mathbf{R}^N)$*

$$dJ(\Omega;V) = \langle g(\Gamma), \gamma_\Gamma(V) \bullet n\rangle_{\mathcal{D}^k(\Gamma\cap D)} \tag{15}$$

*and*

$$G(\Omega) = {}^*(q_L)[G(\Omega)], \quad [G(\Omega)] = {}^*(p_L)g(\Gamma). \ \square \tag{16}$$

REMARK 3.2. In 1907, J. HADAMARD [1] used velocity fields along the normal to the boundary $\Gamma$ of a $C^\infty$ domain to compute the derivative of the first eigenvalue of the plate. Theorem 3.2 and its corollary are generalizations to arbitrary shape functionals of that property to open or closed domains with an arbitrary boundary. The generalization to open domains with a $C^{k+1}$ boundary was done by J.P. ZOLÉSIO [1] in 1979. □

REMARK 3.3. The space $\mathcal{D}^k(\Gamma \cap D)$ is not simple to characterize. However when $\Gamma$ is compact and $D = \mathbf{R}^N$, it coincides with $C^k(\Gamma)$. □

EXAMPLE 3.1. For any measurable subset $\Omega$ of a measurable hold-all D of $\mathbf{R}^N$, consider the volume functional

$$J(\Omega) = \int_\Omega dx. \tag{17}$$

For $\Omega$ with finite volume and $V$ in $\mathcal{D}^1(D, \mathbf{R}^N)$,

$$dJ(\Omega; V) = \int_\Omega \operatorname{div} V dx \tag{18}$$

but for a bounded open domain $\Omega$ with a $C^1$ boundary $\Gamma$

$$dJ(\Omega; V) = \int_\Gamma V \bullet n \quad d\Gamma \tag{19}$$

which is continuous on $\mathcal{D}^0(D, \mathbf{R}^N)$. Here the smoothness of the boundary decreases the order of the distribution $G(\Omega)$. This raises the following question: is it possible to characterize the family of all domains $\Omega$ of $D$ for which the map

$$V \mapsto \int_\Omega \operatorname{div} V \, dx : \mathcal{D}^0(D, \mathbf{R}^N) \to \mathbf{R} \tag{20}$$

is continuous? The answer is yes. It is the family of *finite perimeter sets* with respect to $D$. It contains domains $\Omega$ whose characteristic function belongs to $BV(D)$, the space of $L^1$ functions on $D$ with a distributional gradient in the space of (vectorial) bounded measures. Roughly speaking they are the sets with finite volume and perimeter. □

## 4. SHAPE HESSIAN.

We first study the second order Eulerian semiderivative $d^2 J(\Omega; V; W)$ of a functional $J(\Omega)$ for two time-dependent vector fields $V$ and $W$. A first theorem shows that under some natural continuity hypotheses, $d^2 J(\Omega; V; W)$ is the sum of two terms: the *canonical term* $d^2 J(\Omega; V(0); W(0))$ plus the first order Eulerian semiderivative $dJ(\Omega; \dot{V}(0))$ at $\Omega$ in the direction $\dot{V}(0)$ of the time-partial derivative $\partial_t V(t, x)$ at $t = 0$. As in the study of first order Eulerian semiderivatives, this first theorem reduces the study of second order Eulerian semiderivatives to the time-invariant case. So we shall specialize to fields $V$ and $W$ in $\mathcal{D}^k(D, \mathbf{R}^N)$ and give the equivalent of Hadamard's structure theorem for the canonical term.

### 4.1. Time-dependent case.

The basic framework introduced in sections 2 and 3 has reduced the computation of the Eulerian semiderivative of $J(\Omega)$ to the computation of the derivative

$$j'(0) = dJ(\Omega; V(0)) \tag{1}$$

of the function

$$j(t) = J(\Omega_t(V)). \tag{2}$$

For $t \geq 0$, we naturally obtain

$$j'(t) = dJ(\Omega_t(V); V(t)). \tag{3}$$

This suggests the following definition.

DEFINITION 4.1. *Let $V$ and $W$ belong to $\mathcal{L}_D$ and assume that for all $t \in [0, \tau], dJ(\Omega_t(W); V(t))$ exists for $\Omega_t(W) = T_t(W)(\Omega)$. The functional $J$ is said to have a second order Eulerian semiderivative at $\Omega$ in the directions $(V, W)$ if the following limit exists*

$$\lim_{t \searrow 0} [dJ(\Omega_t(W); V(t)) - dJ(\Omega; V(0))]/t. \tag{4}$$

*When it exists, it is denoted $d^2J(\Omega; V; W)$.* □

REMARK 4.1. This last definition is compatible with the second order expansion of $j(t)$ with respect to $t$ around $t = 0$:

$$j(t) \cong j(0) + tj'(0) + \frac{t^2}{2} j''(0), \tag{5}$$

where

$$j''(0) = d^2J(\Omega; V; V). \quad \square \tag{6}$$

REMARK 4.2. It is easy to construct simple examples (see Example 4.2) with time-invariant fields $V$ and $W$ showing that $d^2J(\Omega; V; W) \neq d^2J(\Omega; W; V)$ (cf. DELFOUR-ZOLÉSIO [2]).□

The next theorem is the analogue of Theorem 3.1 and provides the canonical structure of the second order Eulerian semiderivative.

THEOREM 4.1. *Let $\Omega$ be a domain in the fixed open hold-all $D$ of $\mathbf{R}^N$ and let $m \geq 0$ and $\ell \geq 0$ be integers. Assume that*

(i) $\forall V \in \vec{\mathcal{V}}_D^{m+1,\ell}$, $\forall W \in \vec{\mathcal{V}}_D^{m,\ell}$, *$d^2J(\Omega; V; W)$ exists,*

(ii) $\forall W \in \vec{\mathcal{V}}_D^{m,\ell}$, $\forall t \in [0, \tau]$, *$J$ has a shape gradient at $\Omega_t(W)$ of order $\ell$,*

(iii) $\forall\, U \in \mathcal{V}_D^{\ell}$, *the map*

$$W \mapsto d^2J(\Omega; U; W) : \vec{\mathcal{V}}_D^{m,\ell} \to \mathbf{R} \tag{7}$$

*is continuous.*

*Then for all $V$ in $\overset{\rightarrow m+1,\ell}{\mathcal{V}_D}$ and all $W$ in $\overset{\rightarrow m,\ell}{\mathcal{V}_D}$*

$$d^2J(\Omega;V;W) = d^2J(\Omega;V(0);W(0)) + dJ(\Omega;\dot{V}(0)), \tag{8}$$

*where*

$$\dot{V}(0)(x) = \lim_{t\searrow 0}[V(t,x) - V(0,x)]/t. \quad \square \tag{9}$$

This important theorem gives the canonical structure of the second order Eulerian semiderivative: a first term which depends on $V(0)$ and $W(0)$ and a second term which is equal to $dJ(\Omega;\dot{V}(0))$. When $V$ is time-invariant the second term disappears and the semiderivative coincides with $d^2J(\Omega;V;W(0))$ which can be separately studied for time-invariant vector fields in $\mathcal{V}_D^\ell$.

### 4.2. Time-invariant Case.

DEFINITION 4.2. *Let $\Omega$ be a domain in the open hold-all $D$ of $\mathbf{R}^N$.*

(i) *The functional $J(\Omega)$ is said to be* twice shape differentiable *at $\Omega$ if*

$$\forall\, V,\ \forall\, W \text{ in } \mathcal{D}(D,\mathbf{R}^N), \quad d^2J\,(\Omega;V;W) \text{ exists} \tag{11}$$

*and the map*

$$(V,W) \mapsto d^2J(\Omega;V;W) : \mathcal{D}(D,\mathbf{R}^N) \times \mathcal{D}(D,\mathbf{R}^N) \to \mathbf{R} \tag{12}$$

*is bilinear and continuous. We denote by $h$ the map (12).*

(ii) *Denote by $H(\Omega)$ the vector distribution in $(\mathcal{D}(D,\mathbf{R}^N) \otimes \mathcal{D}(D,\mathbf{R}^N))'$ associated with $h$:*

$$d^2J(\Omega;V;W) = \langle H(\Omega), V \otimes W\rangle = h(V,W), \tag{13}$$

*where $V \otimes W$ is the tensor product of $V$ and $W$ defined as*

$$(V \otimes W)_{ij}(x,y) = V_i(x)W_j(y), \quad 1 \le i,\ j \le N, \tag{14}$$

*and $V_i(x)$ (resp. $W_j(y)$) is the i-th (resp. j-th) component of the vector $V$ (resp. $W$) (cf.* L. SCHWARTZ *[2]'s kernel theorem and* GELFAND-VILENKIN *[1]). $H(\Omega)$ will be called the* Shape Hessian *of $J$* at *$\Omega$.*

(iii) *When there exists a finite integer $\ell \ge 0$ such that $H(\Omega)$ is continuous for the $\mathcal{D}^\ell(D,\mathbf{R}^N) \otimes \mathcal{D}^\ell(D,\mathbf{R}^N)$-topology, we say that $H(\Omega)$ is of* order *$\ell$.* $\square$

THEOREM 4.2. *Let $\Omega$ be a domain with boundary $\Gamma$ in the open hold-all $D$ of* $\mathbf{R}^N$ and *assume that $J$ is twice shape differentiable.*

(i) *The vector distribution $H(\Omega)$ has support in*

$$(\Gamma \cap D) \times (\Gamma \cap D).$$

(ii) *If $\Omega$ is an open or closed domain in $D$ and $H(\Omega)$ is of order $\ell \geq 0$, then there exists a continuous bilinear form*

$$[h] : (\mathcal{D}_D^\ell / D_\Gamma^\ell) \times (\mathcal{D}_D^\ell / L_\Omega^\ell) \to \mathbf{R} \tag{15}$$

*such that for all $[V]$ in $\mathcal{D}_D^\ell / D_\Gamma^\ell$ and $[W]$ in $\mathcal{D}_D^\ell / L_\Omega^\ell$*

$$d^2 J(\Omega; V; W) = [h](q_D(V), q_L(W)) \tag{16}$$

*where $q_D : \mathcal{D}_D^\ell \to \mathcal{D}_D^\ell / D_\Gamma^\ell$ and $q_L : \mathcal{D}_D^\ell \to \mathcal{D}_D^\ell / L_\Omega^\ell$ are the canonical quotient surjections and*

$$D_\Gamma^\ell = \left\{ V \in \mathcal{D}^\ell(D, \mathbf{R}^N) : \partial^\alpha V = 0 \text{ on } \Gamma \cap D, \forall |\alpha| \leq \ell \right\}. \quad \square \tag{17}$$

The next and last result is the extension of Hadamard's structure theorem to second order Eulerian semiderivatives. We need the result established in the Corollary to Theorem 3.2. For a domain $\Omega$ with a boundary $\Gamma$ which is $C^{\ell+1}, \ell \geq 0$, the map

$$q_L(W) \mapsto p_L(q_L(W)) = \gamma_\Gamma(W) \bullet n : \mathcal{D}_\Omega^\ell / L_\Omega^\ell \to \mathcal{D}^\ell(\Gamma \cap D) \tag{18}$$

is a well-defined isomorphism. This will be used for the $V$–component. For the $W$–component we need the following lemma.

LEMMA 4.1. *Assume that the boundary $\Gamma$ of $\Omega$ is $C^{\ell+1}, \ell \geq 0$. Then the map*

$$q_D(V) \mapsto p_D(q_D(V)) = \gamma_\Gamma(V) : \mathcal{D}_D^\ell / D_\Gamma^\ell \to \mathcal{D}^\ell(\Gamma \cap D, \mathbf{R}^N) \tag{19}$$

*is a well-defined isomorphism where*

$$p_D : \mathcal{D}_D^\ell \to \mathcal{D}_D^\ell / D_\Gamma^\ell \tag{20}$$

*is the canonical surjection.* $\square$

REMARK 4.3. When $D = \mathbf{R}^N$ and $\Gamma$ is compact, $\mathcal{D}^\ell(\Gamma \cap D, \mathbf{R}^N) = \mathcal{D}^\ell(\Gamma, \mathbf{R}^N)$ coincides with the space of $\ell$–times continuously differentiables maps from $\Gamma$ to $\mathbf{R}^N$. $\square$

THEOREM 4.3. *Assume that the hypotheses of Theorem 4.2(ii) hold and that the boundary $\Gamma$ of the open domain $\Omega$ is $C^{\ell+1}$ for $\ell \geq 0$.*

(i) *The map*

$$\begin{cases} (v,w) \mapsto h_{D\times L}(v,w) = [h](p_D^{-1}v, p_L^{-1}w) \\ : \mathcal{D}^\ell(\Gamma_D,\mathbf{R}^N) \times \mathcal{D}^\ell(\Gamma_D) \to \mathbf{R} \end{cases} \tag{21}$$

*is bilinear and continuous and for all $V$ and $W$ in $\mathcal{D}^\ell(D,\mathbf{R}^N)$*

$$d^2J(\Omega;V;W) = h_{D\times L}(\gamma_\Gamma V, ((\gamma_\Gamma W)\bullet n)), \tag{22}$$

*where $\Gamma_D = \Gamma \cap D$.*

(ii) *This induces a vector distribution $h(\Gamma_D \otimes \Gamma_D)$ on $\mathcal{D}^\ell(\Gamma_D,\mathbf{R}^N)\otimes\mathcal{D}^\ell(\Gamma_D)$ of order $\ell$*

$$h(\Gamma_D\otimes\Gamma_D) : \mathcal{D}^\ell(\Gamma_D,\mathbf{R}^N)\otimes\mathcal{D}^\ell(\Gamma_D) \to \mathbf{R} \tag{23}$$

*such that for all $V$ and $W$ in $\mathcal{D}^\ell(D,\mathbf{R}^N)$*

$$\langle h(\Gamma_D\otimes\Gamma_D), (\gamma_\Gamma V)\otimes((\gamma_\Gamma W)\bullet n)\rangle = d^2J(\Omega;V;W), \tag{24}$$

*where $(\gamma_\Gamma V)\otimes((\gamma_\Gamma W)\bullet n)$ is defined as the tensor product*

$$((\gamma_\Gamma V)\otimes((\gamma_\Gamma W)\bullet n))_i(x,y) = (\gamma_\Gamma V_i)(x)((\gamma_\Gamma W)\bullet n)(y), \quad x,y\in\Gamma_D \tag{25}$$

*$V_i(x)$ is the i-th component of $V(x)$ and*

$$(\gamma_\Gamma(W)\bullet n)(y) = (\gamma_\Gamma W)(y)\bullet n(y), \quad \forall y\in\Gamma_D. \;\square \tag{26}$$

REMARK 4.4. Finally under the hypotheses of Theorem 4.1 and 4.3

$$\begin{aligned} d^2J(\Omega;V;W) = {} & \langle h(\Gamma_D\otimes\Gamma_D), (\gamma_\Gamma V(0))\otimes((\gamma_\Gamma W(0))\bullet n)\rangle \\ & + \langle (g(\Gamma_D), (\gamma_\Gamma \dot{V}(0))\bullet n\rangle \end{aligned} \tag{27}$$

for all $V$ in $\vec{V}_D^{m+1,\ell}$ and $W$ in $\vec{V}_D^{m,\ell}$. $\square$

EXAMPLE 4.1. Consider Example 3.1. Recall that for $V$ in $\mathcal{D}^1(D,\mathbf{R}^N)$

$$dJ(\Omega;V) = \int_\Omega \operatorname{div} V\,dx. \tag{28}$$

Now for $V$ in $\mathcal{D}^2(D,\mathbf{R}^N)$ and $W$ in $\mathcal{D}^1(D,\mathbf{R}^N)$

$$d^2J(\Omega;V;W) = \int_\Omega \operatorname{div}\left[(\operatorname{div}V)W\right]dx \tag{29}$$

and if $\Gamma$ is $C^1$

$$d^2J(\Omega;V;W)=\int_\Gamma \operatorname{div}V\; W\bullet n\quad d\Gamma \tag{30}$$

which is continuous for pairs $(V;W)\in\mathcal{D}^1(D,\mathbf{R}^N)\times\mathcal{D}^0(D,\mathbf{R}^N)$ or $\mathcal{D}^1(\Gamma,\mathbf{R}^N)\times\mathcal{D}^0(\Gamma,\mathbf{R}^N)$. □

Another interesting observation is that the shape Hessian is, in general, not symmetrical as can be seen from the following example in DELFOUR-ZOLÉSIO [2].

EXAMPLE 4.2. We use the functional (28) and expression (30) in Example 4.1. Choose the following two vector fields

$$V(x,y)=(1,0)\quad\text{and}\quad W(x,y)=(x^2/2,0).$$

Then

$$\operatorname{div}V=0,\quad\text{and}\quad W|_\Gamma=x=\cos\theta$$

and

$$V\bullet n=n_x=\cos\theta \text{ on } \Gamma.$$

As a result $d^2J(\Omega;V;W)=0$ and

$$d^2J(\Omega;W;V)=\int_\Gamma \operatorname{div}W\;(V\bullet n)d\Gamma=\int_0^{2\pi}\cos^2\theta d\theta>0.\quad\square$$

### 4.3. Comparison with methods of perturbation of the identity.

At this juncture it is instructive to compare first and second order Eulerian semiderivatives obtained by the Velocity (Speed) Method with those obtained by first and second order perturbations of the identity: that is, when the transformations $T_t$ are specified a priori by

$$T_t(X)=X+tU(X)+\frac{t^2}{2}A(X),\; X\in\mathbf{R}^N, \tag{31}$$

where $U$ and $A$ are transformations of $\mathbf{R}^N$ verifying the hypotheses of Theorem 2.2. The transformation $T_t$ in (31) is a *second order* perturbation when $A\neq 0$ and a *first order* perturbation when $A=0$. According to Theorem 2.2, first and second order Eulerian semiderivatives associated with (31) can be equivalently obtained by applying the Velocity (Speed) Method to the time-varying velocity fields $V_{UA}$ given by (2.9) and

$$dJ(\Omega;V_{UA})=dJ(\Omega;V_{UA}(0))=dJ(\Omega;U) \tag{32}$$

where we have used Remark 2.1 which says that

$$V_{UA}(0)=U \text{ and } \dot{V}_{UA}(0)=A-[DU]U. \tag{33}$$

Similarly if $V_{WB}$ is another velocity field corresponding to

$$T_t(X) = X + tW(X) + \frac{t^2}{2}B(X), \ X \in \mathbf{R}^N, \tag{34}$$

where $W$ and $B$ verify the hypotheses of Theorem 2.2, then

$$d^2J(\Omega; V_{UA}; V_{WB}) = d^2J(\Omega; V_{UA}(0); V_{WB}(0)) + dJ(\Omega; \dot{V}_{UA}(0)) \tag{35}$$

and

$$d^2J(\Omega; V_{UA}; V_{WB}) = d^2J(\Omega; U; W) + dJ(\Omega; A - [DU]U). \tag{36}$$

Expressions (32) and (35) are to be compared with the following expressions obtained by the Velocity (Speed) Method for two time-invariant vector fields $U$ and $W$

$$dJ(\Omega; U) \text{ and } d^2J(\Omega; U; W). \tag{37}$$

For the Shape gradient the two expressions coincide; for the Shape Hessian we recognize the bilinear term in (36) and (37) but the two expressions differ by the term

$$dJ(\Omega; A - [DU]U). \tag{38}$$

Even for a first order perturbation ($A = 0$), we have a quadratic term in $U$. This situation is analogous to the classical problem of defining second order derivatives on a manifold. The term (38) would correspond to the connexion while the bilinear term $d^2J(\Omega; V; W)$ would be the candidate for the *canonical* second order shape derivative. In this context we shall refer to the corresponding distribution $H(\Omega)$ as the *canonical Shape Hessian*. All other second order shape derivatives will be obtained from $H(\Omega)$ by adding the gradient term $G(\Omega)$ acting as the appropriate acceleration field (connexion).

REMARK 4.5. The method of perturbation of the identity can be made *more canonical* by using the following family of transformations

$$T_t(X) = X + tU(X) + \frac{t^2}{2}(A + [DU]U) \tag{39}$$

which yields

$$dJ(\Omega; U) \text{ for the gradient} \tag{40}$$

and

$$d^2J(\Omega; U; W) + dJ(\Omega; A) \text{ for the Hessian,} \tag{41}$$

where for a first order perturbation ($A = 0$) the second term disappears.□

REMARK 4.6. When $\Omega^*$ is an appropriately smooth domain which minimizes a twice shape differentiable functional $J(\Omega)$ without constraints on $\Omega$, the classical necessary conditions would be (at least formally)

$$dJ(\Omega^*; V) = 0, \ \forall V, \tag{42}$$

$$d^2J(\Omega^*; W; W) \geq 0, \ \forall W, \tag{43}$$

or equivalently for "smooth velocity fields $V$ and $W$"

$$dJ(\Omega^*; V(0)) = 0, \ \forall V \tag{44}$$

$$d^2J(\Omega^*; W(0); W(0)) + dJ(\Omega^*; \dot{V}(0)) \geq 0, \ \forall W. \tag{45}$$

But in view of (44), condition (45) reduces to the following condition on the canonical Shape Hessian

$$d^2J(\Omega^*; W(0); W(0)) \geq 0, \ \forall W. \square \tag{46}$$

# REFERENCES


S. AGMON, A. DOUGLIS, L. NIRENBERG [1], *Estimates near the boundary for solutions of elliptic partial differential equations satisfying general boundary conditions, I.*, Comm. Pure Appl. Math. **12** (1959), 623-727.

[2], *Estimates near the boundary for solutions of elliptic partial differential equations satisfying general boundary conditions, II.*, Comm. Pure Appl. Math. **17** (1964), 35-92.

G. ARUMUGAM, O. PIRONNEAU [1], *On the problems of riblets as a drag reduction device*, Optimal Control Applications and Methods **10** (1989), 93-112.

J. P. AUBIN, A. CELLINA [1], "Differential inclusions," Springer–Verlag, Berlin, 1984.

J. P. AUBIN, H. FRANKOWSKA [1], "Set-valued analysis," Birkäuser, Basel, Berlin, 1990.

V.M. BABIČ [1], *Sur le prolongement des fonctions (in Russian)*, Uspechi Mat. Nauk **8** (1953), 111 - 113.

J. CÉA [1], *Problems of Shape Optimal Design*, in "Optimization of Distributed Parameter Structures,vol II," E.J. Haug and J. Céa, eds., Sijhoff and Noordhoff, Alphen aan den Rijn, The Netherlands, 1981, pp. 1005-1048.

[2], *Numerical Methods of Shape Optimal Design*, in "Optimization of Distributed Parameter Structures,vol II," E.J. Haug and J. Céa, eds., Sijhoff and Noordhoff, Alphen aan den Rijn, The Netherlands, 1981, pp. 1049-1087.

M.C. DELFOUR, J.P. ZOLÉSIO [1], *Shape Sensitivity Analysis via MinMax Differentiability*, SIAM J. on Control and Optimization **26** (1988), 834-862.

[2], *Anatomy of the shape Hessian*, Annali di Matematica Pura et Applicata (to appear).

[3], *Computation of the shape Hessian by a Lagrangian method*, in "Fifth Symp. on Control of Distributed Parameter Systems," A. El Jai and M. Amouroux, eds., Pergamon Press, to appear, pp. 85 - 90.

[4], "Shape Hessian by the Velocity Method: a Lagrangian approach," Proc. CONCOM Conference, Montpellier, France, January 1989, Springer Verlag (to appear).

[5], *Structure of Shape Derivatives for Nonsmooth Domains*, CRM Report 1669, Université de Montréal (April 1990).

N. FUJII [1], *Domain optimization problems with a boundary value problem as a constraint*, in "Control of Distributed Parameter Systems," Pergamon Press, Oxford, New York, 1986, pp. 5-9.

[2], *Second variation and its application in a domain optimization problem*, in "Control of Distributed Parameter Systems 1986," Pergamon Press, Oxford, New York, 1986, pp. 431-436.

M. GUELFAND, N.Y. VILENKIN [1], "Les distributions, Applications de l'analyse harmonique(trad. par G. Rideau)," Dunod, Paris, 1967.

J. HADAMARD [1], *Mémoire sur le problème d'analyse relatif à l'équilibre des plaques élastiques encastrées*, in "Œuvres de J. Hadamard,vol II," (original reference: Mem. Sav. Etrang. 33 (1907), mémoire couronné par l'Académie des Sciences)., C.N.R.S., Paris, 1968, pp. 515-641.

M. NAGUMO [1], *Über die Loge der Integralkurven gewöhnlicher Differentialgleichungen*, Proc. Phys. Math. Soc. Japan **24** (1942), 551–559.

J. NEČAS [1], "Les méthodes directes en théorie des équations elliptiques," Masson (Paris) et Academia (Pragues), 1967.

L. SCHWARTZ [1] "Théorie des distributions," Hermann, Paris, 1966.

[2], *Théorie des noyaux*, in "Proccedings of the International Congress of Mathematicians, Vol I," 1950, pp. 220-230.

J. SIMON [1], *Second variations for domain optimization problems*, "Control of Distributed Parameter Systems (Proc. 4th Int. Conf. in Vorau)," Birkhäuser Verlag, July 1988 (to appear).

J. P. ZOLÉSIO [1], "Identification de domaines par déformation, Thèse de doctorat d'état," Université de Nice, France, 1979.

[2], *The Material Derivative (or Speed) Method for Shape Optimization*, in "Optimization of Distributed Parameter Structures,vol II," E.J. Haug and J. Céa, eds., Sijhofff and Nordhoff, Alphen aan den Rijn, 1981, pp. 1089-1151.

A PROBLEM OF EXACT CONTROLLABILITY OF DISTRIBUTED SYSTEM :

BOUNDARY CONTROL OBTAINED AS LIMIT OF INTERNAL CONTROL.

Caroline Fabre *

INTRODUCTION - We consider a problem of exact controllability of the following model :

$$(P)\quad \begin{cases} u'' + \Delta^2 u = h \\ u(0) = u^0 \ , \ u'(0) = u^1 \ ; \ u = \dfrac{\partial u}{\partial \nu} = 0 \text{ on } \Sigma \end{cases},$$

where $\Omega$ is a bounded domain of $\mathbb{R}^N$ with a $C^2$-boundary $\Gamma$ and $\Sigma = \Gamma \times (0, T)$.

When N = 2, this system models the motion of a vibrating plate, in a very simplified way. Enrike Zuazua has solved, using J.L.Lions' H.U.M, the exact controllability problem of this system when the control is distributed and acts on an $\varepsilon$ - neighborhood of a suitable part of the boundary (see [4]). We present here, a study of the passage to the limit when $\varepsilon \to 0$. We prove that in one dimension of space, (that is, for the beams' problem ), we obtain at the limit the boundary control given by H.U.M which acts on the normal derivative. In space dimension > 1, the question is still open but to point out the difficulty we will state the problem in the general case.

We could consider other boundary conditions and for example $u = \Delta u = 0$ on the boundary.

We also have similar problems for other equations and, for example, one can refer to [1] and [2] concerning the wave equation . For the Schroedinger equation, we already have some results but they are not complete.

PRESENTATION OF THE PROBLEM -Let $\Omega$ be a bounded open set in $\mathbb{R}^N$ with a $C^2$ - boundary $\Gamma$, and let $\nu(y)$ be the unit exterior normal at a point y of $\Gamma$.

Let $\Gamma^0$ be a subset of $\Gamma$. We will say that $\Gamma^0$ satisfy (1.1) if there exists $x^0$ in $\mathbb{R}^N$ with $\Gamma^0 = \{y \in \Gamma \,/\, (y - x^0).\nu(y) > 0\}$.

For T > 0, we write $\Sigma = \Gamma \times (0, T)$, $\Sigma^0 = \Gamma^0 \times (0, T)$, $\omega_\varepsilon = \Omega \cap O_\varepsilon$ where $O_\varepsilon = \bigcup \left(B(x, \varepsilon) \,/\, x \in \Gamma^0\right)$ and $Q_\varepsilon = \omega_\varepsilon \times ]\,0, T[$ . If $\Gamma^0$ satisfy (1.1), for $y^0 \in H_0^2(\Omega)$ and $y^1 \in L^2(\Omega)$, by J.L.Lions ' H.U.M, applied to this problem by E.Zuazua ( see [4]), there exists a control $v_\varepsilon \in L^2(Q)$ such that the solution of

$$\begin{cases} \psi_\varepsilon'' + \Delta^2 \psi_\varepsilon = v_\varepsilon \, \chi_{Q_\varepsilon} \\ \psi_\varepsilon(0) = y^0 \ , \ \psi_\varepsilon'(0) = y^1 \ ; \ \psi_\varepsilon = \dfrac{\partial \psi_\varepsilon}{\partial \nu} = 0 \text{ on } \Sigma \end{cases},$$

where $\chi_{Q_\varepsilon}$ denotes the charasteristic function of $Q_\varepsilon$, satisfies $\psi_\varepsilon(T) = \psi_\varepsilon'(T) = 0$. We recall some results given by the construction of this control $v_\varepsilon$ .

From a solution $\varphi$ of the homogeneous equation :

$$(H)\begin{cases}\varphi'' + \Delta^2\varphi = 0\\ \varphi^0 \in L^2(\Omega),\ \varphi^1 \in H^{-2}(\Omega)\ ;\ \varphi = \dfrac{\partial\varphi}{\partial\nu} = 0 \text{ on } \Sigma\end{cases},$$

we define $\psi$ as the solution of the backward equation,

$$(C)\begin{cases}\psi'' + \Delta^2\psi = \varphi\,\chi_{Q_\varepsilon}\\ \psi(T) = 0,\ \psi'(T) = 0\ ;\ \psi = \dfrac{\partial\psi}{\partial\nu} = 0 \text{ on } \Sigma\end{cases}.$$

Then we consider the operator $\Lambda_\varepsilon$ defined by $\Lambda_\varepsilon(\varphi^0, \varphi^1) = (\psi'(0), -\psi(0))$ from $L^2(\Omega)\times H^{-2}(\Omega)$ to $L^2(\Omega)\times H_0^2(\Omega)$. One can see that

$$\left\langle \Lambda_\varepsilon(\varphi^0, \varphi^1)\ ;\ (\psi'(0), -\psi(0)) \right\rangle = \langle \varphi^1, \psi(0)\rangle_{H^{-2}, H_0^2} - (\varphi^0, \psi'(0))_{L^2} = \int_{Q_\varepsilon} \varphi^2(x,t)\,dx\,dt\,.$$

Suppose that $\Lambda_\varepsilon$ is invertible and consider $\left(\widetilde{\varphi}_\varepsilon^0, \widetilde{\varphi}_\varepsilon^1\right) = \Lambda_\varepsilon^{-1}\left(y^1, -y^0\right)$. Denote by $\widetilde{\varphi}_\varepsilon$ the solution of (H) with initial data $\left(\widetilde{\varphi}_\varepsilon^0, \widetilde{\varphi}_\varepsilon^1\right)$ and by $\psi_\varepsilon$ the solution of (C) with $\widetilde{\varphi}_\varepsilon$ in the right hand side. Then by definition of $\Lambda_\varepsilon$, $\psi_\varepsilon(0) = y^0$ and $\psi_\varepsilon'(0) = y^1$, and $\psi_\varepsilon(T) = \psi_\varepsilon'(T) = 0$ so this solves the control problem when the control acts in an $\varepsilon$ - neighborhood of $\Gamma^0$. To prove that the operator $\Lambda_\varepsilon$ is invertible, we use the Lax-Milgram theorem. For this, we establish the equivalence between the $L^2(\Omega)\times H^{-2}(\Omega)$ norm of the initial data of solutions of (H) and the $L^2$ - norm in $Q_\varepsilon$ of these solutions. This has been done by E.Zuazua in [4] and more exactly he proved the two following results :

*Theorem 1 - There exists a constant C depending only of the geometry of* $\Omega$ *and* T *such that for every solution of* (H) *with initial data* $\left(\theta^0, \theta^1\right) \in H_0^2 \times L^2$, *we have*

$$\|\theta^0\|_{H_0^2}^2 + \|\theta^1\|_{L^2}^2 \le C\left[\frac{1}{\varepsilon^5}\int_{Q_\varepsilon}\left(\theta'^2(x,t) + \theta^2(x,t)\right)dx\,dt\right].$$

By a compactness argument, he deduces from this theorem the following one :

*Theorem 2 - There exists a constant* $C_\varepsilon$ *depending on* $\Omega$ ,T *and* $\varepsilon$ *such that for every solution of* (H) *with initial data* $\left(\varphi^0, \varphi^1\right) \in L^2 \times H^{-2}$, *we have*

$$\| \varphi^0 \|^2_{L^2} + \| \varphi^1 \|^2_{H^{-2}} \leq C_\varepsilon \int_{Q_\varepsilon} \varphi^2(x, t) \; dx \, dt.$$

REMARK - We have no longer any estimate on $C_\varepsilon$.

This proves that $\Lambda_\varepsilon$ is invertible and it allows us to control in an $\varepsilon$ - neighborhood of $\Gamma^0$.

The problem is now to find what happens when $\varepsilon \to 0$ that means to study the convergence of the problem

$$(CE_\varepsilon) \begin{cases} \psi_\varepsilon'' + \Delta^2 \psi_\varepsilon = \widetilde{\varphi}_\varepsilon \chi_{Q_\varepsilon} \\ \psi_\varepsilon(T) = 0 \,,\; \psi_\varepsilon'(T) = 0 \,;\; \psi_\varepsilon = \dfrac{\partial \psi_\varepsilon}{\partial \nu} = 0 \text{ on } \Sigma \end{cases},$$

where,

$$(H_\varepsilon) \begin{cases} \widetilde{\varphi}_\varepsilon'' + \Delta^2 \widetilde{\varphi}_\varepsilon = 0 \\ \widetilde{\varphi}^0_\varepsilon \in L^2(\Omega) \,,\; \widetilde{\varphi}^1_\varepsilon \in H^{-2}(\Omega) \,;\; \widetilde{\varphi}_\varepsilon = \dfrac{\partial \widetilde{\varphi}_\varepsilon}{\partial \nu} = 0 \text{ on } \Sigma \end{cases},$$

and,

$$\langle \widetilde{\varphi}^1_\varepsilon, y^0 \rangle_{H^{-2}, H^2_0} - (\widetilde{\varphi}^0_\varepsilon, y^1)_{L^2} = \int_{Q_\varepsilon} \widetilde{\varphi}^2_\varepsilon (x, t) \, dx \, dt.$$

To study this question, we need estimates on the right hand side of $(CE_\varepsilon)$ and for this we have to get estimates on the problem $(H_\varepsilon)$. We will see later on that the "good" functions to consider are not $\widetilde{\varphi}_\varepsilon$ (they are not bounded in $L^2(Q_\varepsilon)$ ) but $\varphi_\varepsilon = \varepsilon^5 \widetilde{\varphi}_\varepsilon$. Indeed, for N = 1, we will show that $(\varphi^0_\varepsilon , \varphi^1_\varepsilon)$ is bounded in $L^2(\Omega) \times H^{-2}(\Omega)$ and that

$$(1.2) \qquad \frac{1}{\varepsilon^5} \int_{Q_\varepsilon} \varphi^2_\varepsilon (x, t) \; dx \, dt \leq M,$$

where M is independent of $\varepsilon$.

To get this last estimate, we will need results concerning the solution of (1) and precisely :

(i) we'll have to describe exactly the behaviour near the boundary of solutions with finite energy,

(ii) we will see that in problems like $(H_\varepsilon)$, if the initial data are bounded and if the condition (1.2) is fulfilled, then the solution of the limit problem has a finite energy,

(iii) we 'll also have to study the convergence of linear forms of the type :

$$(u^0, u^1, h) \in H^2_0(\Omega) \times L^2(\Omega) \times L^1(0, T; L^2(\Omega)) \to \frac{1}{\varepsilon^5} \int_{Q_\varepsilon} \varphi_\varepsilon (x, t) \, u(x, t) \, dx \, dt ,$$

where u is defined by (P).

Having those 3 points, we can show that problem $(CE_\varepsilon)$ converges and exhibit the limit.

As (iii) uses again (i) and (ii), we begin by these two last points : at the moment we know how to solve (i) in one dimension of space but not for $N \geq 2$ and the difficulty is here in the sense that we could solve the problem if we had this result.

(i) BEHAVIOUR NEAR THE BOUNDARY OF THE SOLUTIONS OF (P) WITH FINITE ENERGY.

In this section $\Omega = [0,\ 1]$ and $\Gamma^0 = \Gamma = \{0\} \cup \{1\}$.
We are interested here in solutions of finite energy of (P) that is for $u^0 \in H_0^2(\Omega)$, $u^1 \in L^2(\Omega)$ and $h \in L^1(0, T; L^2(\Omega))$, we consider u solution of

$$(P)\begin{cases} u'' + \Delta^2 u = h \\ u(0) = u^0 \ ,\ u'(0) = u^1 \ ;\ u = \dfrac{\partial u}{\partial \nu} = 0 \text{ on } \Sigma \end{cases}.$$

We recall that J.L.Lions proved that, in this context, $\Delta u \in L^2(\Sigma)$ ( see [3]) and that the linear mapping $( u^0, u^1, h) \in H_0^2(\Omega) \times L^2(\Omega) \times L^1(0, T; L^2(\Omega)) \rightarrow \Delta u \in L^2(\Sigma)$ is continuous

We consider the expression $\frac{1}{\varepsilon^5}\int_{Q_\varepsilon} u^2(x, t)\, dx\, dt$ : it may look strange but as we will see, it appears naturally in our problem and on an other side, for very regular solutions "we have" (this is false of course but it gives an idea of the meaning of this expression ) :

for $x \in\ ]0, \varepsilon[$, $$u^2(x, t) \approx x^4 \left(\frac{\partial^2 u}{\partial x^2}(0, t)\right)^2$$

so that, $$\frac{1}{\varepsilon^5}\int_{Q_\varepsilon} u^2(x, t)\, dx\, dt \approx \frac{1}{5}\int_0^T \left(\Delta u(0, t)\right)^2 d\sigma\, dt\ .$$

This explains (with the regularity result of J.L.Lions ) the statement of the following theorem :

*Theorem 3 - There exists a constant* C *depending only on* T *such that for every solution of* (P) *with finite energy, we have :*

$$\frac{1}{\varepsilon^5}\int_{Q_\varepsilon} u^2(x, t)\, dx\, dt \le C\left(\| u^0 \|^2_{H_0^2} + \| u^1 \|^2_{L^2} + \| h \|^2_{L^1(0, T; L^2(\Omega))}\right) \cdot$$

Of course, we hope to have the same result in dimension 2 but there still remain some technical difficulties.
To have an idea of the proof, one can refer to [1] where we show a similar result for the wave equation.
This theorem is essential for our problem and we will use it many times.

(ii) REGULARITY OF THE LIMIT OF NON REGULAR PROBLEMS

In this section, $N \geq 1$ and $\Gamma^0$ is any subset of $\Gamma$ with measure > 0. Suppose that we have

$$(H_\varepsilon)\begin{cases}\varphi_\varepsilon'' + \Delta^2\varphi_\varepsilon = 0\\ \varphi_\varepsilon^0 \in L^2(\Omega)\,,\ \varphi_\varepsilon^1 \in H^{-2}(\Omega)\,;\ \varphi_\varepsilon = \dfrac{\partial\varphi_\varepsilon}{\partial\nu} = 0 \text{ on } \Sigma\end{cases},$$

with $(\varphi_\varepsilon^0, \varphi_\varepsilon^1)$ bounded in $L^2(\Omega) \times H^{-2}(\Omega)$.

Then one can say that the functions $\varphi_\varepsilon$ converge (after extraction of a subsequence) for the weak - * topology of $L^\infty(0, T\,;\, L^2(\Omega))$ to $\varphi$ solution of

$$(H)\begin{cases}\varphi'' + \Delta^2\varphi = 0\\ \varphi^0 \in L^2(\Omega)\,,\ \varphi^1 \in H^{-2}(\Omega)\,;\ \varphi = \dfrac{\partial\varphi}{\partial\nu} = 0 \text{ on } \Sigma\end{cases}$$

where $\varphi^0$ and $\varphi^1$ are the weak limit of $\varphi_\varepsilon^0$ and $\varphi_\varepsilon^1$. As we have already mentionned, our functions $\varphi_\varepsilon$ will also satisfy (1.2) and this condition gives the following regularity for the limit :

*Theorem 4 - If* $(\varphi_\varepsilon^0, \varphi_\varepsilon^1)$ *are bounded in* $L^2(\Omega) \times H^{-2}(\Omega)$ *and if furthermore*

$$\sup_\varepsilon \frac{1}{\varepsilon^5}\int_{Q_\varepsilon} \varphi_\varepsilon^2(x, t)\, dx\, dt \leq M,$$

*then the weak - * limit* $\varphi$ *of the functions* $\varphi_\varepsilon$ *satisfies* $\Delta\varphi \in L^2(\Sigma^0)$ . *If* $\Gamma^0$ *satisfies* (1.1) *then* $\varphi^0 \in H_0^2(\Omega)$, $\varphi^1 \in L^2(\Omega)$ *and* $\varphi$ *has a finite energy.*

REMARK - The last point is easily given by the following results of J.L.Lions and E.Zuazua (see [3] , [4]) :

*Theorem 5 - If* $\Gamma^0$ *satisfies* (1.1), *the* $H_0^2(\Omega) \times L^2(\Omega)$ *-norm of the initial data of* (H) *is equivalent to the* $L^2(\Sigma^0)$ *-norm of* $\Delta\varphi$ *on the boundary .*

(iii) STUDY OF THE LIMIT OF SOME LINEAR FORMS

In this section, $\Omega = [0, 1]$ and $\Gamma^0$ satisfies (1.1), which means, for example, that $\Gamma^0 = \{0\}$.
We introduce the following linear forms :

$$L_\varepsilon : H_0^2(\Omega) \times L^2(\Omega) \times L^1(0, T ; L^2(\Omega)) \to \mathbb{R}$$

$$(u^0, u^1, h) \to \frac{1}{\varepsilon^5} \int_{Q_\varepsilon} \varphi_\varepsilon(x, t)\, u(x, t)\, dx\, dt$$

where u is the solution of (P) and $\varphi_\varepsilon$ satisfies $\varphi_\varepsilon'' + \Delta^2 \varphi_\varepsilon = 0$, $\varphi_\varepsilon(0) = \varphi_\varepsilon^0 \in L^2(\Omega)$, $\varphi'_\varepsilon(0) = \varphi_\varepsilon^1 \in H^{-2}(\Omega)$, $\varphi_\varepsilon = \frac{\partial \varphi_\varepsilon}{\partial \nu} = 0$ on $\Sigma$ with initial data bounded in $L^2(\Omega) \times H^{-2}(\Omega)$ .

We want to study their convergence and to exhibit their limit. If we assume that $\varphi_\varepsilon$ satisfies (1.2), using theorem 3 and Holder 's inequality, one can easily see that $(L_\varepsilon)_\varepsilon$ is bounded in $H^{-2}(\Omega) \times L^2(\Omega) \times L^\infty(0, T ; L^2(\Omega))$ . So we already know that under these assumptions and after extraction of a subsequence, $(L_\varepsilon)_\varepsilon$ converges for the weak - * topology of $H^{-2}(\Omega) \times L^2(\Omega) \times L^\infty(0, T ; L^2(\Omega))$ to $L \in H^{-2}(\Omega) \times L^2(\Omega) \times L^\infty(0, T ; L^2(\Omega))$.
We will see that these hypotheses are those which emerge naturally from the control problem.
Having the convergence, it remains to exhibit the limit which is given by the following

*Theorem 6 - Under the above assumptions, the linear forms* $L_\varepsilon$ *converge for the weak - * topology of* $H^{-2}(\Omega) \times L^2(\Omega) \times L^\infty(0, T ; L^2(\Omega))$ *to* L *with*

$$L(u^0, u^1, h) = \frac{1}{5} \int_{\Sigma^0} \Delta\varphi\, \Delta u\, d\sigma\, dt$$

*where* $\varphi$ *is (after extraction of a subsequence) the limit, in a weak sense, of* $\varphi_\varepsilon$.

One can see that if we had theorem 3 in any dimension of space, we would also have this theorem and by the way all the results from the beginning of this paper.

We are now in the position to get the estimate on the controls $\tilde{\varphi}_\varepsilon$.

(iv) ESTIMATE ON THE CONTROLS

From now on, $\Omega = ]0, 1[$ and $\Gamma^0 = \{0\}$.
We begin by some recalls : $y^0$ and $y^1$ are fixed in $H_0^2$ and $L^2$. $\tilde{\varphi}_\varepsilon$ is solution of

$$(H_\varepsilon) \begin{cases} \tilde{\varphi}_\varepsilon'' + \Delta^2 \tilde{\varphi}_\varepsilon = 0 \\ \tilde{\varphi}_\varepsilon^0 \in L^2(\Omega)\,,\ \tilde{\varphi}_\varepsilon^1 \in H^{-2}(\Omega)\,;\ \tilde{\varphi}_\varepsilon = \dfrac{\partial \tilde{\varphi}_\varepsilon}{\partial \nu} = 0 \text{ on } \Sigma \end{cases},$$

with ,

$$(1.3)\quad \left\langle \Lambda_\varepsilon(\varphi^0,\varphi^1)\,;\,(\psi'(0)\,,-\psi(0))\right\rangle = \langle \widetilde{\varphi}^1_\varepsilon, y^0\rangle_{H^{-2},H_0^2} - (\widetilde{\varphi}^0_\varepsilon, y^1)_{L^2} = \int_{Q_\varepsilon} \widetilde{\varphi}^2_\varepsilon(x,t)\,dx\,dt\,.$$

To get an estimate on the controls, we have to describe how the constant $C_\varepsilon$ of Theorem 2 depends on $\varepsilon$. This is given by

*Theorem 7 - There exists a constant C depending only on T such that for every solution of* (H) *with initial data* $(\varphi^0, \varphi^1) \in L^2 \times H^{-2}$, *we have*

$$(1.4)\quad \|\varphi^0\|^2_{L^2} + \|\varphi^1\|^2_{H^{-2}} \leq C\left[\frac{1}{\varepsilon^5}\int_{Q_\varepsilon} \varphi^2(x,t)\,dx\,dt\right],$$

*or equivalently, there exists a constant C depending only on T such that for every solution of* (H) *with initial data* $(\theta^0, \theta^1) \in H_0^2 \times L^2$, *we have*

$$(1.5)\quad \|\theta^0\|^2_{H_0^2} + \|\theta^1\|^2_{L^2} \leq C\left[\frac{1}{\varepsilon^5}\int_{Q_\varepsilon} \theta'^2(x,t)\,dx\,dt\right].$$

This theorem uses (i), (ii) and (iii) and it would be true in any dimension of space if we had (i). The proof of this result will be given later on.

Assume Theorem 7, we write $\varphi_\varepsilon = \varepsilon^5\,\widetilde{\varphi}_\varepsilon$, $\varphi_\varepsilon$ is solution of an homogeneous beam 's equation, and using (1.3), one can show that $(\varphi^0_\varepsilon, \varphi^1_\varepsilon)$ is bounded in $L^2 \times H^{-2}$ and that the condition (1.2) is satisfied.

We can now study the limit of $(CE_\varepsilon)$ .

(v) PASSAGE TO THE LIMIT IN THE PROBLEM OF CONTROL

We rewrite the system $(CE_\varepsilon)$ as follows : $\begin{cases} \psi_\varepsilon'' + \Delta^2\psi_\varepsilon = \dfrac{1}{\varepsilon^5}\,\varphi_\varepsilon\,\chi_{Q_\varepsilon} \\ \psi_\varepsilon(T) = 0\,,\ \psi_\varepsilon'(T) = 0\,;\ \psi_\varepsilon = \dfrac{\partial\psi_\varepsilon}{\partial\nu} = 0 \text{ on } \Sigma \end{cases}$

By definition of $\varphi_\varepsilon$, we have $\psi_\varepsilon(0) = y^0$ and $\psi_\varepsilon'(0) = y^1$ which are fixed in $H_0^2$ and $L^2$.

The hypotheses on the functions $\varphi_\varepsilon$ are

$$\begin{cases} \varphi_\varepsilon'' + \Delta^2 \varphi_\varepsilon = 0 \\ \varphi_\varepsilon = \dfrac{\partial \varphi_\varepsilon}{\partial \nu} = 0 \text{ on } \Sigma \\ (\varphi_\varepsilon^0, \varphi_\varepsilon^1) \text{ is bounded in } L^2 \times H^{-2} \\ \sup\limits_\varepsilon \left( \dfrac{1}{\varepsilon^5} \displaystyle\int_{Q_\varepsilon} \varphi_\varepsilon^2(x,t)\, dx\, dt \right) < \infty \end{cases} .$$

To pass to the limit in ($CE_\varepsilon$), we consider $\psi_\varepsilon$ as defined by transposition from solutions of (P) with finite energy so we have for all $(u^0, u^1, h) \in H_0^2(\Omega) \times L^2(\Omega) \times L^1(0, T; L^2(\Omega))$,

$$\langle \psi_\varepsilon , h \rangle_{L^\infty(0,T;L^2(\Omega)); L^1(0,T;L^2(\Omega))} = -\langle y^0, u^1 \rangle_{L^2} + (y^1, u^0)_{H^{-2}; H_0^2} + \\ + \frac{1}{\varepsilon^5} \int_{Q_\varepsilon} \varphi_\varepsilon(x,t)\, u(x,t)\, dx\, dt$$

In the right hand side of this equality, we recognize exactly the linear forms $L_\varepsilon$ that we have already introduced. We have seen in theorem 6 that under the assumptions on $\varphi_\varepsilon$, $L_\varepsilon$ converges in a weak sense and we have exhibited the limit. This proves that $(\psi_\varepsilon)_\varepsilon$ is bounded in $L^\infty(0, T; L^2(\Omega))$ and if we note $\psi$ its limit for the weak - * topology of $L^\infty(0, T; L^2(\Omega))$ (after extraction of a subsequence), we have

*Theorem 8 - $\psi$ is solution of*

$$\begin{cases} \psi'' + \Delta^2 \psi = 0 \\ \psi(T) = 0 ,\ \psi'(T) = 0 \\ \psi = 0 \text{ on } \Sigma \\ \dfrac{\partial \psi}{\partial \nu} = \dfrac{1}{5} \Delta\varphi \text{ on } \Sigma^0 = \{0\} \times (0,T) \\ \dfrac{\partial \psi}{\partial \nu} = 0 \text{ on } \Sigma - \Sigma^0 = \{1\} \times (0,T) \\ \psi(0) = y^0 \text{ and } \psi'(0) = y^1 \end{cases} ,$$

*where the limit $\varphi$ of $\varphi_\varepsilon$ satisfies (by Theorem 4) $\Delta\varphi(0, t) \in L^2(0, T)$ and has a finite energy.*

*We obtain at the limit the boundary control given by H.U.M which acts on the normal derivative (see* [3] *for* T *big enough and* [4] *for all* T > 0*).*

REMARK - J.L Lions ' H.U.M gives a boundary control in $L^2(0, T)$ for initial data in $L^2(0, 1)$ and $H^{-2}(0, 1)$. To reach these spaces, we approach the initial data $y^0$ and $y^1$ in $L^2(0, 1)$ and $H^{-2}(0, 1)$ by sequences of regular initial data $y_n^0$ and $y_n^1$, then we control in $\varepsilon$-neighborhoods of $\Gamma^0$, and we pass to the limit first when $\varepsilon$ tends to zero and then when n tends to infinity.

PROOF OF THEOREM 7

We have N = 1 and $\Gamma^0$ satisfy (1.1).
For $(\varphi^0, \varphi^1) \in L^2 \times H^{-2}$ and $\varphi$ solution of (H), we will denote by

$$\theta(x, t) = \int_0^t \varphi(x, s)\, ds + \chi \quad \text{where} \quad \begin{cases} \Delta^2 \chi = -\varphi^1 \\ \chi \in H_0^2 \end{cases}.$$

$\theta$ is a solution of (H) with finite energy, $\theta' = \varphi$ and $\|\varphi^0\|_{L^2}^2 + \|\varphi^1\|_{H^{-2}}^2$ is equivalent to $\|\theta^0\|_{H_0^2}^2 + \|\theta^1\|_{L^2}^2$. This justifies the equivalence between (1.4) and (1.5).
We are going to prove (1.5) by a counter - argument : suppose that (1.5) is false, then there exists a sequence $(\varepsilon_n)_n$ of real non negative numbers which tends to zero and sequences $(\theta_n^0)_n \subset H_0^2$, $(\theta_n^1)_n \subset L^2$ such that

$$(1.6)\quad E^2(\theta_n) = \|\theta_n^0\|_{H_0^2}^2 + \|\theta_n^1\|_{L^2}^2 = 1$$

and

$$(1.7)\quad \frac{1}{\varepsilon_n^5} \int_0^T \int_0^{\varepsilon_n} \theta_n'^2(x, t)\, dx\, dt \to 0.$$

By (1.6), $\theta_n^0 \to \theta^0 \in H_0^2$ and $\theta_n^1 \to \theta^1 \in L^2$ respectively for the weak topologies of $H_0^2$ and $L^2$ (and after extraction of subsequences). By continuity with respect to the initial data, $\theta_n$ converges in $L^\infty(0, T; H_0^2(0, 1))$ weak - * to the solution $\theta$ of (H) with $(\theta^0, \theta^1)$ as initial data and $\theta_n$ converges in $L^\infty(0, T; L^2(0, 1))$ weak - * to $\theta'$.

*Lemma 1 - We have* $\theta^0 \in H^4 \cap H_0^2$ *and* $\theta^1 \in H_0^2$

*proof* - we apply theorem 4 to the functions $\theta'_n$ and we get $\frac{\partial \theta'}{\partial x}(0, t) \in L^2(0, T)$ so $\theta'$ has a finite energy .

*Lemma 2 -* $\theta^0 = 0$ *and* $\theta^1 = 0$ *hence* $\theta = 0$.

*Proof* - We consider the following linear continuous forms : (the continuity is given by Theorem 3)

$$\Lambda_n : H_0^2 \times L^2 \times L^1(0, T ; L^2) \to \mathbb{R}$$

$$(u^0, u^1, h) \in H_0^2 \times L^2 \times L^1(0, T ; L^2) \to \frac{1}{\varepsilon_n^5} \int_0^T \int_0^{\varepsilon_n} \theta'_n(x, t)\, u(x, t)\, dx\, dt$$

where u is solution of (P) .

From theorem 6, $\Lambda_n$ converges in a weak - * sense to

$$\Lambda(u^0, u^1, h) = \frac{1}{5} \int_0^T \frac{\partial \theta'}{\partial x}(0, t) \frac{\partial u}{\partial x}(0, t)\, dt.$$

But, from (1.7) and theorem 3,

$$\| \Lambda_n \| \leq c \left( \frac{1}{\varepsilon_n^5} \int_0^T \int_0^{\varepsilon_n} \theta'^2_n(x, t)\, dx\, dt \right)^{1/2} \to 0,$$

hence we have

$$\forall (u^0, u^1, h) \in H_0^2 \times L^2 \times L^1(0, T ; L^2)) \ , \qquad \int_0^T \frac{\partial \theta'}{\partial x}(0, t) \frac{\partial u}{\partial x}(0, t)\, dt = 0.$$

Taking $u = \theta'$ (which has a finite energy) we easily deduce that $\theta = 0$.

*Lemma 3 - There exists a constant c depending only on T such that*

$$\forall n \text{ and } \forall t \in (0, T) \qquad \frac{1}{\varepsilon_n^5} \int_0^{\varepsilon_n} \theta_n^2(x, t)\, dx \leq c\, E^2(\theta_n) = c\, .$$

*Proof* - We use again theorem 3 to get

$$\forall n \qquad \frac{1}{\varepsilon_n^5} \int_0^T \int_0^{\varepsilon_n} \theta_n^2(x, t)\, dx \leq c\, E^2(\theta_n) = c\, .$$

Now,

$$\theta_n(x, t) = \theta_n(x, s) + \int_s^t \theta'_n(x,r)\, dr$$

thus

$$\theta_n^2(x, t) = \theta_n^2(x, s) + \left( \int_s^t \theta'_n(x,r)\, dr \right)^2 + 2\, \theta_n(x, s) \int_s^t \theta'_n(x,r)\, dr\, .$$

Taking into account the sign of the middle term, we have for every $0 < \gamma < 1$,

$$\forall s, \forall n \quad \frac{T}{\varepsilon_n^5} \int_0^{\varepsilon_n} \theta_n^2(x, s)\, dx \leq c + \gamma T \frac{1}{\varepsilon_n^5} \int_0^{\varepsilon_n} \theta_n^2(x, s)\, dx + \frac{1}{\gamma} \frac{T^2}{\varepsilon_n^5} \int_0^T \int_0^{\varepsilon_n} \theta_n'^2(x, t)\, dx\, dt \quad ,$$

where c depends only on T. By (1.7), we deduce Lemma 3 .

We introduce $\psi_n$ obtained from $\theta_n$ by integration in time just as described above. $\psi_n^0 = \chi_n \in H^4 \cap H_0^2$ and $\psi_n^1 = \overset{0}{\theta}_n \in H_0^2$ and they both strongly converge to zero in $H_0^2$ and $L^2$ .

*Lemma 4 - There exists a real positive number p and a subsequence* $(n_k)_k$ *such that*

$$\forall t \in (0, T) \qquad \frac{1}{\varepsilon_{n_k}^5} \int_0^{\varepsilon_{n_k}} \psi_{n_k}^2(x, t)\, dx \underset{k\, \infty}{\longrightarrow} p$$

*( the subsequence is the same for all t) .*

*Proof* - By a similar method than in Lemma 3, one can easily show that there exists c depending only on T such that

$$\forall t \in (0, T), \forall n \qquad \frac{1}{\varepsilon_n^5} \int_0^{\varepsilon_n} \psi_n^2(x, t)\, dx \leq c .$$

For t = 0, we can find a subsequence $(n_k)_k$ such that $\frac{1}{\varepsilon_{n_k}^5} \int_0^{\varepsilon_{n_k}} \psi_{n_k}^2(x, 0)\, dx \underset{k\, \infty}{\longrightarrow} p$ where $p \geq 0$ .

Then, we consider the following linear forms :

$$L_{n_k}^t : H_0^2 \times L^2 \times L^1(0, T ; L^2) \to \mathbb{R}$$

$$(u^0, u^1, h) \in H_0^2 \times L^2 \times L^1(0, T ; L^2)) \to \frac{1}{\varepsilon_{n_k}^5} \int_0^t \int_0^{\varepsilon_{n_k}} \psi_{n_k}(x, s)\, u(x, s)\, dx\, ds ,$$

where u is solution of (P). From Lemma 3 and Theorem 3, these forms are continuous and they are uniformly bounded in t . As the initial data of $\psi_{n_k}$ converge strongly to zero in $H_0^2$ and $L^2$ , we have $L_{n_k}^t(\psi_{n_k}^0, \psi_{n_k}^1, 0) \to 0$. But, by integration by part in time, we obtain

$$2\, L_{n_k}^t(\psi_{n_k}^0, \psi_{n_k}^1, 0) = \frac{1}{\varepsilon_{n_k}^5} \int_0^{\varepsilon_{n_k}} \psi_{n_k}^2(x, t)\, dx - \frac{1}{\varepsilon_{n_k}^5} \int_0^{\varepsilon_{n_k}} \psi_{n_k}^2(x, 0)\, dx ,$$

which proves that $\frac{1}{\varepsilon_{n_k}^5} \int_0^{\varepsilon_{n_k}} \psi_{n_k}^2(x, t)\, dx$ also converges to p and this for all t .

*Lemma 5 - The constant* $p = 0$ .

*Proof* - Denote by $f_k(t) = \frac{1}{\varepsilon_{n_k}^5} \int_0^{\varepsilon_{n_k}} \psi_{n_k}^2(x, t)\, dx$

We have $f_k \to p$ everywhere , $\forall t$, $|f_k(t)| \leq c$ where c is independent of t, k. Then, by Lebesgue 's theorem, $\int_0^T f_k(t)\, dt \to p\, T$.

But, by another way,

$$\int_0^T f_k(t)\,dt = \frac{1}{\varepsilon_{n_k}^5}\int_0^T\int_0^{\varepsilon_{n_k}} \psi_{n_k}^2(x, s)\, dx\, ds \le c\left(\| \psi_{n_k}^0 \|_{H_0^1}^2 + \| \psi_{n_k}^1 \|_{L^2}^2\right) \to 0$$

This proves that $p = 0$ and Lemma 5 .

*End of the proof* - We consider $\Lambda_{n_k}(\psi_{n_k}^0, \psi_{n_k}^1, 0)$.

On one hand, we have $\left|\Lambda_{n_k}(\psi_{n_k}^0, \psi_{n_k}^1, 0)\right| \to 0$ by (1.7) and Lemma 4, and on an other hand, we have by integration by part in time :

$$\Lambda_{n_k}(\psi_{n_k}^0, \psi_{n_k}^1, 0) = \frac{1}{\varepsilon_{n_k}^5}\int_0^{\varepsilon_{n_k}} \psi_{n_k}(x, T)\, \theta_{n_k}(x, T)\, dx - \frac{1}{\varepsilon_{n_k}^5}\int_0^{\varepsilon_{n_k}} \psi_{n_k}(x, 0)\, \theta_{n_k}(x, 0)\, dx -$$

$$- \frac{1}{\varepsilon_{n_k}^5}\int_0^T\int_0^{\varepsilon_{n_k}} \theta_{n_k}^2(x, t)\, dx\, dt \; .$$

Using Lemma 3, 4, 5 and Holder 's inequality, one can see that the boundary terms in time tend to zero when k goes to infinity . We then deduce that

$$\frac{1}{\varepsilon_{n_k}^5}\int_0^T\int_0^{\varepsilon_{n_k}} \theta_{n_k}^2(x, t)\, dx\, dt \to 0$$

and we conclude with E. Zuazua 's Theorem 1 which contradicts (1.6).

Bibliography

[1] C.Fabre et J.P.Puel - Comportement au voisinage du bord des solutions de l'équation des ondes . *C.R. Acad. Sci. Paris,* tome 310, série 1, p 621 - 625, 1990.

[2] C.Fabre - Equation des ondes avec second membre singulier et application à la contrôlabilité exacte *C.R. Acad. Sci. Paris,* tome 310, série 1, p 813 - 818, 1990.

[3] J.L.Lions - Contrôlabilité exacte . Masson . 1988 .

[4] E.Zuazua - Exact controllability of distributed systems for arbitrarily small time. 26th IEEE CDC, Los Angeles, December 1987 .

* Département de mathématique et d'informatique de l'université d'Orléans
et C M A P, Ecole Polytechnique, 91128 Palaiseau Cedex.

# OPTIMAL CONTROL PROBLEMS FOR DISTRIBUTED PARAMETER SYSTEMS GOVERNED BY SEMILINEAR PARABOLIC EQUATIONS IN $L^1$ AND $L^\infty$ SPACES.*

H. O. Fattorini

University of California, Department of Mathematics
Los Angeles, California 90024, USA

**Abstract.** We consider the infinite dimensional nonlinear programming problem of minimizing a real valued function $f_0(u)$ defined in a metric space $V$ subject to the constraint $f(u) \in Y$, where $f(u)$ is defined in $V$ and takes values in a Banach space $E$ and $Y$ is a subset of $E$. We use an extension of a theorem of Kuhn – Tucker type due to Frankowska to obtain Pontryagin's maximum principle for distributed parameter systems described by semilinear parabolic equations in spaces of bounded measurable functions and spaces of regular measures, under general assumptions on the nonlinear term.

**1. Introduction.** Optimal control problems for semilinear distributed parameter systems were considered in [FA2] as particular cases of optimization problems for input-output maps and in [FF1], [FF2] as infinite dimensional nonlinear programming problems. To motivate what follows we explain briefly this treatment.

Consider a semilinear distributed parameter system modelled as a differential equation in a Banach space E,

$$y'(t) = Ay(t) + f(t, y(t), u(t)), \tag{1.1}$$

$$y(0) = y^o, \tag{1.2}$$

where the operator $A$ is the infinitesimal generator of a strongly continuous semigroup $S(t)$, $t \geq 0$. The initial condition $y^o$ is fixed. The control $u(t)$ takes values in a second Banach space $F$ and satisfies a constraint

$$u(t) \in U, \tag{1.3}$$

* This work was supported by the National Science Foundation under grant DMS–8701877

where the set U is a subset of F. The cost functional of the problem is

(1.4) $$y_o(t, u) = \int_0^t f_o(s, y(s, u), u(s))ds$$

where y(t, u) denotes the solution of (1.1)–(1.2) corresponding to the control u (precise conditions on the functions f and $f_o$ will be given later). The control problem is the usual one of minimizing the cost functional $y_o(t, u)$ among all u such that the trajectory y(t, u) satisfies the target condition

(1.5) $$y(t, u) \in Y$$

(Y a given subset of E). The endtime t of the control interval $0 \leq t \leq t$ may be free or fixed.

For the fixed endtime control problem, we consider the functions

(1.6) $$f(u) = y(t, u), \qquad f_o(u) = y_o(t, u)$$

(not to be confused with the functions f(t, y, u) and $f_o(t, y, u)$ in (1.1) and (1.4)) and we define V = V(0, t; U) as the space of all admissible controls (controls subject to measurability conditions to be precised later and satisfying (1.3)). The control problem is a particular case of the following

**Abstract nonlinear programming problem.** Let V be a metric space, E a Banach space, Y a subset of E, $f : V \to E$, $f_o : V \to \mathbf{R}$ = {real numbers}. Characterize the solutions of

(1.7) $$\text{minimize } f_o(u)$$

(1.8) $$\text{subject to } f(u) \in Y$$

The requirement that V be just a metric space (instead of, say, a Banach space) is due to the fact that the most convenient topology for V(0, t; U) is that given by the metric

(1.9) $$d(u, v) = \text{meas } \{t;\ u(t) \neq v(t)\}.$$

introduced by Ekeland [E] in his proof of Pontryagin's maximum principle. A similar formulation can be used for the free time problem taking V as a space of pairs (t, u) endowed with some convenient metric. For the time optimal problem a different model is more useful. Here we use the fact that if **t** is the optimal time and **u** is the optimal control then $y(\mathbf{t}, \mathbf{u}) \in Y$ whereas if $\{t_n\}$ is a sequence with $t_n < t$ we have $y(t_n, u) \notin Y$ for any admissible control; moreover, by continuity of trajectories, in particular of the optimal trajectory, we have $y(t_n, u) \to y(t, u) \in Y$ if $t_n \to t$. Thus, considering the functions

$$f_n(u) = y(t_n, u) \tag{1.10}$$

the time optimal problem is a particular case of the

**Abstract time optimal problem.** Let $\{V_n\}$ be a sequence of metric spaces, $E$ a Banach space, $Y$ a subset of $E$, $f_n: V_n \to E$ such that

$$f_n(V_n) \cap Y = \varnothing . \tag{1.11}$$

Characterize the sequences $\{u_n\}$ such that

$$\operatorname{dist}(f(u_n), Y) \to 0 . \tag{1.12}$$

A sequence $\{u_n\}$, $u_n \in V_n$ that satisfies (1.12) will be called an **optimal sequence** for the abstract time optimal problem.

Both optimal problems were studied in [FF1] and [FF2] in the Hilbert space setting, the results applied to optimal problems for distributed parameter systems, and in a less general formulation in [FA2], where also applications to boundary control systems are considered. There are several motivations to extend the results to the general Banach space setting. The Banach space generalization allows us to handle for instance target conditions of pointwise type $|y(t, u)(x) - y(x)| \leq C$. More importantly, it allows us to include in the abstract treatment problems with state constraints (see [FF3], which cannot be fitted in the Hilbert space setting. Problems with state constraints will not be treated in this paper.

Recently, Frankowska [FR2] [FR2] has generalized the results on the nonlinear programming problem (1.7)–(1.8) to the Banach space case. We state in this paper a generalization of this result under weaker conditions and apply it to parabolic distributed parameter systems. The results in this paper are relevant not only to solutions of (1.7)–(1.8) but to certain approximate or suboptimal solutions, which allows applications to relaxed solutions of optimal control problems [FA7]. Complete proofs of the results in this paper will appear elsewhere ([FA6]).

**2. The Kuhn – Tucker theorem.** We denote by $B(x, r)$ the ball of center $x$ and radius $r \geq 0$ in an arbitrary metric space. Let $u \in V$ and let $\{D_n\}$ be a sequence of subsets of $E$. We denote by $\liminf_{n \to \infty} D_n$ the set of all $y$ such that $\lim_{n \to \infty} \operatorname{dist}(y, D_n) = 0$.

Let $g$ be a function from $V$ into $E$, $\mathbf{u} \in V$. The vector $\xi \in E$ is a *variation* of $g$ at $\mathbf{u}$ if there exists a sequence $\{h_k\} \subset \mathbf{R}_+$ = {positive real numbers} with $h_k \to 0$ and a sequence $\{u_k\} \subset V$ with $d(u_k, \mathbf{u}) \leq h_k$ and such that

$$\frac{g(u_k) - g(u)}{h_k} \to \xi \quad \text{as} \quad k \to \infty .$$

The definition extends in an obvious way to the case where $g$ is defined only in a domain $D(g)$ of $V$. We denote by $\partial g(\mathbf{u})$ (the *variation set* of $g$ at $\mathbf{u}$) the set of all such $\xi$.

We consider the abstract nonlinear programming problem (1.7)–(1.8) for a function $f$ defined in a domain $D(f)$ and a function $f_o$ defined in a domain $D(f_o)$. We denote by $(f_o, f)\colon V \to \mathbf{R} \times E$ the function $(f_o, f)(u) = (f_o(u), f(u))$, whose domain $D((f_o, f)) = D(f_o) \cap D(f)$ we assume nonempty. The hypotheses are the following:

(a) The space $V$ is **complete.**

(b) The target set $Y$ is **closed.**

(c) For each $y \in E$ and every $\mu \in \mathbf{R}$ the functions

$\Phi(u, y) = |f(u) - y|$ if $u \in D(f)$, $\Phi(u, y) = +\infty$ if $u \notin D(f)$

$\Phi_o(u, y) = |f_o(u) - \mu|$ if $u \in D(f_o)$, $\Phi_o(u, y) = +\infty$ if $u \notin D(f_o)$

are **lower semicontinuous.**

The result below applies not only to solutions of (1.7)–(1.8) but also to certain types of approximate solutions. A sequence $\{u^n\} \subset V$ is called an **approximate** or **suboptimal** solution of (1.7)–(1.8) if

$$\limsup_{n \to \infty} f_o(u^n) \leq m, \quad \lim_{n \to \infty} \operatorname{dist}(f(u^n), Y) = 0, \tag{2.1}$$

where $m$ is the minimum of (1.7) subject to (1.8). A sequence $\{y^n\} \subset Y$ will be said to be **associated** with the suboptimal solution $\{u^n\}$ if and only if

$$|f(u^n) - y^n| = \varepsilon_n \to 0. \tag{2.2}$$

**Theorem 2.1** Let $\{u^n\} \subset V$ be a suboptimal solution of (1.7)–(1.8) and let $\{y^n\}$ be a sequence associated with $\{u^n\}$. Then there exists a sequence $\{\delta_n\} \subset \mathbf{R}_+$, $\delta_n \to 0$, a sequence $\{u^n\} \subset V$ and a sequence $\{y^n\} \subset Y$ with

(2.3) $$d(u^n, u^n) + |y^n - y^n| \leq \delta_n$$

and such that: for every sequence $\{D_n\}$ ($D_n$ a convex subset of $\partial(f_o, f)$) and every $\rho > 0$ there exists a sequence $\{\mu_n\} \subset \mathbf{R}$ and a sequence $\{z_n\} \subset E^*$ with

(2.4) $$\max(|\mu_n|, |z_n|) = 1, \ \mu_n \geq 0,$$

(2.5) $$\mu_n \eta^n + \langle z_n, \xi^n - w^n \rangle \geq -\delta_n(1 + \rho)$$

for $(\eta^n, \xi^n) \in \partial(f_o, f)(u^n)$ and $w^n \in C_Y(y^n) \cap B(0, \rho)$. ($C_Y(y)$ the tangent cone to $Y$ at $y$). Moreover, for every limit point $(\mu, z)$ of $\{(\mu_n, z_n)\}$ in the weak $(E^*, E)$ – topology of $E^*$ we have

(2.6) $$\mu\eta + \langle z, \xi \rangle \geq 0$$

for every $(\eta, \xi) \in \liminf_{n \to \infty} D_n$. Finally,

(2.7) $$z \in (\liminf_{n \to \infty} C_Y(y^n))^- ,$$

where $^-$ indicates negative polar.

Under the hypotheses of Theorem 2.1, the multiplier $(\mu, z)$ may be zero. The following condition ([FR1], [FR2]) prevents this.

**Theorem 2. 2** Let $\{D_n\}$ be the sequence of convex sets in Theorem 2.1 and assume there exists $\rho > 0$ and a compact set $Q$ such that the intersection of all the sets

(2.8) $$\overline{\Pi(D_n) - C_Y(y^n) \cap B(0, \rho) + Q}$$

contains an interior point, where $\Pi$ is the canonical projection of $\mathbf{R} \times E$ into $E$. and the bar indicates closure. Then $(\mu, z)$ in (2.6) is not zero.

If the space $E$ has Gâteaux differentiable norm off the origin we may take $D_n = \partial(f_o, f)(u^n)$ or its closed convex hull in Theorems 2.1 and 2.2. If the norm is Fréchet differentiable off the origin and $f(u^n) \to y$ then the vector $z$ belongs to the normal cone $N_Y(y) \subset E^*$.

The Kuhn – Tucker condition (2.6) for solutions of the nonlinear programming problem (1.7)–(1.8), as well as the condition on nontriviality of the multiplier were proved in [FR1] under the assumptions that the norm of $E$ is Gâteaux differentiable off zero and that $f$ is continuous, and in [FR2] for a general Banach space and f, $f_0$ Lipschitz continuous. The generalized version presented here and proved in [FA6], which applies as well to suboptimal solutions is a direct descendant of its Hilbert space version [FF1] [FF2] and the method of proof is similar. An ancestor of this Hilbert space version (where the setup and the hypotheses are much less general) was proved in [FA2]; the case where E is finite dimensional is closely related with the results of [EK1]. We note that allowing the maps f, $f_0$ to be defined only in subsets of V is decisive in the treatment of the point target problem ($Y = \{y\}$) for distributed parameter systems and boundary control systems.

The treatment of the abstract time optimal problem is similar. The result corresponding to Theorem 2.2 is

**Theorem 2.3** Let $\{u^n\}$ be an optimal sequence for the time optimal problem and let $\{y^n\} \subset Y$ be a sequence associated with $\{u^n\}$ (that is, satisfying (2.3)). Then there exists a sequence $\{u^n\}$, $u^n \in V_n$ and a sequence $\{y^n\} \subset Y$ with

$$d(u^n, u^n) + |y^n - y^n| \le \varepsilon_n^{1/2} \tag{2.9}$$

and such that: for every sequence $\{D_n\}$ ($D_n$ a convex subset of $\partial(f)$) and for every $\rho > 0$ there exists a sequence $\{z_n\} \subset E^*$ with

$$|z_n| = 1, \tag{2.10}$$

$$\langle z_n, \xi^n - w^n \rangle \ge -\varepsilon_n^{1/2}(1 + \rho) \tag{2.11}$$

for $\xi^n \in \partial(f)(u^n)$ and $w^n \in C_Y(y^n) \cap B(0, \rho)$. Moreover, for every limit point z of $\{z_n\}$ in the weak $(E^*, E)$ – topology of $E^*$ we have

$$\langle z, \xi \rangle \ge 0 \tag{2.12}$$

for every $(\eta, \xi) \in \liminf_{n \to \infty} D_n$. Finally, we have

$$z \in (\liminf_{n \to \infty} C_Y(y^n))^- . \tag{2.13}$$

**Theorem 2.4** Let $\{D_n\}$ be the sequence of convex sets in Theorem 2.3 and assume there exists $\rho > 0$ and a compact set Q such that the intersection of all the sets

$$\overline{D_n - C_Y(y^n) \cap B(0, \rho) + Q} \tag{2.14}$$

contains an interior point. Then the multiplier z in Theorem 2.3 is not zero.

If the space E has Gâteaux differentiable norm off the origin we may take $D_n = \partial(f)(u^n)$ or its closed convex hull in Theorems 2.3 and 2.4. If the norm is Fréchet differentiable and $f(u^n) \to y$ then z belongs to the normal cone $N_Y(y)$.

**3. Distributed parameter systems described by elliptic differential equations.** Let $\Omega$ be a bounded domain of class $C^{(2)}$ with boundary $\Gamma$ in m-dimensional Euclidean space $\mathbf{R}^m$ , and let A be a uniformly elliptic partial differential operator of class $C^{(2)}$,

$$Ay = \sum_{j=1}^{m}\sum_{k=1}^{m} \frac{\partial}{\partial x_j}\left(a_{jk}(x)\frac{\partial y}{\partial x_k}\right) + \sum_{j=1}^{m} b_j(x)\frac{\partial y}{\partial x_j} + c(x)y$$

with a boundary condition $\beta$ on $\Gamma$. This boundary condition may be either of Dirichlet type or of variational type $Dy = \gamma(x)y$ (D the conormal derivative). The operator A and the boundary condition $\beta$ generate a strongly continuous semigroup $S(t, A, \beta)$ in the space C(K) of continuous functions in K = closure of $\Omega$, the space C(K) endowed with the supremum norm (for the Dirichlet boundary condition the space C(K) is replaced by its subspace $C_o(K)$ consisting of all functions vanishing on $\Gamma$).

The control system is described by the semilinear initial value problem in the space E = C(K),

$$y'(t) = A(\beta)y(t) + f(t, y(t), u(t)) , \tag{3.1}$$

$$y(0) = y^o , \tag{3.2}$$

where $A(\beta)$ is the infinitesimal generator of $S(t, A, \beta)$. There are various reasons to consider the equation (3.1) in a the space C = C(K) rather than, say, in a space $L^p(\Omega)$. One is physical, to wit, the "natural" norm in which temperatures are measured in heat propagation processes is the supremum norm. Othe reason is purely mathematical: in C(K) we may dispense with growth conditions on the nonlinear term which would be unavoidable in a space $L^p(\Omega)$. On the other hand, all the results may be directly applied (in some cases with simplifications) to spaces $L^p(\Omega)$ with $1 < p < \infty$, for instance, to the Navier –Stokes equations in $L^2(\Omega)$. Another way to treat the $L^p(\Omega)$ case is to go from C(K) results to $L^p(\Omega)$ results via Sobolev imbeddings (see [FA3]).

Controls u(t) take values in a closed, bounded subset U (called the *control set*) of a Banach space F and are either strongly or weakly measurable (as precised below). The space V(0, t; U) of all (admissible) controls in an interval

$0 \le t \le \bar{t}$ will be equipped with the distance (1.9). In the assumptions below, $E_\alpha$ is the domain of the fractional power $(-\mathbf{A}(\beta))^\alpha$ with its graph norm $|y|_\alpha$ and the function $f(t, y, u)$ is defined in $[0, T] \times E_\alpha \times U$ for some $\alpha$, $0 \le \alpha < 1$. We denote by $C(0, \bar{t}; E_\alpha)$ the space of all continuous $E_\alpha$–valued functions defined in the interval $0 \le t \le \bar{t}$ with its usual supremum norm.

(a) The function $t \to f(t, y(t), u(t))$ is $(L^1(\Omega), L^\infty(\Omega))$ – weakly measurable for each $y \in C(0, \bar{t}; E_\alpha)$ and each admissible control $u(t)$. For every $C > 0$ there exists a constant $K = K(C)$ such that

$$|f(t, y, u)| \le K \quad (0 \le t \le T,\ y \in E_\alpha, |y|_\alpha \le C, u \in U) \tag{3.3}$$

(b) $f(t, y, u)$ has a Fréchet derivative $\partial_y f(t, y, u)$ with respect to $y$. The function $t \to \partial_y f(t, y(t), u(t))z$ is $(L^1(\Omega), L^\infty(\Omega))$ – weakly measurable for each $y \in C(0, \bar{t}; E_\alpha)$, each admissible control $u(t)$ and every $z$ in $C(K)$. For every $C > 0$ there exists a constant $L = L(C)$ such that

$$|\partial_y f(t, y, u)| \le L \quad (0 \le t \le T,\ y \in E_\alpha, |y|_\alpha \le C, u \in U) \tag{3.4}$$

We note that the requirement of weak $(L^1(\Omega), L^\infty(\Omega))$ measurability (rather than strong measurability) is motivated by the need to include such natural control terms as $f(t, y, u) = f(t, y) + u$. Here, the control functions are elements of the space $L^\infty(\Omega \times [0, T])$; these controls, considered as $L^\infty(\Omega)$ – valued functions are not strongly measurable but only $(L^1(\Omega), L^\infty(\Omega))$ – weakly measurable functions (we note that the present definition of weakly measurable is different from the standard one, where $L^1(\Omega)$ would be replaced by the dual of $L^\infty(\Omega)$). The treatment of (3.1)–(3.2) under these measurability assumptions is slightly nonstandard (see [FA2] for the linear case, [FA7] for the general case) but local existence and uniqueness of $C(K)$–valued solutions $y(t) = y(t, u)$ is proved by successive approximations in the usual way. In general, these solutions do not extend to the entire interval $0 \le t \le T$, but this causes no essential difficulty since we work in a neighborhood of an optimal control, assumed to exist. All that is needed is stability of the global existence property in the distance of the space $V(0, \bar{t}; U)$, which is covered by the following result.

**Lemma 3.1** Let $\mathbf{u}(t) \in V(0, \bar{t}; U)$ be a control whose trajectory $y(t, \mathbf{u})$ exists in $0 \le t \le \bar{t}$. Then there exists $\delta > 0$ such that if $u(t) \in V(0, \bar{t}; U)$ and $d(u, \mathbf{u}) \le \delta$ then the trajectory $y(t) = y(t, u)$ also exists in $0 \le t \le \bar{t}$. Moreover, the function

$$u \to y(\bar{t}, u)$$

from $V(0, t; U)$ into $C(0, t; E)$ is Hölder continuous with exponent $1 - \alpha$ in the ball $B(u, \delta)$.

**4. Construction of convex cones of (limits of) variations.** Construction of variations will be based on the "multiple spike perturbations" classical in control theory [P]. To define these perturbations, we consider a p–dimensional vector $s = (s_1, \ldots s_p)$ such that $0 < s_1 < \ldots < s_p < t$, a p–dimensional nonnegative vector $\alpha = (\alpha_1, \ldots \alpha_p)$, $\alpha_j \geq 0$, and a p– dimensional vector $v = (v_1, \ldots v_p)$ of elements of the control set $U$. Given a control $u \in V(0, t; U)$ and $h > 0$, define

$$
\begin{aligned}
&u_{s,\alpha,h,v}(t) = v_j \quad (s_j - \alpha_j h \leq t \leq s_j,\ j = 1,2,\ldots, p) \\
&u_{s,\alpha,h,v}(t) = u(t) \quad \text{elsewhere}
\end{aligned}
\tag{4.1}
$$

for $h$ so small that the spikes do not overlap.

**Lemma 4.1** There exists a set $e$ of full measure in $0 \leq t \leq t$ such that, for every $s_j \in e$ we have

$$
\lim_{h \to 0+} \frac{y(t, u_{s,\alpha,h,v}) - y(t, u)}{h} = \xi(t, s, \alpha, u, v) = \sum_{j=1}^{p} \alpha_j \xi(t, s_j, u, v_j), \tag{4.2}
$$

in $0 \leq t \leq t$, where $\xi(t, s, u, v)$ is defined by

$$
\begin{aligned}
&\xi(t, s, u, v) = 0 \quad (t < s) \\
&\xi(t, s, u, v) = S(t, s, u)\{f(t, y(s, u), v) - f(t, y(s, u), u(s))\} \quad (t \geq s),
\end{aligned}
\tag{4.3}
$$

$S(t, s, u)$ the solution operator of the linearized equation

$$
z'(s) = (A(\beta) + \partial_y f(t, y(s, u), u(s)))z(s) \quad (0 \leq s \leq t). \tag{4.4}
$$

Convergence in (4.2) is uniform outside of the intervals $(s_j + \delta, s_j + \delta)$ for any $\delta > 0$; moreover, $h^{-1}| y(t, u_{s,\alpha,h,v}) - y(t, u)|$ is bounded by a family of functions with equicontinuous integrals.

The assumptions on the kernel $f_0(t, y, u)$ of the cost functional (1.4) are

(a) $f_0(t, y, u)$ is measurable for each $y \in C(0, t; E_\alpha)$ and each $u \in V(0, t; E)$. For every $C > 0$ there exists a constant $K = K(C)$ such that

$$
|f_0(t, y, u)| \leq K \quad (0 \leq t \leq T,\ y \in E_\alpha, |y|_\alpha \leq C, u \in U) \tag{4.5}
$$

(b) $f_o(t, y, u)$ has a Fréchet derivative $\partial_y f_o(t, y, u)$ with respect to $y$. The function $t \to \partial_y f_o(t, y(t), u(t))$ is measurable for each $y \in C(0, t; E_\alpha)$ and each $u \in V(0, t; E)$. For every $C > 0$ there exists a constant $L = L(C)$ such that

$$|\partial_y f_o(t, y, u)| \le L \quad (0 \le t \le T,\ y \in E_\alpha, |y|_\alpha \le C, u \in U)$$

**Lemma 4.2** There exists a set $e$ of full measure in $0 \le t \le t$ such that, for every $s_j \in e$ we have

$$\text{(4.6)} \qquad \lim_{h \to 0+} \frac{y_o(t, u_{s,\alpha,h,v}) - y_o(t, u)}{h} = \xi_o(t, s, \alpha, u, v) = \sum_{j=1}^{p} \alpha_j \xi_o(t, s_j, u, v_j),$$

where

$$\text{(4.7)} \qquad \begin{aligned} &\xi_o(t, s, u, v) = 0 \quad (t < s) \\ &\xi_o(t, s, u, v) = f_o(s, y(s, u), v) - f_o(s, y(s, u), u(s)) + \\ &\quad + \int_s^t \langle \partial_y f_o(\sigma, y(\sigma, u), u(\sigma)), \xi(\sigma, s, u, v)\rangle d\sigma \quad (t \ge s) \end{aligned}$$

The sets $D_n$ required in Theorem 2.3 will consist of all elements of the form

$$\text{(4.8)} \qquad \xi(t, s, \alpha, u^n, v)) \in E$$

with arbitrary $\alpha$ and $\mathbf{v}$ and $\mathbf{s}$ with $s_j \in e$, where $e$ is the set in Lemma 4.1, while the sets $D_n$ in Theorem 2.2 consists of all the elements of the form

$$\text{(4.9)} \qquad (\xi_o(t, s, \alpha, u^n, v), \xi(t, s, \alpha, u^n, v)) \in \mathbf{R} \times E$$

with arbitrary $\alpha$ and $\mathbf{v}$ and $\mathbf{s}$ with $s_j \in e_n$, where $e_n$ is the set in Lemma 4.1 and Lemma 4.2 corresponding to $u^n$ (we may assume they are the same). Obviously, the $D_n$ are convex. To figure out the elements of $\liminf_{n\to\infty} D_n$ needed in Theorems 2.1 and 2.2 we use the following result:

**Lemma 4.3** Let $\{t_n\}$ be a sequence with $t_n < t$, $\{u^n\}$ a sequence in $V(0, t; U)$. Assume that

$$\Sigma\,(t - t_n) < \infty, \qquad \Sigma\, d_n(u^n,u) < \infty,$$

where $d_n$ is the distance (1.9) in the space $V(0, t_n; U)$. Then there exists a subset $e$ of $\cap\, e_n$ of full measure in $0 \le t \le t$ such that, for $s \in e$ we have

$$\xi(t_n, s, u^n, v) \to \xi(t, s, u, v), \quad \xi_o(t_n, s, u^n, v) \to \xi_o(t, s, u, v).$$

The proof of Lemma 4.3 is essentially the same as that of the corresponding Hilbert space result in [FF2]. It follows from Lemma 4.3 that elements of the form (4.8) (resp. 4.9) with $u^n = u$ and $s_j \in e$ belong to $\liminf_{n \to \infty} D_n$ and can thus be used in (2.12) (resp. in (2.6)).

**5. The maximum principle** Let $u(t)$ be an optimal control for the control problem with cost functional (1.4). We apply Theorem 2.1 with functions $f, f_o$ defined by (1.6) to the sequence $\{u^n\} = \{u\}$, which obviously is an approximate solution (with $\varepsilon_n = 0$) of the nonlinear programming problem (1.7)–(1.8). Theorem 2.1 produces a multiplier $(\mu, z) \in \mathbf{R} \times E$ satisfying inequality (2.6) for the convex cones of (limits of) variations constructed in the previous section. Manipulations similar to the ones in [FA2] then produce the maximum (or, rather, minimum) principle

(5.1) $$\mu f_o(s, y(s, u), u(s)) + \langle z(s), f(s, y(s, u), u(s))\rangle =$$

$$= \min_{v \in U} \{\mu f_o(s, y(s, u), v) + \langle z(s), f(s, y(s, u), v)\rangle\}$$

for $s$ almost everywhere in the control interval $0 \le t \le t$, where $z(s)$ is the solution of the adjoint backwards initial value, or final value problem:

(5.2) $$z'(s) = -(A(\beta)^* + \partial_y f(t, y(s, u), u(s))^*)z(s) - \mu\partial_y f_o(t, y(s, u), u(s)) \quad (0 \le s \le t).$$

(5.3) $$z(t) = z.$$

There are some technicalities in this final value problem. Note first that the adjoint $A(\beta)^*$ is an operator in the dual of the space $C(\Omega)$, which can be identified with the space $\Sigma(K)$ of all regular Borel measures defined in $K$, the closure of $\Omega$. In this space, $A(\beta)^*$ is not a semigroup generator, since it is not even densely defined. The treatment of (5.2) parallels closely that of (3.1), the role of the space $C(\Omega)$ played by $L^1(\Omega)$ and the role of $L^\infty(\Omega)$ played by $\Sigma(K)$; in particular, solutions of the backwards equation (5.2) belong to $L^1(\Omega)$ for $t < t$, thus both sides of (5.1) make sense.

The key question is whether nontriviality of the multiplier $(\mu, z)$ (and thus of the maximum principle (5.1)) can be insured. By virtue of (4.2) and of a limiting argument, the closure of the set $\Pi(D_n) \subset E$ will contain all elements of the form

$$\int_0^t S(t, s, u)\{f(s, y(s, u), v(s)) - f(s, y(s, u), u(s))\}ds\,,$$

for every $v \in V(0, t; U)$, that is, the reachable space of the linearized system

$$z'(s) = (A(\beta) + \partial_y f(t, y(s, u), u(s)))z(s) + g(t) \quad (0 \le s \le t) \tag{5.4}$$

$$z(0) = 0 \tag{5.5}$$

where the class of admissible controls consists of all $g$ of the form

$$g(t) = f(t, y(t, u), v(s)) - f(t, y(t, u), u(t)).$$

However, due to the smoothing properties of parabolic equations, this reachable set is very "thin" in the space $C(\Omega)$ or, for that matter, in any space $L^p(\Omega)$; typically, it will be contained in the domain of some fractional power of the infinitesimal generator $A(\beta)$ (see [FA1] for the linear case). Thus, one must rely on $C_Y(y^n) \cap B(0, \rho)$ to cause the sets in (2.8) to contain a common open set. A situation where this happens is that where the target set is a convex set with nonempty interior, (or more generally a set satisfying an open cone condition) or a $C^{(1)}$ manifold of finite codimension.

The treatment of the time optimal problem is similar: the maximum principle is (5.1) with $\mu = 0$.

**6. Final remarks.** In view of the observations in the previous section, the point target case $Y = \{y\}$ is not amenable to the treatment. It can be studied in other ways, roughly speaking working in the domain of $A(\beta)$ rather than in the whole space E. This has been done in [FA1] in the linear case and in [FA2] in the semilinear case (see also the references in [FA2]). Some of the results can be extended to the present setting, but the treatment of the time optimal problem only extends in spaces $L^p(\Omega)$ for $1 < p < \infty$, since the proof of the maximum principle depends on $L^p$ estimations on the derivative $y'(t)$ of the solution of the abstract differential equation $y'(t) = Ay(t) + f(t)$ ($A$ the infinitesimal generator of an analytic semigroup) in terms of the $L^p$ norm of $f(t)$. These estimations depend in turn on results for vector valued singular integrals [BO], [BU], [DV], which require conditions verified in $L^p$ spaces for $1 < p < \infty$ but not in such spaces as $C(K)$. The same observation holds for results concerning invariance of the Hamiltonian proved in [FA5] in Hilbert spaces.

The methods in this paper can be used with minor modifications for the treatment of the equation (3.1) in the space $L^1(\Omega)$; as for the equation (5.2), the role of $C(\Omega)$ is played by $L^1(\Omega)$ and the role of $L^\infty(\Omega)$ is played by $\Sigma(K)$. The nonlinear term may take values in $\Sigma(K)$ but the solutions take values in $L^1(\Omega)$. This sort of setting is natural when modelling diffusion processes and allows for control terms such as as the "travelling delta" $u(t)\delta(x - x(t))$.

## References


[AE] J-P. AUBIN and I. EKELAND, **Applied Nonlinear Analysis,** Wiley- Interscience, New York (1984)

[BO] J. BOURGAIN, Some remarks on Banach spaces in which martingale difference sequences are unconditional, Ark. Mat. 21 (1983) 163-168

[BU] D. L. BURKHOLDER, A geometric characterization of Banach spaces that implies the existence of certain singular integrals of Banach–space valued functions, *Conference on Harmonic Analysis in honor of A. Zygmund,* Wadsworth (1983) 270–286

[C] F. CLARKE, **Optimization and Nonsmooth Analysis,** Wiley - Interscience, New York (1983)

[DV] G. DORE and A. VENNI, On the closedness of the sum of two closed operators, Math. Zeitschrift 196 (1987) 189-201

[E] I. EKELAND, Nonconvex minimization problems, Bull. Amer. Math. Soc. 1 (NS) (1979) 443-474

[FA1] H. O. FATTORINI, The time optimal control problem in Banach spaces, Appl. Math. Optimization 1 (1974/75) 163-188

[FA2] H. O. FATTORINI, A unified theory of necessary conditions for nonlinear nonconvex control systems, Applied Math. Optim. 15 (1987) 141-185

[FA3] H. O. FATTORINI, Optimal control of nonlinear systems: convergence of suboptimal controls, I, Lecture Notes in Pure and Applied Mathematics vol. 108, Marcel Dekker, New York (1987) 159-199

[FA4] H. O. FATTORINI, Optimal control of nonlinear systems: convergence of suboptimal controls, II, Springer Lecture Notes in Control and Information Sciences vol. 97, Berlin (1987) 230-246

[FA5] H. O. FATTORINI, Constancy of the Hamiltonian in infinite dimensional systems, to appear in Proceedings of *4th. International Conference on Control of Distributed Parameter Systems,* Vorau, July 1988

[FA6] H. O. FATTORINI, Optimal control problems in Banach spaces, to appear.

[FA7] H. O. FATTORINI, Existence and the maximum principle for relaxed solutions of control problems in infinite dimensional spaces, to appear.

[FF1] H. O. FATTORINI and H. FRANKOWSKA, Necessary conditions for infinite dimensional control problems, Proceecedings of *8th. International Conference on Analysis and Optimization of Systems,* Antibes-Juan Les Pins, June 1988

[FF2] H. O. FATTORINI and H. FRANKOWSKA, Necessary conditions for infinite dimensional control problems, to appear in Mathematics of Control, Signals and Systems

[FF3] H. O. FATTORINI and H. FRANKOWSKA, Explicit convergence estimates for suboptimal controls I, II, to appear.

[FF4] H. O. FATTORINI AND H. FRANKOWSKA, Infinite dimensional control problems with state constraints, to appear in *Proceedings of IFIP-IIASA Conference on Modelling and Inverse Problems of Control for Distributed Parameter Systems,* Laxenburg, July 1989

[FR1] H. FRANKOWSKA, A general multiplier rule for infinite dimensional optimization problems with constraints, to appear.

[FR2] H. FRANKOWSKA, Some inverse mapping theorems, to appear.

[P] L. S. PONTRYAGIN, V. G. BOLTYANSKII, R. V. GAMKRELIDZE and E. F. MISCHENKO, **The Mathematical Theory of Optimal Processes** (Russian), Goztekhizdat, Moscow (1961)

# SHAPING THE REFERENCE INPUT RESPONSE OF LINEAR DISTRIBUTED PARAMETER SYSTEMS VIA OUTPUT FEEDBACK

Dieter Franke

Universität der Bundeswehr Hamburg
Fachbereich Elektrotechnik
Holstenhofweg 85, D - 2000 Hamburg 70

## 1. Introduction

In recent years a new representation of dynamic systems has been proposed by several authors. It has been restricted to lumped parameter systems so far and is characterized by generalized Fourier series expansions of the input, the state and the output, using a suitable orthogonal basis on a finite time interval. To this end, Paraskevopoulos et al. [1] and Vlassenbroeck et al. [2] cut off the time axis at some point $t = T$ and consider the system on the time interval $[0,T]$. In several papers the new representation is utilized for system analysis and identification [3], [4]. Franke [5] uses a different approach which avoids cutting off the time axis by introducing a nonlinearly distorted time coordinate

$$\tau = 1 - 2e^{-\alpha t}, \quad \alpha > 0, \tag{1}$$

thus mapping the interval $0 \leq t < \infty$ on the interval $-1 \leq \tau \leq 1$. The state equations

$$\dot{\underline{x}} = \underline{A}\,\underline{x} + \underline{B}\,\underline{u}, \quad \underline{x}(0) = \underline{x}_0, \tag{2}$$

$$\underline{y} = \underline{C}\,\underline{x}, \tag{3}$$

$\underline{x} \in \mathbb{R}^n$, $\underline{u} \in \mathbb{R}^p$, $\underline{y} \in \mathbb{R}^q$, $t \in [0,\infty)$, rewritten versus the $\tau$-coordinate take the form

$$\alpha(1-\tau)\hat{\underline{x}}' = \underline{A}\,\hat{\underline{x}} + \underline{B}\,\hat{\underline{u}}, \quad \hat{\underline{x}}(-1) = \underline{x}_0, \tag{4}$$

$$\hat{\underline{y}} = \underline{C}\,\hat{\underline{x}}, \tag{5}$$

where $\tau \in [-1, 1]$. Then by inserting the series expansions

$$\hat{\underline{u}}(\tau) = \sum_k \underline{u}_k^* p_k(\tau), \quad \hat{\underline{x}}(\tau) = \sum_k \underline{x}_k^* p_k(\tau), \quad \hat{\underline{y}}(\tau) = \sum_k \underline{y}_k^* p_k(\tau), \tag{6}$$

into (4), where $p_k(\tau)$ are Legendre polynomials, and by applying Galerkin's method, one obtains the following algebraic relations between Fourier coefficients (for the case $\underline{x}_o = \underline{O}$):

$$\underline{x}_k^* = [(k+1)\alpha\underline{I} - \underline{A}]^{-1} \cdot \left\{ (2k+1)\alpha(-1)^{k+1} \sum_{j=0}^{k-1} (-1)^j \underline{x}_j^* + \underline{B}\,\underline{u}_k^* \right\}, \tag{7}$$

$$\underline{y}_k^* = \underline{C}\,\underline{x}_k^*, \qquad k = 0, \ldots, N-1. \tag{8}$$

It can be shown that this Galerkin approximation minimizes the mean square state equation error for whatever choice of $\alpha > 0$ and N, provided the over-all system is stable.

Based on (7) and (8), a novel access to linear feedback control has been proposed in [6], [7] which is oriented at direct shaping of the reference input response by balancing the generalized Fourier coefficients in the closed loop. In the present paper the method will be extended to a class of linear distributed parameter systems (DPS).

## 2. A new representation of linear DPS

Let the infinite-dimensional system be given by its state equations

$$\partial\underline{x}(t,z)/\partial t = \underline{A}_z\underline{x}(t,z) + \underline{B}(z)\underline{u}(t), \tag{9}$$

$$\underline{y}(t) = \int_\Omega \underline{C}(z)\underline{x}(t,z)d\Omega, \tag{10}$$

where $\underline{u} \in \mathbb{R}^p$, $\underline{x} \in L_2(\Omega)$ and $\underline{y} \in \mathbb{R}^p$ are the control, the state and the output, respectively; $0 \le t < \infty$, $z \in \Omega$, where $\Omega$ is a simply connected finite spatial region. $\underline{A}_z$ is a linear matrix differential operator with respect to z, and $\underline{B}(z)$, $\underline{C}(z)$ are given space dependent matrices. Let the initial state be $\underline{x}(0,z) = \underline{x}_o(z)$ and the boundary conditions be formally homogeneous.

In the following we do not use Laplace transform methods. However, there will arise a relation to Green's function methods. Therefore, let $\underline{G}(s,z,\zeta)$ be the Green's matrix corresponding to $(s\underline{I} - \underline{A}_z)$. Hence, the complex valued input-output equations of the above system are

$$\underline{X}(s,z) = \int_\Omega \underline{G}(s,z,\zeta)\underline{B}(\zeta)d\Omega \cdot \underline{U}(s), \tag{11}$$

$$\underline{Y}(s) = \int_\Omega \underline{C}(z)\underline{X}(s,z)d\Omega. \tag{12}$$

Now (9) and (10) will be subject to time transformation (1) which yields

$$\alpha(1-\tau)\partial\hat{\underline{x}}(\tau,z)/\partial\tau = \underline{A}_z\hat{\underline{x}}(\tau,z) + \underline{B}(z)\hat{\underline{u}}(\tau), \tag{13}$$

$$\hat{\underline{y}}(\tau) = \int_\Omega \underline{C}(z)\hat{\underline{x}}(\tau,z)d\Omega, \qquad -1 \leq \tau \leq 1, \tag{14}$$

with formally homogeneous boundary conditions and initial state $\hat{\underline{x}}(-1, z) = \underline{x}_o(z)$.

Fourier series expansions are quite common in the analysis of DPS, however they are usually applied with respect to spatial coordinates. Here we use expansions with respect to $\tau$,

$$\hat{\underline{u}}(\tau) = \sum_k \underline{u}_k^* p_k(\tau) = \underline{U}^*\underline{p}(\tau), \tag{15}$$

$$\hat{\underline{x}}(\tau,z) = \sum_k \underline{x}_k^*(z) p_k(\tau) = \underline{X}^*(z)\underline{p}(\tau), \tag{16}$$

$$\hat{\underline{y}}(\tau) = \sum_k \underline{y}_k^* p_k(\tau) = \underline{Y}^*\underline{p}(\tau), \tag{17}$$

where again $p_k(\tau)$ are Legendre polynomials. By inserting these series into (13) and by minimizing the mean square equation error via Galerkin's method to be applied with respect to $\tau$, one obtains a boundary value problem for each Fourier coefficient $\underline{x}_k^*(z)$. The solution of this problem is quite analogous to (7), namely (for the case $\underline{x}_o(z) = \underline{0}$)

$$\underline{x}_k^*(z) = [(k+1)\alpha\underline{I} - \underline{A}_z]^{-1}\Big\{(2k+1)\alpha(-1)^{k+1} \cdot \sum_{j=0}^{k-1} (-1)^j \underline{x}_j^*(z) + \underline{B}(z)\underline{u}_k^*\Big\}, \tag{18}$$

$$\underline{y}_k^* = \int_\Omega \underline{C}(z)\underline{x}_k^*(z)d\Omega, \qquad k = 0, \ldots, N-1. \tag{19}$$

In (18), $(^{-1})$ denotes the inverse operator which in view of (11) can be rewritten using Green's matrix:

$$\underline{x}_k^*(z) = \int_\Omega \underline{G}[(k+1)\alpha, z, \zeta] \cdot \Big\{(2k+1)\alpha(-1)^{k+1} \cdot \sum_{j=0}^{k-1} (-1)^j \underline{x}_j^*(\zeta) + \underline{B}(\zeta)\underline{u}_k^*\Big\} d\Omega, \qquad k=0, \ldots, N-1. \tag{20}$$

However, in contrast to (11), all equations are real valued now. It should be remarked that of course $(k+1)\alpha$, $k = 0, \ldots, N-1$, are required to be in the resolvent set of operator $\underline{A}_z$. This can be met by suitable choice of $\alpha$.

It can be ovserved that eqs. (20) as well as eqs. (7) have a triangular structure which allows evaluation in a recursive manner. If for example, the control is given by its Fourier coefficients $\underline{u}_k^*$, then the $\underline{x}_k^*(z)$ can be obtained from (20) and the $\underline{y}_k^*$ from (19); and vice versa, if the output is prescribed by its Fourier coefficients $\underline{y}_k^*$, then the $\underline{u}_k^*$ and $\underline{x}_k^*(z)$ can also be computed from (19), (20) in a recursive manner.

An important question to be treated next is how many Fourier coefficients to be considered. To this end we define the relative degree of DPS.

## 3. Infinite-dimensional systems with finite relative degree

For the finite-dimensional system (2), (3) the relative degree $d_i$ with respect to output $y_i$ is defined as

$$d_i = \min\left\{k \,\middle|\, \underline{c}_i^T \underline{A}^{k-1} \underline{B} \neq \underline{0}^T, \quad k = 1, \ldots, n\right\}. \tag{21}$$

Now for the infinite-dimensional system (9), (10) assume that for some finite integer k

$$\int_\Omega \underline{c}_i^T(z) \underline{A}_z^{k-1} \underline{B}(z) d\Omega \neq \underline{0}^T.$$

Then the relative degree $d_i$ with respect to output $y_i$ will be defined as [8]

$$d_i = \min\left\{k \,\middle|\, \int_\Omega \underline{c}_i^T(z) \underline{A}_z^{k-1} \underline{B}(z) d\Omega \neq \underline{0}^T, \; k = 1, 2, \ldots\right\}. \tag{22}$$

The simple meaning of $d_i$ in (21) and (22) is that there is at least one component of $\underline{u}$ acting directly on the $d_i$-th derivative of $y_i(t)$.

For example, we have $d_i = 1$, if

$$\int_\Omega \underline{c}_i^T(z) \underline{b}_j(z) d\Omega \neq 0 \quad \text{for some j.}$$

Obviously, finite $d_i$ requires colocated or overlapping spatial supports for actuators and sensors.

Example: Euler-Bernoulli beam equation

$$\frac{\partial^2 x}{\partial t^2} + \frac{\partial^4 x}{\partial z^4} = u_1(t)\delta(z-z_1) + u_2(t)\delta(z-z_2), \quad 0 < z < 1, \tag{23}$$

with forces $u_1$ and $u_2$ acting pointwise at $z_1$ and $z_2$, respectively.

Boundary conditions:

$$x(t,0) = x(t,1) = 0, \tag{24}$$

$$\left.\frac{\partial^2 x}{\partial z^2}\right|_{z=0} = \left.\frac{\partial^2 x}{\partial z^2}\right|_{z=1} = 0. \tag{25}$$

Let the velocity $\partial x/\partial t$ be measured in colocated points:

$$y_1(t) = \left.\frac{\partial x}{\partial t}\right|_{z_1}, \quad y_2(t) = \left.\frac{\partial x}{\partial t}\right|_{z_2}. \tag{26}$$

Then by introducing the state variables

$$x_1(t,z) = \frac{\partial^2 x}{\partial z^2}, \quad x_2(t,z) = \frac{\partial x}{\partial t}, \tag{27}$$

the system matrices are

$$\underline{A}_z = \begin{bmatrix} 0 & \partial^2/\partial z^2 \\ -\partial^2/\partial z^2 & 0 \end{bmatrix}, \quad \underline{B}(z) = \begin{bmatrix} 0 & 0 \\ \delta(z-z_1) & \delta(z-z_2) \end{bmatrix},$$

$$\underline{C}(z) = \begin{bmatrix} 0 & \delta(z-z_1) \\ 0 & \delta(z-z_2) \end{bmatrix},$$

and therefore the relative degrees here are

$$\underline{d_1 = d_2 = 1.} \tag{28}$$

## 4. Controller design by balancing generalized Fourier coefficients

In the following we restrict ourselves to infinite-dimensional systems with finite relative degree. If the relative degree is finite, it turns out to be a small integer in most situations. This is favourable for the controller design by balancing a small number of Fourier coefficients in the closed loop.

Let the system (9), (10) be augmented by a linear feedback control law, in the simplest case

$$\underline{u}(t) = \underline{K}_o \underline{w}(t) - \underline{K}\, \underline{y}_M(t), \tag{29}$$

where

$$\underline{y}_M(t) = \int_\Omega \underline{C}_M(z)\underline{x}(t,z)d\Omega \in \mathbb{R}^q \tag{30}$$

is a vector of measured variables, $\underline{w}(t)$ is the reference input, and $\underline{K}$, $\underline{K}_o$ are constant matrices to be determined.

The design procedure aims at direct shaping of the reference input response. To this end the following steps are proposed:

(a) Rewrite the controller equations, similar to the plant equations, in terms of Fourier coefficients:

$$\underline{u}_k^* = \underline{K}_o\underline{w}_k^* - \underline{K}\,\underline{y}_{Mk}^*, \tag{31}$$

$$k = 0, \ldots, N-1.$$

$$\underline{y}_{Mk}^* = \int_\Omega \underline{C}_M(z)\underline{x}_k^*(z)d\Omega. \tag{32}$$

(b) Select reference input $\underline{w}(t)$, e.g. unit step function, hence $\underline{W}^*$.
(c) Select $\alpha > 0$.
(d) Prescribe $\underline{Y}^*$ by prescribing $\underline{y}(t)$. (Number of Fourier coefficients oriented at the relative degrees $d_i$, see [7] and exemplary discussion in the next section).
(e) Calculate $\underline{U}^*(\alpha)$ and $\underline{X}^*(z,\alpha)$ from plant equations (19), (20).
(f) Calculate $\underline{Y}_M^*(\alpha)$ from (32).
(g) Calculate controller parameters $\underline{K}_o$, $\underline{K}$ from (31).

It should be emphasized that the above procedure requires nothing else than solving linear equations, and this overcomes a main drawback of Riccati design or eigenvalue assignment.

It should also be pointed out that the design procedure does not necessarily imply stability, although it has a stabilizing tendency whenever $\underline{y}(t)$ is prescribed well-damped in step (d), see the example in the next section. In any case, stability should be examined in a final step, e.g. via Ljapunov's direct method.

## 5. Numerical example

As an example we consider the active damping of the Euler-Bernoulli beam, eqs. (23) - (28), by feedback.

Using the abbreviations

$$\dot{\underline{y}}(t) = \begin{bmatrix} \partial x/\partial t|_{z_1} \\ \partial x/\partial t|_{z_2} \end{bmatrix}, \quad \underline{y}(t) = \begin{bmatrix} x(t,z_1) \\ x(t,z_2) \end{bmatrix}, \tag{33}$$

the following alternative feedback laws will be discussed:

$$\text{I)} \quad \underline{u}(t) = \underline{K}_O\underline{w}(t) - \underline{K}\,\dot{\underline{y}}(t), \tag{34}$$

$$\text{II)} \quad \underline{u}(t) = \underline{K}_O\underline{w}(t) - \underline{K}_P\underline{y}(t) - \underline{K}_D\dot{\underline{y}}(t). \tag{35}$$

Eq. (35) can be regarded as a multivariable PD-Controller with feedback of both deflection and velocity.

The design objectives are

- Stabilization with guaranteed spillover prevention,
- Matching of a simple closed-loop transfer model including noninteraction.

## 5.1 Stability

It can be shown by Ljapunov's direct method:
Controller (34) stabilizes the beam equation asymptotically if

(i) $\varphi_i(z_k) \neq 0, \quad i = 1, 2, 3, \ldots$
$\quad k = 1, 2,$

where $\varphi_i(z)$ are the eigenfunctions of the beam equation, and

(ii) $\underline{K}$ is any symmetric and positive definite matrix.

For control law (35) to stabilize the beam equation asymptotically, condition (ii) has to be replaced by

(iii) $\underline{K}_P$ and $\underline{K}_D$ are any symmetric and positive definite matrices.

## 5.2 Model matching

According to (28) the relative degrees with respect to velocities $\dot{y}_1$ and $\dot{y}_2$ are $d_1 = d_2 = 1$, hence the relative degrees with respect to deflections $y_1$ and $y_2$ are $\tilde{d}_1 = \tilde{d}_2 = 2$. This motivates prescription of a simple and well damped response to unit step input

$$\underline{w} = \begin{bmatrix} 1 \\ 0 \end{bmatrix} : \qquad \hat{\underline{y}}(\tau) \overset{!}{=} \begin{bmatrix} 0.25(1+\tau)^2 \\ 0 \end{bmatrix}, \; -1 \le \tau \le 1. \tag{36}$$

The polynomial $0,25(1+\tau)^2$ is the simplest one which on the one hand meets the system's transient abilities, characterized by $\tilde{d}_1 = 2$, and on the other hand meets the steady state requirement

$$\hat{\underline{y}}(+1) = \underline{w}, \quad \text{hence } \underline{y}(t\to\infty) = \underline{w}. \tag{37}$$

By inserting the time transformation (1) into (36) one obtains the original function

$$\underline{y}(t) = \begin{bmatrix} (1 - e^{-\alpha t})^2 \\ 0 \end{bmatrix} \qquad 0 \le t < \infty. \tag{38}$$

From (38) it can be seen that the input-output behaviour to be matched is a second order lumped parameter model. Moreover, the time scaling parameter $\alpha$ to be selected has a very simple meaning: $(-\alpha)$ is the dominant pole of the finite-dimensional model to be matched.

Due to the second order polynomial, $\hat{\underline{y}}(\tau)$ contains only three Fourier coefficients:

$$\underline{y}_0^* = \begin{bmatrix} 1/3 \\ 0 \end{bmatrix}, \quad \underline{y}_1^* = \begin{bmatrix} 1/2 \\ 0 \end{bmatrix}, \quad \underline{y}_2^* = \begin{bmatrix} 1/6 \\ 0 \end{bmatrix}. \tag{39}$$

In the same way, if the reference input is selected as

$$\underline{w} = \begin{bmatrix} 0 \\ 1 \end{bmatrix},$$

the Fourier coefficients of the output are prescribed as

$$\underline{y}_0^* = \begin{bmatrix} 0 \\ 1/3 \end{bmatrix}, \quad \underline{y}_1^* = \begin{bmatrix} 0 \\ 1/2 \end{bmatrix}, \quad \underline{y}_2^* = \begin{bmatrix} 0 \\ 1/6 \end{bmatrix}. \tag{40}$$

This model includes noninteraction.

Balancing the Fourier coefficients (39), (40) in the closed loop yields a set of linear equations for the controller matrices (see step (g) in section 4). These equations are overdetermined, and therefore they are solved approximately in the least square sense.

5.3 Numerical results

The following points of measurement and control will be selected,

$$z_1 = 1 - \sqrt{2}/2 \approx 0.293, \qquad z_2 = \sqrt{2}/2 \approx 0.707, \tag{41}$$

which makes the problem symmetric with respect to z.

First, controller (34) will be computed by selecting $\alpha = 5$:

$$\underline{K}(5) = \begin{bmatrix} 57.1 & -46.6 \\ -46.6 & 57.1 \end{bmatrix} \quad \underline{K}_0(5) = \begin{bmatrix} 225.2 & -186.8 \\ -186.8 & 225.2 \end{bmatrix}.$$

Obviously $\underline{K}(5)$ is positive definite which guarantees asymptotic stability of the closed-loop system. Simulation results can be seen from Figures 1, 2 and 3. In Fig. 2 the response of the actual system is compared to the response of the model versus the $\tau$-coordinate. The

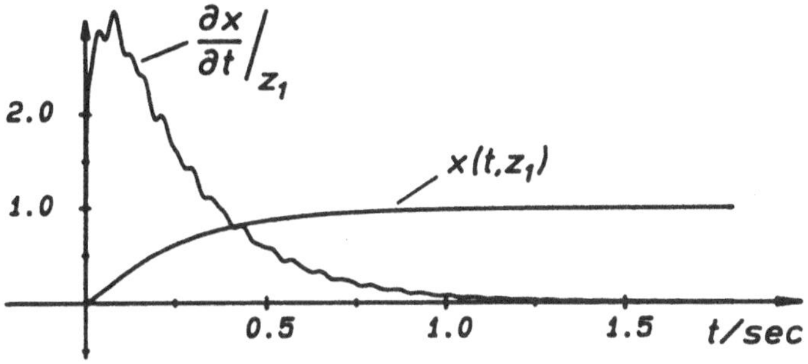


Fig. 1. Response to $\underline{w}^T = (1,0)$ at point $z_1$ with feedback law (34), $\alpha = 5$

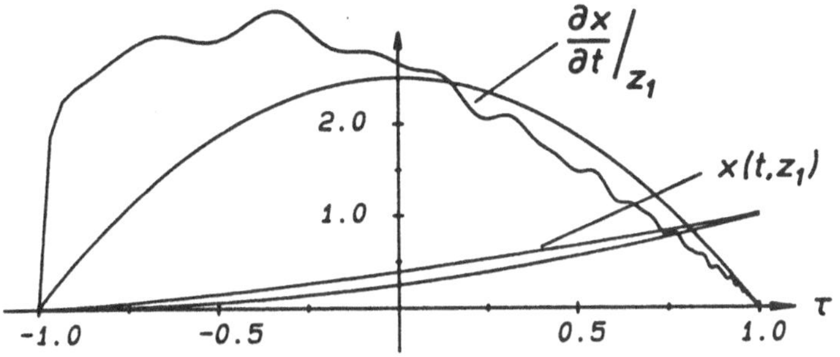


Fig. 2. Transformation of Fig. 1 on the $\tau$-coordinate

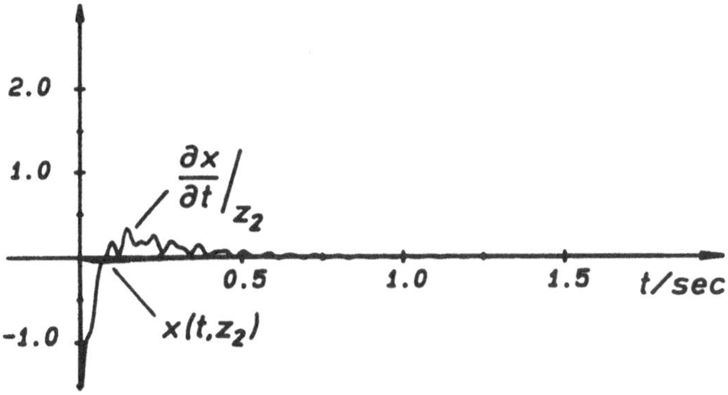


Fig. 3. Response to $\underline{w}^T = (1,0)$ at point $z_2$ with feedback law (34), $\alpha = 5$

error between model and actual system is due to the approximations made. From Fig. 3 it can be seen that the noninteraction is quite favourable.

Feedback law (35) is underlying the simulations in Figures 4, 5 and 6. The error between model and actual system is now smaller than in the previous case, because the controller has more parameters to meet the specified Fourier coefficients. Again, $z_1$ and $z_2$ have been selected according to (41), however $\alpha = 50$ now. The gain matrices ob-

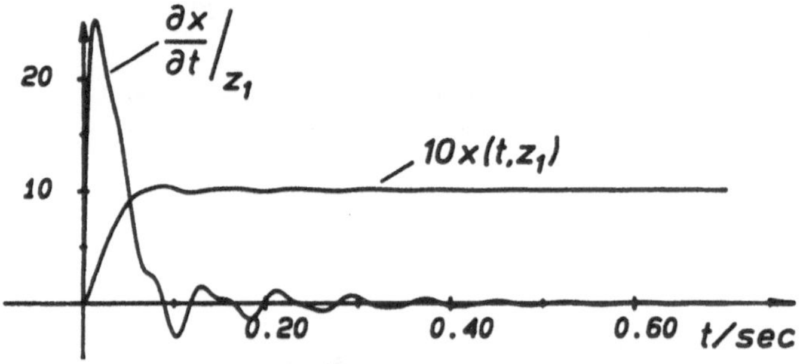


Fig. 4. Response to $\underline{w}^T = (1,0)$ at point $z_1$ with feedback law (35), $\alpha = 50$

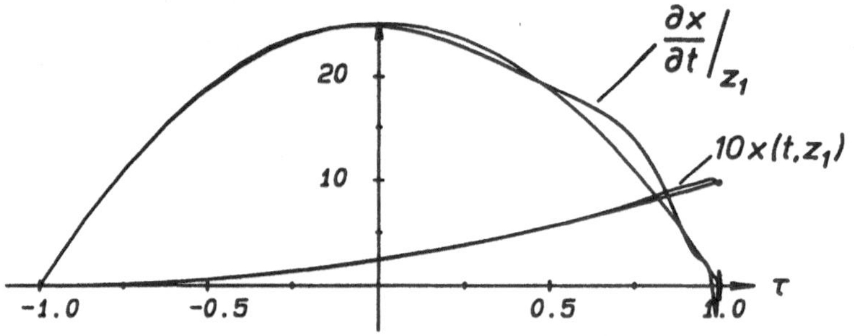


Fig. 5. Transformation of Fig. 4 on the $\tau$-coordinate

tained by balancing Fourier coefficients are

$$\underline{K}_P(50) = \begin{bmatrix} 1098 & 443 \\ 443 & 1098 \end{bmatrix}, \quad \underline{K}_D(50) = \begin{bmatrix} 34.9 & 7.7 \\ 7.7 & 34.9 \end{bmatrix},$$

$$\underline{K}_o(50) = \begin{bmatrix} 1324 & 257 \\ 257 & 1324 \end{bmatrix}.$$

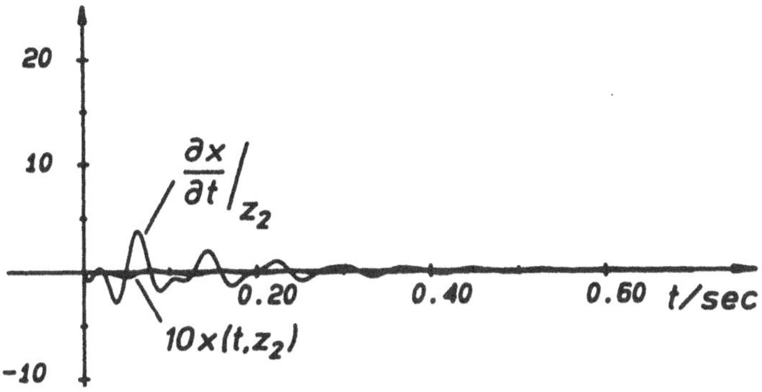


Fig. 6. Response to $\underline{w}^T = (1,0)$ at point $z_2$ with feedback law (35), $\alpha = 50$

Asymptotic stability is guaranteed, since $\underline{K}_P$ and $\underline{K}_D$ are positive definite.

## References.


[1] Paraskevopoulos, P.N., Sparis, P.D. and Mouroutsos, S.G.: The Fourier series operational matrix of integration. Int. J. Systems Sci. 10 (1985), pp. 171-176

[2] Vlassenbroeck, J. and Van Dooren, R.: A Chebyshev Technique for Solving Nonlinear Optimal Control Problems. IEEE Trans. Automat. Control, vol. AC-33, pp. 333-340, April 1988

[3] Paraskevopoulos, P.N.: Chebyshev Series Approach to System Identification, Analysis and Optimal Control. Journal of the Franklin Institute, Vol. 316 (1983), pp. 135-157

[4] Liu, C.-C. and Shih, Y.-P.: System analysis, parameter estimation and optimal regulator design of linear systems via Jacobi series. Int. J. Control, Vol. 42 (1985), pp. 211-224

[5] Franke, D.: A generalized Fourier series approach for the representation of dynamical systems. Automatisierungstechnik 36 (1988), pp. 68-73 (in German)

[6] Franke, D.: Linear controller design by balancing generalized Fourier coefficients. Automatisierungstechnik 36 (1988), pp. 133-138 (in German)

[7] Franke, D.: A data condensing root locus method for multivariable control systems. Automatisierungstechnik 36 (1988), pp. 480-486 (in German)

[8] Franke, D.: Feedback control of infinite-dimensional systems with finite relative order. Proc. Intern. AMSE Conference "Signals and Systems", Miami, Florida (USA), 1989, AMSE Press, Vol. 1, pp. 131-140

# SECOND ORDER OPTIMALITY CONDITIONS FOR NONLINEAR PARABOLIC BOUNDARY CONTROL PROBLEMS

Helmuth Goldberg, Fredi Tröltzsch
Technische Universität Chemnitz, Sektion Mathematik
DDR-9010 Chemnitz, PSF 964

## 1. Introduction

In this paper we shall derive sufficient second order optimality conditions for a nonlinear parabolic boundary control problem with constraints on the control and the state. By means of a semigroup technique we extend the results of /4/ and /8/ to the case of a domain of arbitrary dimension and additional state-constraints.

Let $D \subset R^n$ be a bounded domain with sufficiently smooth, say $C_\infty$-, boundary $\Gamma$, such that D is locally at one side of $\Gamma$. Moreover, we are given real numbers $T > 0$, $\mu \geq 0$, $u_1 < u_2$, $t_i$, $c_i$, $i = 1,\dots,k$, $t_i \in (0,T]$, real functions $q \in W_p^\sigma(D)$, $f_i \in W_p^\sigma(D)$, $i=1,\dots,k$ (p and $\sigma$ will be specified later), and a real function $b = b(t,r,x,u)$: $[0,T] \times R \times R \times [u_1,u_2] \longrightarrow R$. We assume that b is twice continuously differentiable with respect to (x,u) and fulfils a Carathèodory type condition: The continuity of b and of its partial derivatives with respect to (x,u) is uniform with respect to (t,r), and these functions are measurable with respect to (t,r) for all fixed (x,u).

We are going to investigate the following *optimal control problem*:

Minimize

$$\int_D (w(T,r) - q(r))^2 dr + \mu \int_0^T u(t,r)^2 dS_r dt$$

subject to

$$\begin{aligned} & w_t(t,r) = \Delta w(t,r) - w(t,r) && \text{in } D \\ & w(0,r) = 0 && \text{in } D \\ & \partial w / \partial n(t,r) = b(t,r,w(t,r),u(t,r)) && \text{on } \Gamma, \end{aligned} \tag{1.1}$$

$t \in (0,T]$ (by $\partial/\partial n$ the outward normal derivative is denoted), and to

$$u_1 \leq u(t,r) \leq u_2 \quad \text{a.e. on } [0,T] \times \Gamma,$$

$$\int_D f_i(r) w(t_i,r) dr \leq c_i \;, \; i = 1,\dots,k.$$

The control u is supposed to be measurable, thus $u \in U_\infty = L_\infty((0,T)\times\Gamma)$, and the state w is defined as mild solution of (1.1) (see section 2) with $w \in C([0,T],W_p^\sigma(D))$.
We assume that b is linear with respect to u,

$$b(t,r,x,u) = b_1(t,r,x) + b_2(t,r,x)u. \tag{1.2}$$

For many boundary conditions with background in mathematical physics this is no serious restriction. For instance, boundary conditions describing different phenomena of heat exchange of the type

$$\partial w/\partial n = a(t,r,w)(u - w),$$
$$\partial w/\partial n = u(\vartheta - w), \text{ or}$$
$$\partial w/\partial n = a(u^4 - w^4)$$

can be covered by (1.2) (take $u := u^4$ as a new control in the last case).

## 2. Transformation to a mathematical programming problem

According to the assumptions on b the mapping

$$(w(r),v(r)) \longmapsto b(t,r,w(r),v(r)) =: \mathcal{B}(t,w,v)(r)$$

is twice continuously Fréchet-differentiable from $C(\bar{D})\times L_p(\Gamma)$ to $L_p(\Gamma)$ for almost all t, $1 \le p \le \infty$. Moreover, for abstract functions $x \in X := C([0,T],C(\Gamma))$, $u \in U_{\nu,p} := L_\nu((0,T),L_p(\Gamma))$, by

$$(x(t),u(t)) \longmapsto \mathcal{B}(t,x(t),u(t)) =: B(x,u)(t)$$

a twice continuously Fréchet-differentiable mapping from $X\times U_{\nu,p}$ to $U_{\nu,p}$ is defined, $1 \le \nu < \infty$ , $1 \le p < \infty$ , an abstract Nemytskii operator.
We shall use the first and second order derivatives $B'(x_o,u_o)h$ and $B''(x_o,u_o)[h_1,h_2]$ at $(x_o,u_o) \in X\times U_\infty$ in the direction $h = (v,z) \in X\times U_\infty$. It holds

$$B'(x_o,u_o)h = B_x v + B_u z$$
$$(B_x v)(t,r) = b_x(t,r,x_o(t,r),u_o(t,r))v(t,r) =: b_x^o(t,r)v(t,r) \tag{2.1}$$
$$(B_u z)(t,r) = b_u(t,r,x_o(t,r),u_o(t,r))z(t,r) =: b_x^o(t,r)z(t,r).$$

Thus the image of $B_x$, $B_u$ is obtained by multiplication with a certain bounded and measurable function. Therefore, $B_x$ and $B_u$ can be extended continuously to continuous linear operators acting in $L_\nu((0,T),L_p(\Gamma))$. Analogously,

$$(B''(x_o,u_o)[h,h])(t,r) = (B_{xx}[v,v] + 2B_{xu}[v,z] + B_{uu}[z,z])(t,r)$$
$$= b_{xx}^o(t,r)v(t,r)^2 + 2b_{xu}^o(t,r)v(t,r)z(t,r) \tag{2.2}$$

has bounded and measurable coefficients $b_{xx}^o$ and $b_{xu}^o$ (note that $b_{uu}^o = 0$ by (1.2)).

At next we shall introduce the notion of a mild solution to (1.1). Therefore, we take p and $\sigma \in R$ such that $n/p < \sigma < 1 + 1/p$ and define an operator A acting in $L_p(D)$ by

$$D(A) = \{ w \in W_p^2(D) \mid \partial w / \partial n = 0 \quad \text{on } \Gamma \},$$
$$Aw = -\Delta w + w, \quad w \in D(A).$$

It is known that A is densely defined and closed in $L_p(D)$, and that -A is the infinitesimal generator of a strongly continuous and analytic semigroup S(t) of linear continuous operators in $L_p(D)$.
Moreover, we need the "Neumann" operator N, which assigns to $g \in L_p(\Gamma)$ the solution w of the elliptic boundary value problem

$$\Delta w - w = 0 \qquad \text{in } D,$$
$$\partial w / \partial n = g \qquad \text{on } \Gamma.$$

N is continuous from $L_p(\Gamma)$ to $W_p^s(D)$ for $s < 1 + 1/p$.
A mild solution w of (1.1) is any function of $C([0,T], W_p^{\sigma}(D))$, which solves the Bochner integral equation

$$w(t) = \int_0^t AS(t-s)N\mathcal{B}(s, \tau w(s), u(s))ds, \quad t \in [0,T], \tag{2.3}$$

where $\tau$ denotes the trace operator. Note that $\sigma > n/p$ implies $w \in C([0,T], C(\bar{D}))$.
It can be shown by standard methods that (2.3) is uniquely solvable on $[0,T]$ for all $u_1 \leq u \leq u_2$, if $T > 0$ is sufficiently small (see /9/).
For many types of boundary conditions, in particular for conditions describing heat exchange processes, it can be proved by maximum principle arguments that $|w(t,r)|$ is uniformly bounded. Then the solution of (2.3) exists globally.
(2.3) is not the actual state equation we are going to deal with. We introduce a new state function $x(t) = \tau w(t)$ and obtain from (2.3)

$$x(t) = \int_0^t \tau AS(t-s)N\mathcal{B}(s, x(s), u(s))ds, \tag{2.4}$$

which is our actual equation of state. We consider x as element of $C([0,T], C(\Gamma))$. Finally, we define linear operators K, C, $C_i$ by

$$(Kf)(t) = \int_0^t \tau AS(t-s)Nf(s)ds,$$
$$Cf = \int_0^T AS(T-s)Nf(s)ds, \qquad C_i f = \int_0^{t_i} AS(t_i-s)Nf(s)ds.$$

It is known from AMANN /2/ that

$$\|AS(t)N\|_{L_p(\Gamma)\to W_p^{\sigma}(D)} \le c\, t^{-(1+(\sigma-\sigma')/2)} \tag{2.5}$$

for $0<\sigma<\sigma'<1+1/p$. Therefore, K, C, and $C_i$ are continuous in the following spaces: K: $U_{\nu,p}\to C([0,T],C(\Gamma))$, C, $C_i$: $U_{\nu,p}\to W_p^{\sigma}(D)$, if $\nu > 2(\sigma'-\sigma)^{-1}$.

Now we have all prerequisites to formulate the optimal control problem in the form of the following

Mathematical programming problem (P):

Minimize

$$F(x,u) = \|CB(x,u) - q\|_2^2 + \mu\,\|u\|_2^2$$

subject to $x \in X$,

$$x = KB(x,u)$$
$$g_i(x,u) = \langle \phi_i\,,\, C_iB(x,u)\rangle_p - c_i \le 0,\ i = 1,\dots,k,$$
$$u \in U_{ad}.$$

In this setting, $\|\cdot\|_2$ denotes $L_2$-norms of the underlying spaces, $\langle\cdot,\cdot\rangle_p$ is the pairing between $L_{p'}(D)$ and $L_p(D)$, and $\phi_i \in L_p(D)^*$ are defined by

$$\langle \phi_i\,,\, v\rangle_p = \int_D f_i(r)v(r)dr\,.$$

Moreover,

$$U_{ad} = \{u\in U_{\nu,p} \mid u_1 \le u(t,r) \le u_2\}\,.$$

Here we assume $\nu > 2(\sigma'-\sigma)^{-1}$. It should be noticed that $L_\infty((0,T)\times\Gamma)$ can be continuously embedded into $L_\nu((0,T),L_p(\Gamma))$ $(1\le \nu,p<\infty)$, but not into $L_\infty((0,T),L_\infty(\Gamma))$ (cf. FATTORINI /3/). Each function $u\in U_{ad}$ can be represented by $u\in L_\infty((0,T)\times\Gamma)$. In the sequel we shall denote by $g(x,u)$ the column vector $(g_1(x,u),\dots,g_k(x,u))^t$.

## 3. Existence of optimal controls and first order necessary optimality conditions

The proof of existence is the first reason for the assumption of linearity of $b(t,r,x,u)$ with respect to u. We shall require additionally the following natural assumptions:

(A1) For all $u\in U_{ad}$ there is exactly one $x\in X$ with $x = KB(x,u)$.

(A2) The feasible set M,

$$M = \{(x,u) \mid x = KB(x,u),\ g(x,u)\le 0,\ u\in U_{ad}\}$$

is non-void.

(A1) is simply a restriction on T, as it was already pointed out above.

Theorem 1: Under the assumptions (A1), (A2) the control problem (P) admits at least one optimal solution $(x_o, u_o)$.

Proof: The method is already standard. If $(x_n, u_n)$ M is a minimizing sequence, then $u_n \rightarrow u$ weakly in $U_{\nu,p}$ can be assumed. Compactness of K ensures strong convergence $x_n \rightarrow x$ in X. The linearity of b with respect to u yields $x = KB(x,u)$, and from the weak lower semicontinuity of F the optimality of (x,u) is derived. #

In an optimal pair $(x_o, u_o)$ the component $u_o$ is said to be an optimal control and $x_o$ its corresponding state.
In order to define the notion of regularity, which is essential for any optimality condition, we introduce the so-called linearizing cone. An element $h = (v,z) \in X \times U_\infty$ belongs to L(M) iff

$$v = KB_x v + KB_u z \tag{3.1}$$

$$z \in \bigcup_{\lambda > 0} (U_{ad} - \{u_o\})$$

$$g'(x_o, u_o)h + rg(x_o, u_o) \leq 0, \quad r \geq 0.$$

Note that $B_x$, $B_u$ are the partial derivatives of B at $(x_o, u_o)$.
The set L(M) is said to be the linearizing cone of M at $(x_o, u_o)$. The optimal control $u_o$ is said to be regular, if it satisfies the assumption

(A3) There is a $\bar{h} = (\bar{v}, \bar{z}) \in X \times U_\infty$, such that $\bar{h}$ solves (3.1), $\bar{z} \in U_{ad} - \{u_o\}$, and

$$g_i(x_o, u_o) + g_i'(x_o, u_o)\bar{h} < 0, \quad i = 1, \ldots, k. \tag{3.2}$$

The Lagrange function L is defined by

$$L(x,u,y_1,y_2) = F(x,u) + \int_0^T \langle y_1(t), x(t) - (KB(x,u))(t) \rangle_p \, dt + y_2^t g(x,u),$$

where $y_1 \in L_{\nu'}((0,T), L_{p'}(\Gamma))$ and $y_2 \in R_+^k$ $(1/\nu + 1/\nu' = 1, 1/p' + 1/p = 1)$.
By definition, the equation $x = KB(x,u)$ is regarded in $X = C([0,T], C(\Gamma))$, thus the general theory of necessary optimality conditions would lead to a Lagrange multiplier $y_1 \in X^*$. However, this space can be avoided by an embedding technique, which mainly exploits the "smoothing property" $K: U_{\nu,p} \rightarrow X$ (see /10/, thm. 1.3.2). In this way we arrive at the

Theorem 2: Let $u_o$ be a regular optimal control with corresponding state $x_o$. Then there are $y_1 \in L_{\nu'}((0,T), L_{p'}(\Gamma))$, $y_2 \in R_+^k$ such that

$$L_x(x_o, u_o, y_1, y_2) = 0 \tag{3.3}$$

$$L_u(x_o, u_o, y_1, y_2)(u - u_o) = 0 \quad \forall u \in U_{ad} \tag{3.4}$$

$$y_2^t \, g(x_o,u_o) = 0 \tag{3.5}$$

Proof: The result follows from /10/, thm. 1.3.2 and the smoothing property of K. #

Equation (3.3) is the adjoint equation

$$y_1(t) = B_x(t)\left\{ -C^*(w_o(T)-q) + \sum_{i=1}^{k} y_2^i C_i^* \Phi_i + K^* y_1 \right\}(t), \tag{3.6}$$

where C, $C_i$, and K are regarded as operators with image in $L_p(D)$ and $U_{\nu',p'}$ respectively, thus $C^*$, $C_i^*$ : $L_{p'}(D) \rightarrow U_{\nu',p'}$, $K^*$ : $U_{\nu',p'} \rightarrow U_{\nu',p'}$. $B_x(t)$ is the extension of $B_x$ to $U_{\nu',p'}$. It is acting just by the multiplication of $\{\ldots\}(t,r)$ with $b_x(t,r,x_o(t,r),u_o(t,r))$. Moreover, $w_o(T) := CB(x_o,u_o)$.
Actually, we have even more regularity of $y_1$:

Theorem 3: The assumption $q$, $f_i \in W_p^{\sigma}(D)$, $i = 1,\ldots,k$, implies that $y_1 = y_1(t,r)$ is bounded and measurable on $(0,T)\times\Gamma$.

Proof: We show at first that $C^*(w_o(T)-q)(t,r)$ is continuous (to be more precise: it has a continuous representative). Take $v = v(t,r)$ from $U_{\nu,p}$. Then with $\bar{w} := w_o(T) - q \in W_p^{\sigma}(D)$

$$\langle \bar{w} , Cv \rangle_p = \langle \bar{w} , \int_0^T AS(T-s)Nv(s)ds \rangle_p$$

$$= \int_0^T \langle (AS(T-s)N)^* \bar{w} , v(s) \rangle_p \, ds$$

(here we regard $AS(T-s)N$ as operator from $L_p(\Gamma)$ to $L_p(D)$, hence $(AS(T-s)N)^*$: $L_{p'}(D) \rightarrow L_{p'}(\Gamma)$)

$$= \int_0^T \langle \tau S_{p'}(T-s)\bar{w} , v(s) \rangle_p \, ds$$

after an integration by parts, where $S_{p'}(t)$ denotes the semigroup generated by $A_{p'}$, which is the counterpart of A defined in $L_{p'}(D)$. The restriction of $S_{p'}$ to $W_p^{\sigma}(D)$ is a strongly continuous semigroup, too.
Hence

$$\Psi(t) = S_{p'}(T-t)\bar{w}$$

belongs to $C([0,T],W_p^{\sigma}(D)) \hookrightarrow C([0,T],C(\bar{D}))$. Thus $(C^*\bar{w})(t) = \tau\Psi(t)$ is contained in $C([0,T],C(\Gamma))$.
Completely analogous we find

$$(C_i^* \Phi_i)(t) = \begin{cases} S_{p'}(t_i-t)f_i, & 0 \le t \le t_i \\ 0 , & t_i < t \le T, \end{cases}$$

and $f_i \in W_p^{\sigma}(D)$ ensures that $(C_i^* \Psi_i)(t)$ is piecewise continuous on $[0,T]$ with values in $C(\Gamma)$, thus $(C_i^* \phi_i)(t,r)$ belongs to $L_\infty((0,T)\times\Gamma)$.
$K^*$ admits the form

$$(K^* y_1)(t) = \int_t^T \tau A_{p'} S_{p'}(s-t) N_{p'} y_1(s)ds = \int_t^T k(t,s) y_1(s)ds,$$

and it can be shown that the restriction of $k(t,s)$ to $L_p(\Gamma)$ satisfies

$$\|k(t,s)\|_{L_p(\Gamma)\to C(\Gamma)} \le c(t-s)^{-\alpha} ,$$

where $\alpha \in (0,1)$. (3.6) can be written

$$y_1(t) = \varphi(t) + B_x(t) \int_t^T k(t,s) y_1(s)ds, \tag{3.7}$$

where $\varphi(t,r)$ is from $L_\infty((0,T)\times\Gamma) \subsetneq L_\nu((0,T),L_p(\Gamma))$ $\forall \nu<\infty$.
If $y_1$ belongs to $U_{\nu,p}$, then the right hand side of (3.7) is contained in $L_\infty((0,T)\times\Gamma)$, if $\nu$ is sufficiently large. Now a Neumann series argument yields $y_1 \in L_\infty((0,T)\times\Gamma)$. #

## 4. Sufficient second order optimality conditions

The theory of second order optimality conditions for problems in function spaces was developed rapidly after the basic investigations by IOFFE /5/. In particular, MAURER /7/ worked out second order conditions for abstract optimal control problems with application to optimal control problems governed by systems of nonlinear ordinary differential equations. Such conditions were applied in many papers to different questions of optimal control of ordinary differential equations. Some corresponding references are quoted in /8/.
In the paper /4/ by GOLDBERG and the author necessary and sufficient second order optimality conditions have been derived for a control problem governed by the one-dimensional heat equation with nonlinear boundary condition subject to constraints on the control. Here we extend these results to the n-dimensional case and allow additional constraints on the state.

In what follows we shall use different norms of $h = (v,z) \in X\times U_\infty$ :
We denote by $\|\cdot\|_{\nu,p}$ the natural norm of $L_\nu((0,T),L_p(\Gamma))$, $\|\cdot\|_2 := \|\cdot\|_{2,2}$, and by $\|\cdot\|_\infty$ the norm of X. We put

$$\|h\|_2 = \max(\|v\|_2, \|z\|_2), \quad \|h\|_{\infty,\nu,p} = \max(\|v\|_\infty, \|z\|_{\nu,p}).$$

We shall impose the following second order conditions on $(x_o,u_o)$:

(SOC) There is a $\delta > 0$ such that

$$L''(x_o,u_o,y_1,y_2)[h,h] \geq \delta \|h\|_2^2 \tag{4.1}$$

for all $h \in L(M)$.

<u>Remark</u>: In (4.1) we can substitute $\|z\|_2$ for $\|h\|_2$, as v and z are connected by the linearized equation (3.1).

<u>Theorem 4</u>: Suppose that $u_o$ is a regular control, such that $(x_o,u_o) \in M$, the first order necessary conditions (3.3-5) and the second order condition (SOC) are satisfied. Assume further that $q \in W_p^{\sigma}(D)$ and $f_i \in W_p^{\sigma}(D)$, $i = 1,\dots,k$. Then there are constants $\alpha > 0$ and $\varepsilon > 0$, such that

$$F(x,u) - F(x_o,u_o) \geq \alpha\,(\|x - x_o\|_2^2 + \|u - u_o\|_2^2) \tag{4.2}$$

for all $(x,u) \in M$ with $\|(x-x_o,u-u_o)\|_{\infty,\nu,p} < \varepsilon$.

<u>Proof</u>: To prove theorems of this type, the so-called two-norm discrepancy plays a decisive role. In $X \times U_\infty$ we shall work with $\|\cdot\| := \|\cdot\|_{\infty,\nu,p}$ and the $L_2$-norm $\|\cdot\|_2$. Moreover, we shall denote by $r_j(h,E)$ the j-th order remainder term of a certain differentiable mapping E (at $(x_o,u_o)$).
According to MAURER /7/ we have to verify the following conditions:
If $\|h\| \to 0$ $(h = (v,z) \in X \times U_\infty)$, then

(a) $|r_1(g,h)|/\|h\|_2 \to 0$,

(b) $|r_2(L,h)|/\|h\|_2^2 \to 0$,

and there is a $c > 0$ such that

(c) $|L''(x_o,u_o,y_1,y_2)[h_1,h_2]| \leq c\,\|h_1\|_2\,\|h_2\|_2 \quad \forall h \in X \times U_\infty$.

Then the theorems 3.1 and 3.5 of /7/ yield the statement of the theorem.
From the Taylor formula for $b = b(t,r,x,u)$ we obtain

$$\|r_1(B,h)\|_2 \leq \alpha(h)\|h\|_2 \tag{4.3}$$

$$\|r_2(B,h)\|_1 = \beta(h)\|h\|_2^2 \tag{4.4}$$

where $\alpha(h)$, $\beta(h) \to 0$ for $\|h\| \to 0$, and $\|\cdot\|_1$ is the norm of $L_1((0,T),L_1(\Gamma))$.
For (4.4) the linearity of b with respect to u is essential. This can be seen as follows: We assume for short that $b(t,r,x,u) = b_1(t,r,x)u$, denote by $b^{\vartheta}_{1,xx}(t,r)$ the function $b_{1,xx}(t,r,x_o(t,r) + \vartheta v(t,r))$, and introduce similar expressions for the other derivatives.
Then for $\vartheta \in (0,1)$

$$\|(B''(x_o + \vartheta v, u_o + \vartheta z) - B''(x_o,u_o))[h,h]\|_1 =$$

$$= \| (b^{\vartheta}_{1,xx}(u_o + \vartheta z) - b^o_{1,xx} u_o) v^2 + 2(b^{\vartheta}_{1,x} - b^o_{1,x}) vz \|_1$$
$$\le \| (b^{\vartheta}_{1,xx} - b^o_{1,xx}) u_o \|_\infty \|v\|_2^2 + \| b^{\vartheta}_{1,xx} v \|_\infty \|v\|_2 \|z\|_2 +$$
$$+ 2 \| b^{\vartheta}_{1,x} - b^o_{1,x} \|_\infty \|v\|_2 \|z\|_2 .$$

For $\|v\|_\infty \to 0$ the $\|\cdot\|_\infty$ -factors tend to zero, thus (4.4) follows easily.

To verify (a), we consider

$$r_1(g_i,h) = \langle \Phi_i \, , \int_0^{t_i} AS(t_i - s) N r_1(B,h) ds \rangle_p$$
$$= \int_0^{t_i} \langle (AS(t_i - s)N)^* \Phi_i \, , \, r_1(B,h) \rangle_p ds .$$

Along the lines of the proof of theorem 3 we deduce that the left hand side in the duality brackets is bounded and measurable. Hence by (4.3)

$$|r_1(g_i,h)| \le c\,\alpha(h) \|h\|_2 ,$$

implying (a).

L" is given by

$$L''(x_o,u_o,y_1,y_2)[h_1,h_2] = F''(x_o,u_o)[h_1,h_2] - \langle y_1 \, , \, KB''(x_o,u_o)[h_1,h_2] \rangle_p$$

where $h_i = (v_i,z_i)$ and

$$F''(x_o,u_o)[h_1,h_2] = 2\langle CK(B_x v_1 + B_u z_1) \, , \, CK(B_x v_2 + B_u z_2) \rangle_2$$
$$+ 2\langle w_o(T) - q \, , \, CKB''(x_o,u_o)[h_1,h_2] \rangle_2 .$$

To show (b) we confine ourselves to the part $f(x,u) := \langle y_1 \, , \, KB(x,u) \rangle_p$ of L. We have

$$r_2(f,h) = \int_0^T \langle y_1(t) \, , \int_0^t \tau\, AS(t-s) N r_2(B,h)(s) ds \rangle_p \, dt$$
$$= \int_0^T \langle \int_t^T (\tau\, AS(s-t)N)^* y_1(s) ds \, , \, r_2(B,h)(t) \rangle_p \, dt .$$

$y_1(t)$ is bounded and measurable, hence $y_1 \in L_\nu((0,T),L_p(\Gamma))$, too. Thus the integral in the brackets is continuous, as $\nu$ , p are sufficiently large. Therefore,

$$|r_2(f,h)| \le c \; \|r_2(B,h)\|_1 \le c\, \beta(h) \|h\|_2^2$$

by (4.4). Analogously

$$|f''(x_o,u_o)[h_1,h_2]| = \int_0^T \langle \int_t^T (\tau AS(t-s)N)^* y_1(s) ds, B''(x_o,u_o)[h_1,h_2](t) \rangle_p dt$$

$$\le c \| B''(x_o,u_o)[h_1,h_2] \|_1 \le c \; \|h_1\|_2 \; \|h_2\|_2$$

is obtained. The discussion of the part F(x,u) of L is analogous, but tedious. In the estimations, the continuity of K as operator in

$L_2((0,T),L_2(\Gamma))$ must be used as a basic tool. The technique is mainly along the lines of /8/. #

Remarks:

(i) Thus, $u_o$ is a locally optimal control with respect to the $L_p$-norm, as the mapping $u \mapsto x = x(u)$ is continuous from $U_{\nu,p}$ to X.

(ii) Theorem 4 remains true without the assumption of linearity of b with respect to u, if $\|(x - x_o, u - u_o)\|_{\infty,\infty} < \varepsilon$ is required. This, however, is a very hard restriction (see section 5).

## 5. Second order conditions and approximation

It is known that sufficient second order optimality conditions can be supposed to show the stability of solutions of mathematical programming problems or optimal control problems with respect to perturbations of the given data. In particular, this refers to the numerical approximation of optimal control problems governed by nonlinear ordinary differential equations. Here we mention the paper by ALT /1/, where optimal control is considered as a particular case of a more general class of problems. Moreover, we refer to the investigations on stability by MALANOWSKI /6/.

The integral equation method permits to extend this approach also to control problems governed by parabolic partial differential equations in one-dimensional domains with nonlinear boundary conditions, see TRÖLTZSCH /8/. Similarly, the sufficient second order conditions of section 4 apply to show strong $L_2$-convergence of optimal controls, if the nonlinear parabolic control problem of section 1 is approximated by a suitable numerical method. Then we have to solve (1.1) by a finite difference or finite element technique, and the set of controls must be discretized, too, for instance by piecewise linear or smoother functions.

A discussion of these aspects of approximation would go beyond the scope of this paper. We shall only explain, where the linearity of b is needed: It is the discretization of controls, which is responsible for this assumption.

Let $U^n_{ad} \subset U_\infty$, $n \in N$, be a set of discretized controls, $S_i \subset L_\nu((0,T),L_p(\Gamma))$, i =1,2. We define with a certain norm $\|\cdot\|$

$$d(u,S_1) = \inf_{v \in S_1} \|u - v\|$$

$$d(S_2,S_1) = \sup_{u \in S_2} d(u,S_1).$$

In order to show convergence of approximating controls one has to assume

$$d(u_o, U^n_{ad}) \leq \alpha(n), \qquad (5.1)$$

$$d(U_{ad}, U^n_{ad}) \leq \alpha(n), \qquad (5.2)$$

where $\alpha(n) \rightarrow 0$, $n \rightarrow \infty$. The norm $\|\cdot\|$, which underlies the definition of the distance d, must comply with the differentiability of B(x,u). The choice $\|\cdot\| := \|\cdot\|_\infty$ would not cause restrictions on the nonlinearity of b(t,r,x,u). But then, as a rule, (5.1) can only be satisfied, if $u_o$ is sufficiently smooth (cf. ALT /1/).
For $\|\cdot\| = \|\cdot\|_{\nu,p}$ assumption (5.1) can be fulfilled, but then differentiability with respect to this norm and the assumptions (a) - (b) on the remainder term lead to the requirement of linearity with respect to u. We refer to /8/.

REFERENCES


/1/ ALT, W.: On the approximation of infinite optimization problems with an application to optimal control problems. Appl. Math. Optimization 12 (1984), 15-27.

/2/ AMANN, H.: Parabolic evolution equations with nonlinear boundary conditions. J. Differential Equations 72 (1988), 201-269.

/3/ FATTORINI, H.O.: Optimal control problems for semilinear parabolic distributed parameter systems (1990), to appear.

/4/ GOLDBERG, H., and F. TRÖLTZSCH: Second order optimality conditions for a class of control problems governed by nonlinear integral equations with application to parabolic boundary control. Optimization 20 (1989), 687-698.

/5/ IOFFE. A.D.: Necessary and sufficient conditions for a local minimum 3: Second order conditions and augmented duality. SIAM J. Control Optimization 17 (1979), 266-288.

/6/ MALANOWSKI, K.: Stability and sensitivity of solutions to optimal control problems for systems with control appearing linearly. Appl. Math. Optimization 16 (1987), 73-91.

/7/ MAURER, H.: First and second order sufficient optimality conditions in mathematical programming and optimal control. Math. Programming Study 14 (1981), 163-177.

/8/ TRÖLTZSCH, F.: Approximation of nonlinear parabolic boundary control problems by the Fourier method - convergence of optimal controls. Accepted for publication in Optimization.

/9/ TRÖLTZSCH, F.: On convergence of semidiscrete Ritz-Galerkin schemes applied to the boundary control of parabolic equations with nonlinear boundary condition. Accepted for publication in ZAMM.

/10/ TRÖLTZSCH, F.: Optimality conditions for parabolic control problems and applications. Teubner-Texte zur Mathematik, Vol. 62, Teubner-Verlag, Leipzig 1984.

# ON A WEIGHTING METHOD IMPROVING IDENTIFIABILITY OF DISTRIBUTED PARAMETER SYSTEMS

F. Guyon - J.P. Yvon
UTC - BP 649 - 60206 Compiègne Cedex - France

J. Henry
INRIA - Domaine de Voluceau - BP 105 - 78153 Rocquencourt - France

## 1. INTRODUCTION

### 1.1 Motivations

The output least-square estimation problem of the coefficient of an elliptic or parabolic problem is generally ill-conditionned (Chavent [1], Kunisch [4]). Even with a performing quasi-Newton minimization method like BFGS this leads to a very slow convergence and to highly oscillating estimated coefficents. In this paper we present a method for improving the conditionning of the problem by including in the error function a time dependent weighting factor. This weighting is designed in order to improve the convergence of the minimization algorithm that is to have a condition number of the hessian of the error function as small as possible. The method can also be applied to the estimation problem for ordinary differential equations.

The method is specially intended for evolution equation where the number of observation is large compared to the number of parameter to estimate.

Locally after linearization and discretization, the output least square estimation problem for parabolic equations is an ordinary least square problem. For the sake of simplicity we present our weighting method within this framework.

### 1.2. The basic example : identification of diffusion coefficients

As a reference problem let us consider the following system

$$\begin{cases} \dfrac{\partial u}{\partial t} - \displaystyle\sum_{i=1}^{n} \dfrac{\partial}{\partial x_i}\left[\tau(x)\dfrac{\partial u}{\partial x_i}\right] = f(x,t) & \text{in } \Omega\times(0,T) \\ u(x,t) = 0 \quad \text{on } \Gamma \text{ the boundary of } \Omega \\ u(x,0) = u_0(x) \end{cases} \tag{1.1}$$

where $\tau(.):\overline{\Omega}\to \mathbb{R}$ is a continuous fonction ($\tau(x) \geq \underline{\tau} > 0$), which is the "parameter" (here the diffusion coefficient) to be identified.

If we consider the simple case of a distributed observation, this corresponds to the following criterion

$$J(\tau) = \int_0^T \int_\Omega |u(x,t;\tau) - z(x,t)|^2\, dx\, dt = \int_0^T \|u(\tau) - z\|^2_{L^2(\Omega)}\, dt \tag{1.2}$$

The identification problem consists in minimizing the functionnal (1.2) under the constraint $\tau \geq \underline{\tau}$.

### 1.3. Orthonormalization of sensitivity functions

The main goal of this study is to introduce a weighting term in (1.2) :

$$J_W(\tau) = \int_0^T \langle W(t)(u(t;\tau) - z(t)), u(t;\tau) - z(t) \rangle_{L^2(\Omega)}\, dt \tag{1.3}$$

where $W(t) \in \mathcal{L}(L^2(\Omega);L^2(\Omega))$ such that the hessian of $J_W$ is well conditionned. An ideal case would be to determine a new scalar product in $L^2(0,T;L^2(\Omega))$ such that the sensitivity functions are orthonormal for this product.

We first recall classical facts in least square estimation. Then we turn to problems where the observation is time dependent and we present our method in that case. Then we apply it to the OLS estimation problem for parabolic equations. We end by some numerical experiments.

## 2. WEIGHTED LINEAR LEAST SQUARE PROBLEMS

### 2.1 Least-square estimator

We consider the following problem : let $A \in \mathcal{M}_{p,q}(\mathbf{R})$ be a matrix with rank(A) = q $\leq$ p, the classical *least square problem* consists in finding $\hat{\theta} \in \mathbf{R}^q$ such that

$$|A\hat{\theta}-b| \leq |A\theta-b| \quad , \forall \theta \in \mathbf{R}^q, \tag{2.1}$$

where $b \in \mathbf{R}^p$ represents a vector of "measurements" of the form

$$b = A\bar{\theta} + v \tag{2.2}$$

where $\bar{\theta}$ is the "true" value of parameters and v is a given noise with zero mean and covariance matrix R.

It is well known that the least square estimator is given by

$$\hat{\theta} = (A^T A)^{-1} A^T b,$$

with a covariance matrix of error of estimation

$$\mathbf{E}\,[\,(\theta-\bar{\theta})\,(\theta-\bar{\theta})^T\,] = P = (A^T A)^{-1} A^T R A \,(A^T A)^{-1}$$

The classical weighted form of (2.1) consists in modifying the least square criterion

$$\begin{cases} \text{for } W \in \mathcal{S}_p(\mathbf{R}),\ W > 0, \text{ find } \hat{\theta}_W \text{ such that :} \\ |A\hat{\theta}_W-b|_W \leq |A\theta-b|_W \quad \forall \theta \in \mathbf{R}^q, \end{cases} \tag{2.3}$$

where $|x|_W^2 = x^T W x$, and ($\mathcal{S}_p(\mathbf{R})$ is the space of symmetric matrices of order p)

Then $\hat{\theta}_W$ is given by

$$\hat{\theta}_W = (A^T W A)^{-1} A^T W b.$$

The covariance of the error of this estimator is then

$$P_W = (A^T W A)^{-1} A^T W R W A \,(A^T W A)^{-1} \tag{2.4}$$

### 2.2 The minimum variance estimator

It is classical that if we look for the minimum variance linear estimator of $\bar{\theta}$, then the estimator $\hat{\theta}_{MV}$ is given by

$$\hat{\theta}_{MV} = (A^T R^{-1} A)^{-1} A^T R$$

and the corresponding covariance matrix of the error is

$$P = (A^T R^{-1} A)^{-1}.$$

In that conditions it is clear that if we want to determine the weighting matrix W in order to minimize the variance of the error, the optimal choice for W is

$$W = R^{-1}, \tag{2.5}$$

and in this case both estimators are identical, i.e. : $\hat{\theta}_{R^{-1}} = \hat{\theta}_{MV}$.

## 2.3 Introduction of a "Newton type" weighting

The basic idea is to reduce the weighted least-square problem (2.3) to a simple one in which the projection on Range (A) is trivial. It is clear that a convenient choice is to select W such as

$$A^T W A = I \quad , \text{I identity matrix in } \mathbb{R}^q. \tag{2.6}$$

But there is (at least in the case $q < p$) non uniqueness of such a W. A possible idea consists in adding a condition, for instance we impose to $\hat{\theta}_W$, to realize the minimum variance of the error. From (2.4) we have

$$\text{var} ( \hat{\theta}_W - \bar{\theta} ) = \text{tr } P_w = \text{tr } (A^T W A )^{-1} A^T W R W A (A^T W A )^{-1} = \text{tr } A^T W R W A \tag{2.7}$$

by virtue of (2.6), then the problem'is reduced to

$$\begin{cases} \min\limits_{W>0} \text{tr } (A^T W R W A), \\ A^T W A = I. \end{cases} \tag{2.8}$$

This problem has a solution given by the Lyapunov equation :

$$W A A^T R^{-1} + R^{-1} A A^T W = 2\, R^{-1} A (A^T R^{-1} A)^{-1} A^T R^{-1}. \tag{2.9}$$

Unfortunately in most examples, we do not have any information on the noise, then various choices can be envisaged.

• The first one is to take R = I in (2.8)(2.9), but the resulting problem involves the solution of a Lyapunov equation.

• The second one consists in obtaining the uniqueness of W satisfying (2.6) by minimizing a given norm of W.

This can be done by considering the problem

$$\begin{cases} \min\limits_{W>0} \|W\|_F^2, \\ A^T W A = I, \end{cases} \tag{2.10}$$

(where $\|W\|_F^2 = \text{tr } W^T W$ is the Frobenius norm of W) which has the solution

$$W = A\, (A^T A)^{-2} A^T. \tag{2.11}$$

Then a step of the gradient algorithm applied to the weighted function $J_W(\theta) = \frac{1}{2} \| A\theta - b \|_W^2$ is exactly a step of the Newton algorithm applied to the original function $J(\theta) = \frac{1}{2} \| A\theta - b \|^2$.

This is evidently has no interest to solve the least-square problem, which can be done by an orthogonalisation procedure. In fact in the sequel we will adapt these ideas to a less trivial situation.

Remark

One can notice that W defined by (2.11) satisfies (2.9) with R = I. Furthermore the estimator is the same as the original one : $\theta_W = \theta$.

## 3. LINEAR LEAST-SQUARE PROBLEMS INVOLVING TIME

### 3.1 The least-squares estimator

Now we consider the problem where the observation depends on time. Let S be the family of matrices :

$$t \mapsto S(t) : [0,T] \to \mathcal{M}_{p,q}$$

Therefore the quadratic least-square error is given by

$$J(\theta) = \frac{1}{2} \int_0^T \| S(t)\theta - z(t) \|_{\mathbb{R}^p}^2 dt, \tag{3.1}$$

where $z(t) \in \mathbb{R}^p$, is the vector of measurements of the form

$$z(t) = S(t)\bar{\theta} + v(t) \tag{3.2}$$

where v(t) is a noise.

The estimator $\hat{\theta}$ which minimizes (3.1) is given by the condition

$$\left[ \int_0^T S^T(t)\, S(t)\, dt \right] (\hat{\theta} - \bar{\theta}) = \int_0^T S^T(t)\, v(t)\, dt \tag{3.3}$$

which, in the case where

$$H = \int_0^T S^T(t)\, S(t)\, dt \in \mathcal{S}_q \text{ is invertible,} \tag{3.4}$$

leads to

$$\hat{\theta} = H^{-1} \int_0^T S^T(t)\, z(t)\, dt \,; \quad (\hat{\theta} - \bar{\theta}) = H^{-1} \int_0^T S^T(t)\, v(t)\, dt \,. \tag{3.5}$$

### 3.2. Presentation of two possible time-dependent weightings

Our goal is to adapt the previous ideas by introducing now a family of weighting matrices

$$t \mapsto W(t) : [0,T] \to \mathcal{S}_p$$

and to replace the least square function (3.1) by

$$J_W(\theta) = \frac{1}{2} \int_0^T (S(t)\theta - z(t))^T W(t) (S(t)\theta - z(t))dt. \tag{3.6}$$

The problem of finding a weighting function orthonormalizing the sensitivity functions can be then formulated as

$$\begin{cases} \text{Does there exist a family of symmetric matrices} \\ \quad t \mapsto W(t) \in \mathscr{S}_p \text{ such that} \\ \int_0^T S^T(t)\, W(t)\, S(t)\, dt = I \end{cases} \tag{3.7}$$

If this problem has a solution (this will be studied below), there is no reason that it should be unique. If we follow the same lines as in the previous section two main choices can be made.

<u>*A First choice for W(.)*</u>

The error on the estimator $\hat{\theta}_W$ is given by :

$$\left[\int_0^T S^T(t)\, W(t)\, S(t)\, dt\right] (\hat{\theta}_W - \bar{\theta}) = \int_0^T S^T(t)\, W(t)\, v(t)\, dt$$

which by (3.7) is simply

$$\hat{\theta}_W - \bar{\theta} = \int_0^T S^T(t)\, W(t)\, v(t)\, dt\,. \tag{3.8}$$

As we have the majoration

$$| \hat{\theta}_W - \bar{\theta} |^2 \leq \left( \int_0^T |S^T(t)\, W(t)|^2 dt \right) |v|^2_{L^2(0,T;R^p)}\,, \tag{3.9}$$

it seems quite natural to look for a family of matrices W (t) such as

$$\min_{W(.) \in \mathscr{S}_p} \int_0^T |S^T(t)\, W(t)|^2\, dt \tag{3.10}$$

$$\int_0^T S^T(t)\, W(t)\, S(t)\, dt = I. \tag{3.7}$$

It remains to choose a *norm* in (3.9). For a practical standpoint it is convenient to take an euclidian norm on $\mathscr{S}_p$, one possible choice (which is similar to (2.8)) being then the ***Frobenius norm*** $|\,.\,|_F$ associated to the scalar product $< A,B >_F$ defined by

$$< A,B >_F = \operatorname{tr}(AB^T). \tag{3.11}$$

<u>Formal solution</u>

One way to solve this problem is to introduce the following lagrangian

$$\begin{cases} L(W,\Lambda) = \frac{1}{2} \int_0^T |S^T(t)\, W(t)|^2\, dt + \frac{1}{2} \int_0^T |W(t)\, S(t)\, |^2\, dt \\ \qquad\qquad + < \Lambda, I - \int_0^T S^T(t)\, W(t)\, S(t)\, dt >_F. \end{cases} \tag{3.12}$$

The condition giving the stationnarity of L with respect to W leads to

$$S(t)S^T(t)\, W(t) + W(t)\, S(t)S(t)^T = S(t)\Lambda S(t)^T, \tag{3.13}$$

which is a Lyapunov equation which does not necessarily have a unique solution (the natural condition being $S(t)S(t)^T > 0$ which is clearly too strong). This choice will not be studied any further.

*A second possible choice for W(.).*

It consists simply to seek a W of minimum norm :

$$\begin{cases} \min\limits_{W(.)} \int_0^T \|W(t)\|_F^2\, dt \quad , \quad W(t) \text{ symmetric} , \\ \int_0^T S^T(t)W(t)S(t)\, dt = I \qquad (I : \text{identity in } \mathbb{R}^q) \end{cases} , \tag{3.14}$$

As before we introduce the lagrangian

$$L(W,\Lambda) = \frac{1}{2} \int_0^T \|W(t)\|_F^2\, dt + < \Lambda , I - \int_0^T S^T(t)W(t)S(t)\, dt >_F , \tag{3.15}$$

the stationnarity of L with respect to W gives the following condition

$$\int_0^T < W(t), \delta W(t) >_F dt \; - \int_0^T < S(t)\Lambda S^T(t), \delta W(t) >_F dt = 0 \; ,$$

which yields

$$W(t) = S(t)\Lambda S^T(t). \tag{3.16}$$

If we impose that W satisfies the constraint in (3.7) , this leads to

$$\int_0^T S^T(t)S(t)\Lambda S^T(t)S(t)\, dt = I, \tag{3.17}$$

which is a linear equation with respect to $\Lambda$ and which can be explicited via

$$Q.\, \Lambda = I \tag{3.18}$$

where Q is given via a KRONECKER product [1] by :

---

(1) The KRONECKER product T of A and B, $T = A \otimes B$, is a ***tensor*** defined by $T_{ijk\ell} = a_{ij}\, b_{k\ell}$ then if C is a matrix the product T.C is a matrix D defined by $d_{ik} = \sum\limits_{j,\ell} T_{ijk\ell}\; c_{j\ell}$ .

$$Q = \int_0^T [S^T(t)S(t)] \otimes [S^T(t)S(t)]\, dt .$$

**Remark 3.1**

It is not clear whether the equation (3.18) admits a solution, this point will be made more precise later. If $\Lambda$ is a solution of (3.17) then $\Lambda^T$ is also a solution and, as a consequence, if the solution $\Lambda$ is unique it is symmetric.

Proposition 3.1

The solution $\overline{W}$ of problem (3.14), if it does exist, is given by the set of two equations :

$$\begin{cases} Q.\Lambda = I \\ \overline{W}(t) = S(t)\Lambda S^T(t) \end{cases}, \qquad (3.19)$$

where $\Lambda$ and $\overline{W}$ are symmetric.

Proof. Equations (3.19) represent the set of necessary conditions of problem (3.14).

## 4. ANALYSIS OF THE METHOD

### 4.1. The strong identifiability hypothesis

Let $S:[O;T] \to \mathcal{M}_{p,q}$ be a continuous function ; we want to solve

$$\min_{\theta \in R^q} \frac{1}{2} \int_0^T \| S(t)\theta - z(t) \|^2_{R^p}\, dt. \qquad (4.1)$$

Definition 4.1

The parameters $\theta$ are *identifiable* if the mapping

$$\theta \mapsto S(t)\theta \;:\; R^q \to L^2(0,T;R^p) \qquad (4.2)$$

is injective.

A necessary and sufficient condition of identifiability is that the identifiability grammian

$$\mathcal{J} = \int_0^T \| S(t)^T S(t) \|\, dt$$

has full rank.

In the sequel we will make an hypothesis which is stronger than (4.2).

Strong identifiability hypothesis.

$$\begin{cases} \text{The mapping } \mathcal{C} \in \mathcal{L}(L^2(0;T;\mathcal{S}_p);\mathcal{S}_q) \text{ defined by} \\ \qquad \mathcal{C}Z = \int_0^T S^T(t)\, Z(t)\, S(t)\, dt \\ \text{is onto.} \end{cases} \qquad (4.3)$$

Proposition 4.1

Assumption (4.3) implies identifiability.

Proof. Assume that

$$S(t)\theta \equiv 0, \tag{4.4}$$

then if we define $R = \theta\,\theta^T \in \mathcal{S}_q$, (4.4) implies that

$$\mathcal{C}^*R = S(t)RS^T(t) = S(t)\theta\theta^T S^T(t) = \|S(t)\theta\|^2 = 0,$$

where $\mathcal{C}^*$ is the adjoint of $\mathcal{C}$ but, from (4.3) $\mathcal{C}^*$ is injective, this implies $R = \theta\theta^T = 0$ which implies $\theta = 0$.

Theorem 4.1

Under assumption (4.3) the problem (3.14) admits a unique solution $\overline{W}$ given by the set of equations

$$\begin{cases} \int_0^T S^T(t)S(t)\Lambda S^T(t)S(t)\,dt = I; \\ \overline{W}(t) = S(t)\Lambda S^T(t). \end{cases} \tag{4.5}$$

Proof.

The first equation of (4.5) may be written as

$$\mathcal{C}\mathcal{C}^*\Lambda = I \tag{4.6}$$

and, as $\mathcal{C}$ is onto this equation admits a unique solution $\Lambda$. Then there exists a $\overline{W}$ which satisfies the set of optimality conditions for (3.13). As in that case the function to be minimized is strictly convex and the constraints are linear, the necessary optimality conditions are sufficient, then $\overline{W}$ is the solution of (3.14). ∎

It is possible to show that the weighting W thus computed has the following property : the error on $\theta_W$ due to the lack of exact convergence of the minimization algorithm is minimized and equidistributed on the components of $\theta_W$.

## 4.2. Regularization

As we have mentionned previously the sole assumption of identifiability does not imply (4.3). Furthemore, even if (4.3) is satisfied, equation (4.5) may be ill-conditionned.

To overcome this difficulty, we propose a regularization.

*Regularization of (3.18)*

One can replace equation (3.18) by

$$(Q + \varepsilon I).\,\Lambda = I \qquad \text{where } \varepsilon > 0 \text{ is given}, \tag{3.18$_\varepsilon$}$$

this can be done directly by penalizing the constraint in (3.14) :

$$\begin{cases} \min\limits_{W(.);K} \int_0^T \|W(t)\|_F^2\,dt + \frac{1}{\varepsilon}\| K - I \|_F^2 \\ \\ \int_0^T S^T(t)W(t)S(t)\,dt = K \quad ; \; K \in \mathcal{S}_q \end{cases} \tag{4.7}$$

The corresponding lagrangian is

$$L_\varepsilon(W,K,\Lambda) = \frac{1}{2}\int_0^T \|W(t)\|_F^2 \, dt$$

$$+ \frac{1}{2\varepsilon}\| K - I \|_F^2 - < \Lambda , K - \int_0^T S^T(t)W(t)S(t) \, dt >_F. \tag{4.8}$$

Stationnarity of this lagrangian with respect to W and K leads respectively to

$$W_\varepsilon(t) = S(t)\Lambda_\varepsilon S^T(t) ,$$
$$K_\varepsilon = I - \varepsilon\Lambda .$$

Plugging these relations in the second relation of (4.7) one gets

$$\begin{cases} (Q + \varepsilon I). \Lambda_\varepsilon = I \\ W_\varepsilon(t) = S(t)\Lambda_\varepsilon S^T(t) \end{cases} \tag{4.9}$$

4.3. The time discretized problem

Let us consider the time discretization of problem (3.1). Let $\{t_i\}_{i=1}^N$ be the discretization times : $t_i = i\frac{T}{N}$, $S_i = S(t_i)$.

Problem (3.1) becomes the minimization of :

$$J(\theta) = \frac{1}{2} \sum_{i=1}^{N} | S_i \theta - z_i |^2. \tag{4.10}$$

This problem is of the form (2.1) for the matrix $A \in \mathcal{M}_{p,Nq}$ having $S_i$'s as block rows :

$$A = \begin{bmatrix} S_1 \\ \cdot \\ \cdot \\ \cdot \\ S_i \\ \cdot \\ \cdot \\ \cdot \\ S_N \end{bmatrix}$$

So, given a fully discretized problem, it is natural to consider various row splitting of the matrix A.

Let us assume that $N = \ell N'$ and consider the sub-splitting of A :

$$A = \begin{bmatrix} \Sigma_1 \\ \Sigma_2 \\ \cdot \\ \cdot \\ \cdot \\ \Sigma_{N'} \end{bmatrix} \qquad \Sigma_k \in \mathcal{M}_{\ell p,q} \tag{4.11}$$

$$\Sigma_k = \begin{bmatrix} S_{\ell(k-1)+1} \\ \vdots \\ S_{\ell k} \end{bmatrix} \qquad k : 1, \dots, N'$$

As an analogous of (4.3), we may define the strong identifiability hypothesis for the time discretized problem :

The mapping $\mathscr{B} \in \mathscr{L}((\mathscr{S}_p)^N ; \mathscr{S}_q)$ defined by :

$$\mathscr{B} Z = \sum_{i=1}^{N} S_i^T Z_i S_i \tag{4.12}$$

is onto.

The following result shows how this property depends on the splitting of A.

Theorem 4..2

If the strong identifiability hypothesis (4. 12) is satisfied for the row of splitting A by $S_i$'s, it is also satisfied for the sub splitting by $\Sigma_k$'s. Furthermore, if A has full rank, the property is true for A itself without splitting.

**Proof** : By assumption the mapping $\mathscr{B}^*$ :

$$\Lambda \in \mathscr{S}_q \to \{S_i \Lambda S_i^T\}_{i=1}^{N} \text{ is injective.}$$

Assume that there exists a $\Lambda \in \mathscr{S}_q$ such that :

$$\mathscr{B}_\Sigma^* \Lambda = \{\Sigma_k \Lambda \Sigma_k^T\}_{k=1}^{N'} = 0$$

But for each k, the $\ell$ diagonal blocks of $\Sigma_k \Lambda \Sigma_k^T$ are :

$$S_{\ell(k-1)+j} \Lambda S_{\ell(k-1)+j}^T \quad j = 1, \dots, \ell$$

and they are null. By the injectiveness of $\mathscr{B}^*$, $\Lambda$ is null and this proves that $\mathscr{B}_\Sigma^*$ is also injective.

To prove the strong identifiability without row-splitting, we have to prove that :

$$\Lambda \in \mathscr{S}_q \to A \Lambda A^T \text{ is injective.}$$

As A has full rank :

$$rk(A \Lambda A^T) = rk \Lambda,$$

and so

$$A \Lambda A^T = 0 \Rightarrow \Lambda = 0.$$

The practical interest of the preceding result is due to the fact that in order to minimize the volume of computation of $J_W$ and $\nabla J_W$ one has to use a row splitting of A as fine as possible. This result suggests to test successive refinements till (4. 12) is no more satisfied. Anyhow to satisfy (4.12) it is necessary that :

$$\frac{N}{\ell}\ \frac{\ell p(\ell p+1)}{2} > \frac{q(q+1)}{2}.$$

## 5. THE IDENTIFICATION METHOD

### 5.1. The weighted identification algorithm

Starting from the example of §1.2, after discretization, the problem is reduced to :

$$\begin{cases} \frac{dy}{dt} + A(\theta)\, y = b(t) \quad y(t) \in \mathbb{R}^n \\ y(0) = y_0 \end{cases}, \tag{5.1}$$

where $\theta \in \mathbb{R}^q$, $A(\theta) \in \mathcal{M}_{n,n}$ and $b(t) \in \mathbb{R}^n$. The least square criterion is now

$$J(\theta) = \frac{1}{2} \int_0^T \| Cy(t;\theta) - z(t) \|^2_{\mathbb{R}^p} dt, \tag{5.2}$$

where $C \in \mathcal{M}_{p,n}$ is the observation operator and $z(t) \in \mathbb{R}^p$ represents the measurements.

The linearized problem around a state $\bar{y}(t) = y(t;\bar{\theta})$ is defined by

$$\begin{cases} \frac{d(\delta y)}{dt} + A(\bar{\theta})\, \delta y = B(t;\bar{y}(t))\, \delta\theta \text{ on } (0;T) \\ \delta y(0) = 0 \end{cases}, \tag{5.3}$$

with $B(t;\bar{y}(t))\, \delta\theta \triangleq -[\frac{dA}{d\theta}(\bar{\theta})\, \delta\theta]\, \bar{y}(t)$.

In order to have a closed representation of (5.3) let us introduce the family of operators

$$t \mapsto S(t) : [0,T] \to \mathcal{M}_{p,q}$$

defined by

$$S(t)\, \delta\theta = C\, \delta y(t) \qquad (\delta y \text{ being defined by (5.3)}). \tag{5.4}$$

Therefore the quadratic least-square error (corresponding to the linearized problem) is given by

$$\tilde{J}(\delta\theta) = \frac{1}{2} \int_0^T \| S(t)\, \delta\theta - \bar{z}(t) \|^2_{\mathbb{R}^p} dt, \tag{5.5}$$

with $\bar{z}(t) = z(t) - C\bar{y}(t)$.

A possible algorithm is defined by the following sequence of calculations :

Step 1. For a given value of $\bar{\theta}$ calculate the solution $\overline{W}$ of (3.19) whith S given by (5.4).
Step 2. Update the value of $\theta$ *either* by solving the optimization problem

$$\min_{\theta} \int_0^T [y(t;\theta) - z(t)]^T\, \overline{W}(t)\, [y(t;\theta) - z(t)]\, dt, \tag{5.6}$$

*or* by performing a finite number of steps of an optimization method for (5.6).

**Remark 5.1**

We have, for any $\delta y = S\,\delta\theta$ given by (5.3), the relation

$$\int_0^T \delta y^T(t)\ \overline{W}(t)\ \delta y(t)\,dt = \|\delta\theta\|^2, \tag{5.6)'}$$

which gives for the linearized problem a hessian equal to identity. But we can observe in most situations that the terms neglected in the linearization result in a problem which is highly non convex and the behaviour of the problem (5.6)' is very different from (5.6). Futhermore it is clear that, in general, the operator $\overline{W}(t)$ is not *positive definite* (see appendix) and, as a consequence, the problem may be not well posed, thus it is usually necessary to regularize this problem.

## 5.2. An example of parameter estimation in a parabolic problem satisfying the strong identifiability condition

There are very few results on the identifiability of parabolic equations (Kitamura - Nakagiri [3], Nakagiri [5], Courdesses - Amouroux [2]). The strong identifiability presented in (4.3) on the discretized linearized problem is still more difficult to check. We present here a very simple situation studied in [2] where identifiability and strong identifiability turn out to be equivalent.

Consider the one-dimensionnal problem :

$$\frac{\partial u}{\partial t} - a\frac{\partial^2 u}{\partial x^2} - bu = 0 \qquad \begin{array}{l} x \in \Omega = ]0,1[ \\ t \in ]0,T[ \end{array} \tag{5.8}$$

with boundary conditions :

$$u(0,t) = u(1,t) = 0$$

and initial conditions :

$$u(0,x) = \alpha_1 \sqrt{2}\ \sin \pi x + \alpha_2 \sqrt{2}\ \sin 2\pi x.$$

The parameters to estimate are the constants a and b :

$$\theta = (a\quad b)^T \qquad\qquad 0 < a_1 \leqslant a \leqslant a_2$$

The system is observed at point $x_0$ : $Cu = u(x_0,t)$.

It is shown in [2] that the identifiability of $\theta$ is equivalent to :

$$\alpha_1\,\alpha_2 \sin \pi x_0\ \sin 2\pi x_0 \neq 0 \tag{5.9}$$

Let us consider the linearization of the state around $u(\overline{\theta})$ corresponding to a given value $\overline{\theta}$ of the parameter. We obtain :

$$S(t) = \begin{bmatrix} \frac{\partial u}{\partial a}(x_0,t) \\ \frac{\partial u}{\partial b}(x_0,t) \end{bmatrix} = \sqrt{2}\,t\,\alpha_1 \sin \pi x_0\, e^{(b-\pi^2 a)t} \begin{bmatrix} -\pi^2 \\ 1 \end{bmatrix} + \sqrt{2}\,t\,\alpha_2 \sin 2\pi x_0\, e^{(b-4\pi^2 a)t} \begin{bmatrix} -4\pi^2 \\ 1 \end{bmatrix}$$

Using (5.9) and the linear independance of time functions in the fomula, it is easy to show that if :

$$S^T(t)\,\Lambda\,S(t) = 0 \quad \forall\, t \in ]0,T[ \quad \Lambda \in \mathcal{S}_2$$

then the entries of $\Lambda$ :

$$\Lambda = \begin{bmatrix} \lambda_1 & \lambda_{12} \\ \lambda_{12} & \lambda_2 \end{bmatrix}$$

must satisfy :

$$\begin{bmatrix} \pi^4 & -2\pi^2 & 1 \\ 4\pi^4 & -5\pi^2 & 1 \\ 16\,p^4 & -8\pi^2 & 1 \end{bmatrix} \begin{bmatrix} \lambda_1 \\ \lambda_{12} \\ \lambda_2 \end{bmatrix} = 0$$

which implies $\Lambda = 0$ as this matrix is nonsingular.

## 6. NUMERICAL EXPERIMENTS

Numerical experimentations are based on the example (1.1) in one dimension :

$$\begin{cases} \frac{\partial u}{\partial t}(x;t) - \frac{\partial}{\partial x}\left[\tau(x)\frac{\partial u}{\partial x}(x;t)\right] = f(x;t) \text{ in } ]0,1[\times]0,T[ \\ u(0,t) = \alpha_0 \\ u(1,t) = \alpha_1, u(x,0) = u_0(x) \end{cases} \tag{6.1}$$

with a "true" value of $\tau$ being :

$$\tau(x) = 1 + 5x\,. \tag{6.2}$$

A classical RITZ-GALERKINE approximation with piecewise linear functions is used to reduce problem (6.1) to a finite dimensional system. In the following numerical results the number of spatial nodes is 7, hence there are 6 parameters to estimate. Observation is distributed or punctual at one or several nodes.

On the various figures the convergence is illustrated by considering the evolution of the mean square error on the coefficients with respect to the number of iterations. It must be mentionned that the curves are piecewise straight lines which join the points where the error is actually calculated, thus the curves do not give information on the local rates of convergence.

Figures 1 and 2 :

• curves (a) coresponds to a minimizaton of the original functional by a BFGS algorithm after 2000 iterations in order to show the error on parameters.

• curves (b) represents the results obtained by the following algorithm : for a given value of $\varepsilon$ one computes $W_\varepsilon$ by (4.9) and a complete minimization of the weighted functional is performed this gives a new value for $\theta$, then $\varepsilon$ is divided by a given factor (for instance 100) and the procedure is repeated.

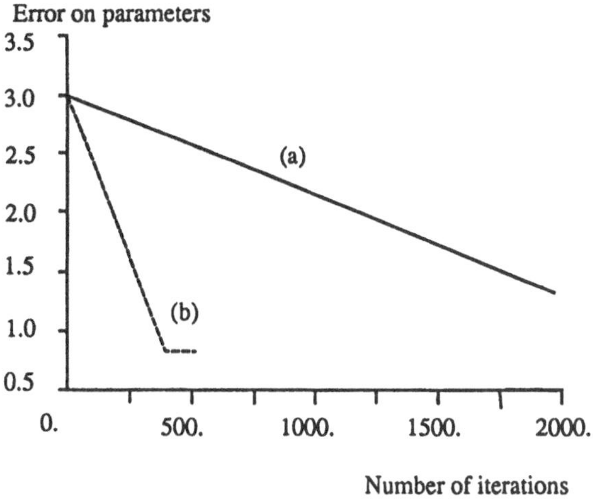


Figure 1

There is only one observation and the identifiability is very poor. The results show simply that the weighted functional gives a better result on the error even if this error does not vanish.

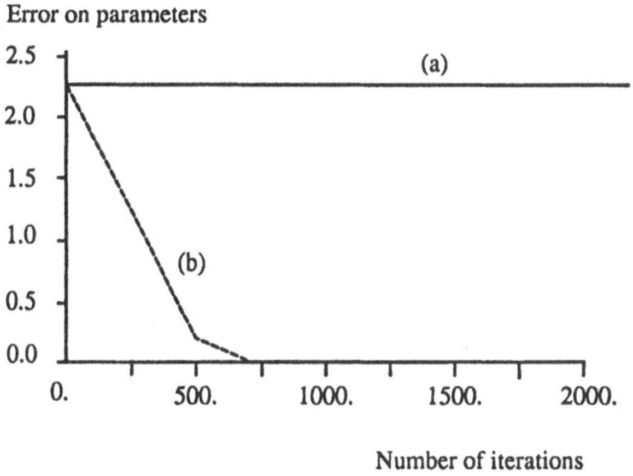


Figure 2

Observation at nodes 1;2,3. The initial guess on θ is (5,...,5). The figure shows the poor convergence of the non weighted functional (a) compared to the weighted one (b).

## CONCLUSION

The proposed method improves the convexity of the criterion. It is clear that the computational effort required by this method may be rewarding only in the case of ill conditionned problems. For that reason the efficiency of the method is particularly illustrated by examples where the classical approach fails. A lot of questions still remain pending in particular the study of the "strong identifiability hypothesis" and the interpretation of the new estimate $\theta_W$.

## REFERENCES

[1] Chavent, G. (1987). New trends in identification of distributed parameter systems, in Proceedings 10th IFAC World Congress Munchen, Pergamon Press.
[2] Courdesses, M. - Amouroux, M. (1982). Identifiabilité et identification des systèmes à paramètres répartis, in Proceedings 3rd Symposium IFAC Control of Distributed Parameter Systems, Toulouse June 1982.
[3] Kitamura, S. - Nakagiri, S. (1977). Identifiability of spatially-varying and constant parameters in distributed systems of parabolic type. SIAM J. Cont. Opt. Vol. 15, n°5, pp. 785-802.
[4] Kunisch, K. (1986). A survey of some recent results on the output least squares formulation of parameter estimation problems, in Proceedings *IFAC Congress on Control of Distributed Parameter Systems*, Pergamon Press, pp. 17-23.
[5] Nakagiri, S. (1983). Identifiability of linear systems in Hilbert spaces. SIAM J. Cont. Opt. Vol. 21, n°4, pp. 501-530.
[6] Sorenson, H.W. (1980). Parameter estimation. Marcel Dekker Inc.

## Appendix

A drawback of the method is that the matrix $\Lambda$ is not nessarily positive definite nor W(t). The next simple example shows the difficulty. The state of the system satisfies

$$\begin{cases} \dot{x}_1(t) = a_1 x_1(t) + \theta_1 e^{a_1 t}, x_1(0) = 0 \\ \dot{x}_2(t) = a_2 x_2(t) + \theta_2 e^{a_2 t}, x_2(0) = 0 \end{cases}$$

and is given by

$$\begin{cases} x_1(t) = t\ \theta_1 e^{a_1 t}, \\ x_2(t) = t\ \theta_2 e^{a_2 t}. \end{cases}$$

The observation is given by

$$y(t) = x_1(t)+x_2(t)$$

and the least square functionnal is given by

$$J(\theta_1,\theta_2) = \int_0^\infty |y(t)-z(t)|^2\, dt .$$

Under these conditions we have

$$S(t) \stackrel{\text{déf}}{=} [s_1(t), s_2(t)] = [\ t\ \ e^{a_1 t}, t\ \ e^{a_2 t}] .$$

The sytem of equation giving the symmetric matrix $\Lambda = \begin{pmatrix} \Lambda_{11} & \Lambda_{12} \\ \Lambda_{12} & \Lambda_{22} \end{pmatrix}$, is

$$\begin{pmatrix} a & 2b & c \\ b & 2c & d \\ c & 2d & e \end{pmatrix} \begin{pmatrix} \Lambda_{11} \\ \Lambda_{12} \\ \Lambda_{22} \end{pmatrix} = \begin{pmatrix} 1 \\ 0 \\ 1 \end{pmatrix}$$

where

$$a = \int_0^\infty s_1^4(t)\,dt\ ,\quad b = \int_0^\infty s_1^3(t)\, s_2(t)\,dt\ ,\ c = \int_0^\infty s_1^2(t)\, s_2^2(t)\,dt\ ,\ d = \int_0^\infty s_1(t)\, s_2^3(t)\,dt\ ,\ e = \int_0^\infty s_2^4(t)\,dt\ .$$

After some calculations it is easy to show that the solution $\Lambda$ is not definite, and furthermore that the function W(t) is not necessarily positive.

# DISCRETIZATION ERROR IN OPTIMAL CONTROL*

William W. Hager
Department of Mathematics
University of Florida
Gainesville, Florida 32611 USA

and

Asen L. Dontchev
Institute of Mathematics
Bulgarian Academy of Sciences
Sofia, Bulgaria

Abstract

Sensitivity analysis is used to estimate the error associated with Euler's discretization to a nonlinear optimal control problem with convex control constraints.

## 1. Introduction.

In this paper we use a result from sensitivity analysis to estimate the error in Euler's approximation to a nonlinear optimal control problem with convex control constraints. This paper presents some of the key ingredients in the analysis while the complete theory appears in [4]. In earlier papers, Budak *et al.* [1] and Cullum [2] prove convergence of the optimal value associated with discrete approximations to state and control constrained problems. Mordukhovich [8] shows that the discrete optimal cost converges to the true optimal cost if and only if a relaxation of the control problem is stable. Estimates for the error in the optimal control associated with higher order discretizations of unconstrained nonlinear problems are derived by Hager [5]. Dontchev [3] obtains an error estimate for Euler's approximation applied to an optimal control problem with convex cost, linear system dynamics, and linear inequality state and control constraints. In this paper, the assumptions of cost convexity and constraint linearity are dropped -- we consider a problem with nonlinear system dynamics and a general convex control constraint. Our method of analysis makes use of the so-called averaged modulus of smoothness, introduced by Sendov and Popov [9].

*This work was supported by the U.S. Army Research Office Contract DAAL03-89-G-0082 and by the Bulgarian Ministry of Science Contract 127. The research was performed while the first author was a visitor at the University of Florida.

## 2. Abstract result.

This section states the abstract sensitivity result that is applied to the optimal control problem. We consider a family of equations, each equation depending on a parameter $p$ contained in a metric space $P$. Associated with each $p \in P$, there is a closed subset $\Omega_p$ of a Banach space $Z_p$, a normed vector space $Y_p$, and a pair of maps $T_p : Z_p \to Y_p$ and $F_p : \Omega_p \to 2^{Y_p}$. We consider the following problem:

$$\text{Find } z \in \Omega_p \text{ such that } T_p(z) \in F_p(z). \tag{1}$$

For convenience, it is assumed that $0 \in P$. The continuity of the solution map $\Sigma$ defined by

$$\Sigma(p) = \{ z \in \Omega_p : T_p(z) \in F_p(z) \}$$

is related to stability properties of the following linearized problem:

$$\text{Find } z \in \Omega_p \text{ such that } L_p(z - z_p) + y \in F_p(z), \tag{2}$$

where $L_p : Z_p \to Y_p$ is linear, $z_p \in Z_p$, and $y \in Y$.

Throughout this paper, $\|\cdot\|$ denotes a norm in the appropriate space. Letting $B_r(z)$ denote the closed ball with center $z$ and radius $r$, we make the following definitions:

$$D_\rho(p) = \sup_{\substack{y,z \in B_\rho(z_p) \cap \Omega_p \\ y \neq z}} \frac{\|T_p(z) - T_p(y) - L_p(z-y)\|}{\|z-y\|}$$

and

$$\delta(p) = \|T_p(z_p) - y_p\|$$

where $y_p \in F_p(z_p)$. (The norms above may depend on $p$ although this dependence is not indicated explicitly.) The following result is a consequence of Corollary 1 in [4]:

THEOREM 1. *Suppose that for some positive $\sigma$ and $\gamma$ and for each $y \in B_\sigma(y_p)$,* (2) *has a unique solution, denoted $\Psi_p(y)$, that satisfies the inequality*

$$\|\Psi_p(y_1) - \Psi_p(y_2)\| \le \gamma \|y_1 - y_2\| \text{ for every } y_1 \text{ and } y_2 \in B_\sigma(y_p). \tag{3}$$

*If $D_\rho(p)$ and $\delta(p)$ tend to zero as $\rho$ and $p$ tend to 0, then for each $\rho$ and $p$ in a neighborhood of zero with $\rho \ge \gamma\delta(p)/(1 - \gamma D_\rho(p))$, equation* (1) *has a unique solution $z$ such that*

$$\|z_p - z\| \le \frac{\gamma}{1 - \gamma D_\rho(p)} \|T_p(z_p) - y_p\|.$$

## 3. Euler's method.

We apply Theorem 1 to Euler's discretization of the following nonlinear control problem with control constraints:

$$\begin{aligned} &\text{minimize } \int_I g(x(t),u(t))\,dt \\ &\text{subject to } \dot{x}(t) = f(x(t),u(t)) \text{ and } u(t) \in U \text{ a. e. } t \in I, \\ &x(0) = a, \; x \in W^{1,\infty}, \; u \in L^\infty, \end{aligned} \tag{4}$$

where $f : R^{n+m} \to R^n$, $g : R^{n+m} \to R$, $U \subset R^m$ is nonempty, closed and convex, $a$ is the given starting condition, $I$ is the interval $[0,1]$, $L^\infty$ is the space of essentially bounded functions, and $W^{1,\infty}$ is the space of Lipschitz continuous functions. We assume that there exists a solution $(x^*,u^*)$ to (4) with $u^*$ Riemann integrable, that there exists a closed set $\Delta \subset R^{n+m}$ where both $f$ and $g$ are twice continuously differentiable, and that there exists $\delta > 0$ such that $(x^*(t),u^*(t)) \in \Delta$ and the distance from $(x^*(t),u^*(t))$ to the boundary of $\Delta$ is at least $\delta$ for every $t \in I$. When we write $\dot{x}^*$, we mean a function whose values on $I$ coincide with those of $f(x^*,u^*)$.

Let $H$ denote the Hamiltonian defined by

$$H(x,u,\lambda) = g(x,u) + \lambda^T f(x,u),$$

and let $\lambda = \lambda^*$ be the solution of the adjoint equation

$$\dot{\lambda}(t) = -\frac{\partial H(x(t),u(t),\lambda(t))}{\partial x} \quad \text{a. e. } t \in I, \; \lambda(1) = 0,$$

associated with $x = x^*$ and $u = u^*$. By the minimum principle [7, p. 134], we have:

$$\frac{\partial H(x^*(t),u^*(t),\lambda^*(t))}{\partial u}(v - u^*(t)) \geq 0 \quad \text{a. e. } t \in I \text{ and for every } v \in U.$$

Given a natural number $N$, let $h = 1/N$ be the mesh spacing, and let $x_i$ and $u_i$ denote approximations to $x(t)$ and $u(t)$ at $t = t_i = ih$. We consider the Euler discretization of (4) given by

$$\begin{aligned} &\text{minimize } \sum_{i=0}^{N-1} h\, g(x_i,u_i) \\ &\text{subject to } x_{i+1} = x_i + h f(x_i,u_i) \text{ and } u_i \in U, \; i = 0, 1, \cdots, N-1, \; x_0 = a. \end{aligned} \tag{5}$$

If $(x^h,u^h)$ denotes a solution to (5), let $\lambda = \lambda^h$ denote the solution of the discrete adjoint equation

$$\lambda_i = \lambda_{i+1} + \frac{h\,\partial H(x_i,u_i,\lambda_{i+1})}{\partial x}, \quad i = N-1, N-2, \cdots, 0, \; \lambda_N = 0, \tag{6}$$

associated with $x = x^h$ and $u = u^h$. By the discrete minimum principle [7, p. 280], we have

$$\frac{\partial H(x_i^h, u_i^h, \lambda_{i+1}^h)}{\partial u_i}(v - u_i^h) \geq 0 \text{ for all } v \in U, \quad i = 0, 1, \cdots, N-1. \tag{7}$$

In order to estimate the distance between $(x^*, u^*)$ and $(x^h, u^h)$, we need a coercivity type assumption for the discrete problem. Define the following matrices:

$$A(t) = \frac{\partial f^*(t)}{\partial x}, \ B(t) = \frac{\partial f^*(t)}{\partial u}, \ Q(t) = \frac{\partial^2 H^*(t)}{\partial^2 x}, \ R(t) = \frac{\partial^2 H^*(t)}{\partial^2 u}, \ S(t) = \frac{\partial^2 H^*(t)}{\partial x \partial u}.$$

Here $f^*(t)$ and $H^*(t)$ stand for $f(x^*(t), u^*(t))$ and $H(x^*(t), u^*(t), \lambda^*(t))$, respectively. Letting $A_i$, $B_i$, $Q_i$, $S_i$, and $R_i$ denote the corresponding time varying matrices evaluated at $t = t_i$, we assume that there exists a scalar $\alpha > 0$, $\alpha$ independent of $N$, such that

$$u^T R_i u \geq \alpha |u|^2, \ 0 \leq i \leq N-1, \text{ whenever } u = v - w \text{ with } v \text{ and } w \in U, \tag{8}$$

and

$$\sum_{i=0}^{N-1} x_i^T Q_i x_i + u_i^T R_i u_i + 2x_i^T S_i u_i \geq \alpha \sum_{i=0}^{N-1} |u_i|^2 \tag{9}$$

whenever $u_i = v_i - w_i$ for some $v_i$ and $w_i \in U$, and

$$x_{i+1} = x_i + hA_i x_i + hB_i u_i, \ i = 0, 1, \cdots, N-1, \ x_0 = 0.$$

Obviously, the discrete condition (8) holds if there exists $\alpha > 0$ such that

$$u^T R(t) u \geq \alpha |u|^2 \text{ for every } t \in I \text{ and for each } u = v - w \text{ with } v \text{ and } w \in U.$$

In Appendix 1 of [4], we show that assumption (9) for the discrete problem can be deduced from an analogous assumption for the continuous problem if $u^*$ is continuous. In analyzing the discrete problem (5), we utilize a discrete $L^p$ norm defined by

$$(\|u\|_{L^p})^p = \sum_{i=0}^{N-1} h|u_i|^p, \ 1 \leq p < \infty, \quad \text{and} \quad \|u\|_{L^\infty} = \text{maximum } \{ |u_i| : 0 \leq i < N \}.$$

If $\phi$ and $v$ satisfy the finite difference system

$$\phi_{i+1} = \phi_i + hA_i\phi_i + hv_i, \ i = 0, 1, \cdots, N-1, \ \phi_0 = 0,$$

then there exists a constant $c$, independent of $h$, such that

$$|\phi_j| \leq c\|v\|_{L^1} \leq c\|v\|_{L^2}. \tag{10}$$

Squaring this inequality, multiplying by $h$, and summing over $j$ yields

$$\|\phi\|_{L^2} \leq c\|v\|_{L^2}.$$

Hence, if the coercivity condition (9) holds relative to the control, then the following joint state-control coercivity condition holds: There exists $\alpha > 0$ such that

$$h \sum_{i=0}^{N-1} x_i^T Q_i x_i + u_i^T R_i u_i + 2x_i^T S_i u_i \geq \alpha(\|x\|_{L^2}^2 + \|u\|_{L^2}^2)$$

whenever $u_i = v_i - w_i$ for some $v_i$ and $w_i \in U$, and

$$x_{i+1} = x_i + hA_i x_i + hB_i u_i, \; i = 0, 1, \cdots, N-1, \; x_0 = 0.$$

Our convergence result for the discrete problem is expressed in terms of a modulus of smoothness introduced by Sendov and Popov [9]. The local modulus of continuity $\omega(u;t,h)$ of the function $u$ is defined by

$$\omega(u;t,h) = \sup\{|u(a)-u(b)| : a, b \in [t-h/2, t+h/2] \cap I\},$$

while the average modulus of smoothness $\tau$ is given by

$$\tau(u;h) = \int_I \omega(u;t,h)\,dt.$$

In [9, pp. 8–11] it is shown that $\tau(u;h) \to 0$ as $h \to 0$ if and only if the bounded function $u$ is Riemann integrable on $I$; moreover, $\tau(u;h) = O(h)$ if and only if $u$ has bounded variation on $I$.

THEOREM 2. *If $u^*$ is Riemann integrable and the coercivity assumptions* (8) *and* (9) *hold, then for all N sufficiently large, there exists a local minimizer $(x^h, u^h)$ of* (5) *such that*

$$\underset{0 \le i \le N-1}{\text{maximum}} |u^*(t_i) - u_i^h| = O(h + \tau(u^*;h)),$$

$$\underset{0 \le i \le N}{\text{maximum}} |x^*(t_i) - x_i^h| = O(h + \tau(u^*;h)),$$

$$\underset{0 \le i \le N}{\text{maximum}} |\lambda^*(t_i) - \lambda_i^h| = O(h + \tau(u^*;h)),$$

$$\underset{0 \le i \le N-1}{\text{maximum}} \left|\dot{x}^*(t_i) - \frac{x_{i+1}^h - x_i^h}{h}\right| = O(h + \tau(u^*;h)).$$

Hence, if $u^*$ has bounded variation, then each of these error estimates is of order $h$.

*Proof.* We apply Theorem 1 to the necessary conditions associated with the discrete problem (5). The parameter $p$ of Theorem 1 is identified with the mesh spacing $h$, the set $\Omega_p$ consists of discrete triples $(x, u, \lambda)$ where $u_i \in U$ for each $i$. Component $i$, $0 \le i \le N-1$, of the operators $T_p$ and $F_p$, denoted $T_i^h$ and $F_i^h$ respectively, is the following:

$$T_i^h(x,u,\lambda) = \begin{bmatrix} \dfrac{\partial H(x_i,u_i,\lambda_{i+1})}{\partial x} + \dfrac{\lambda_{i+1}-\lambda_i}{h} \\ \dfrac{\partial H(x_i,u_i,\lambda_{i+1})}{\partial u} \\ f(x_i,u_i) - \dfrac{x_{i+1}-x_i}{h} \end{bmatrix} \quad \text{and} \quad F_i^h(x,u,\lambda) = \begin{bmatrix} 0 \\ \partial U(u_i) \\ 0 \end{bmatrix},$$

where $\partial U(u_i) = \{ w : w^T(v-u_i) \ge 0 \text{ for every } v \in U \}$ is the normal cone to the set $U$ at $u_i$. In the discrete space $Z_p$ associated with $\Omega_p$, we use the $L^\infty$ norm for each of the 3 components $x$, $u$, and $\lambda$. In the discrete space $Y_p$ associated with the range of $T_p$, we use an $L^1$ norm for the first and last component and the $L^\infty$ norm for the middle component. That is, if $y = (a,b,c) \in Y_p$, then

$$\|y\|_p = \|a\|_{L^1} + \|b\|_{L^\infty} + \|c\|_{L^1}.$$

The point $z_p$ of Theorem 1 is given by $z_p = (x^I, u^I, \lambda^I)$ where

$$x_i^I = x^*(t_i), \quad u_i^I = u^*(t_i), \quad \lambda_i^I = \lambda^*(t_i).$$

Component $i$ of the point $y_p$, denoted $y_i^h$, is the triple

$$y_i^h = \begin{bmatrix} 0 \\ \dfrac{\partial H(x_i^I,u_i^I,\lambda_i^I)}{\partial u} \\ 0 \end{bmatrix}.$$

Observe that with this choice for $y_p$, we have $y_p \in F(z_p)$. The linear operator $L_p$ is defined in the following way: It acts on a discrete triple $(x,u,\lambda)$ to produce a vector whose $i$-th component is

$$L^h(x,u,\lambda)_i = \begin{bmatrix} A_i^T\lambda_{i+1} + Q_ix_i + S_iu_i + \dfrac{\lambda_{i+1}-\lambda_i}{h} \\ R_iu_i + S_i^Tx_i + B_i^T\lambda_{i+1} \\ A_ix_i + B_iu_i - \dfrac{x_{i+1}-x_i}{h} \end{bmatrix}.$$

It can be verified that under the smoothness assumptions and for $\rho$ smaller than $\delta$, we have $D_\rho(h) \to 0$ as $h \to 0$. Now consider the term $T_p(z_p) - y_p$:

$$(T_p(z_p) - y_p)_i = \begin{bmatrix} \dfrac{\partial H(x_i^I, u_i^I, \lambda_{i+1}^I)}{\partial x} + \dfrac{\lambda_{i+1}^I - \lambda_i^I}{h} \\ \dfrac{\partial H(x_i^I, u_i^I, \lambda_{i+1}^I)}{\partial u} - \dfrac{\partial H(x_i^I, u_i^I, \lambda_i^I)}{\partial u} \\ f(x_i^I, u_i^I) - \dfrac{x_{i+1}^I - x_i^I}{h} \end{bmatrix}. \tag{11}$$

The middle component of this vector is $O(h)$ since $\lambda^*$ is Lipschitz continuous. Since the analysis of the first and last component in (11) is similar, we only focus on the last component:

$$\left| f(x_i^I, u_i^I) - \frac{x_{i+1}^I - x_i^I}{h} \right| \le \frac{1}{h} \int_{t_i}^{t_{i+1}} |f(x_i^I, u_i^I) - \dot{x}^*(t)|\, dt \le \tag{12}$$

$$\frac{1}{h} \int_{t_i}^{t_{i+1}} |f(x_i^I, u_i^I) - f(x^*(t), u^*(t))|\, dt \le \frac{c}{h} \int_{t_i}^{t_{i+1}} h + \omega(u^*; t, 2h)\, dt$$

where $c$ denotes a generic constant, independent of $h$. Multiplying (12) by $h$, summing over $i$, and exploiting the inequality $\tau(u; kh) \le k\tau(u; h)$ for each natural number $k$ (see [9, p. 11]), it follows that

$$\|T_p(z_p) - y_p\|_p = O(h + \tau(u^*; h)).$$

Next, we need to analyze the linearization and establish the existence of a constant $\gamma$ satisfying (3). The analysis essentially parallels that of [6] except that continuous norms are replaced by their discrete analogues. We need to examine how the solution to the following system depends on the perturbations $q_i$, $r_i$, and $s_i$:

$$\begin{gathered} A_i^T \lambda_{i+1} + Q_i x_i + S_i u_i + \frac{\lambda_{i+1} - \lambda_i}{h} + q_i = 0, \quad \lambda_N = 0, \\ (R_i u_i + S_i^T x_i + B_i^T \lambda_{i+1} + r_i)(v - u_i) \ge 0 \text{ for every } v \in U, \\ A_i x_i + B_i u_i - \frac{x_{i+1} - x_i}{h} + s_i = 0, \quad x_0 = a, \end{gathered} \tag{13}$$

$i = 0, 1, \cdots, N-1$. Note that the system (13) constitutes the first-order necessary conditions (see [7, p. 280]) associated with the following quadratic program:

$$\text{minimize } h \sum_{i=0}^{N-1} \frac{1}{2} x_i^T Q_i x_i + \frac{1}{2} u_i^T R_i u_i + x_i^T S_i u_i + q_i^T x_i + r_i^T u_i \tag{14}$$

subject to $x_{i+1} = x_i + hA_i x_i + hB_i u_i + hs_i$ and $u_i \in U$, $0 \le i \le N-1$, $x_0 = a$.

By Lemma 2 of [6], there is a one-to-one correspondence between a solution to (14) and a solution to (13) when (9) holds.

Now consider the perturbations $(q^i, r^i, s^i)$ for $i = 1$ and 2. Let $(x^i, u^i, \lambda^i)$ denote the associated solutions to (13). Referring to Section 2 of [6] and replacing continuous norms by the corresponding discrete norms, we have

$$\|u^1 - u^2\|_{L^2} \le c(\|q^1 - q^2\|_{L^1} + \|r^1 - r^2\|_{L^2} + \|s^1 - s^2\|_{L^1}).$$

Utilizing (10), we also conclude that

$$\|x^1 - x^2\|_{L^\infty} + \|\lambda^1 - \lambda^2\|_{L^\infty} \le c(\|q^1 - q^2\|_{L^1} + \|r^1 - r^2\|_{L^2} + \|s^1 - s^2\|_{L^1}). \tag{15}$$

Finally, by (8) and Lemma 1 of [6], we have

$$\|u^1 - u^2\|_{L^\infty} \le c(\|r^1 - r^2\|_{L^\infty} + \|x^1 - x^2\|_{L^\infty} + \|\lambda^1 - \lambda^2\|_{L^\infty}).$$

Combining this with (15) yields

$$\|x^1 - x^2\|_{L^\infty} + \|u^1 - u^2\|_{L^\infty} + \|\lambda^1 - \lambda^2\|_{L^\infty} \le$$
$$c(\|q^1 - q^2\|_{L^1} + \|r^1 - r^2\|_{L^\infty} + \|s^1 - s^2\|_{L^1}).$$

Hence, there exists a constant $\gamma$ such that (3) holds with $\sigma = \infty$.

By Theorem 1, there exists a solution to the discrete necessary conditions (6) and (7) associated with (5) which satisfies the first 3 estimates of Theorem 2. The discrete and continuous state equations along with the previously established error estimates imply that

$$\left|\dot{x}^*(t_i) - \frac{x_{i+1}^h - x_i^h}{h}\right| = |f(x^*(t_i), u^*(t_i)) - f(x_i^h, u_i^h)| = O(h + \tau(u^*; h)),$$

which gives the last estimate of Theorem 2. The fact that $x^h$ and $u^h$ are local minimizers of (5) is established in [4].

## REFERENCES


[1] B. M. Budak, E. M. Berkovich, and E. N. Solov'eva, *The convergence of difference approximations in optimal control problems*, Zh. Vychisl. Math. i Mat. Fiz., 9 (1969), pp. 522–547.

[2] J. Cullum, *Finite-dimensional approximations of state-constrained continuous optimal control problems*, SIAM J. Control, 10 (1972), pp. 649–670.

[3] A. L. Dontchev, *Error estimates for a discrete approximations to constrained control problems*, SIAM J. Numer. Anal., 13 (1981), pp. 500–514.

[4] A. L. Dontchev and W. W. Hager, *Sensitivity in nonlinear optimal control*, July 4, 1990, submitted.

[5] W. W. Hager, *Rate of convergence for discrete approximations to unconstrained control problems*, SIAM J. Numer. Anal., 13 (1076), pp. 449–471.

[6] W. W. Hager, *Multiplier methods for nonlinear optimal control*, SIAM J. Numer. Anal., 27 (1990), pp. 1061–1080.

[7] A. D. Ioffe and V. M. Tihomirov, *Theory of extremal problems*, North Holland, Amsterdam, 1979.

[8] B. Sh. Mordukhovich, *On difference approximations of optimal control systems*, Appl. Math. Mech., 42 (1978), pp. 452–461.

[9] Bl. Sendov and V.A. Popov, *The averaged moduli of smoothness*, J. Wiley & Sons, New York, 1988.

# THE EULER-BERNOULLI PLATE IS EXACTLY CONTROLLABLE VIA BENDING MOMENTS ONLY

Mary Ann Horn[1] and Irena Lasiecka[2]
Department of Applied Mathematics, University of Virginia
Thornton Hall, Charlottesville, VA 22903

## I. Introduction and Statement of the Problem

Let $\Omega$ be a bounded open set in $R^2$ with a sufficiently smooth boundary, $\Gamma$. We consider the following model of the Euler-Bernoulli equation with control action in the bending moments:

$$w_{tt} + \Delta^2 w = 0 \qquad \text{in } Q_T = (0,T)\times\Omega \tag{1.1.a}$$

$$\left.\begin{array}{l} w(0,\cdot) = w_0 \in H_0^1(\Omega) \\ w_t(0,\cdot) = w_1 \in H^{-1}(\Omega) \end{array}\right. \qquad \text{in } \Omega \tag{1.1.b}$$

$$w|_{\Sigma} = 0 \qquad \text{in } \Sigma_T = (0,T)\times\Gamma \tag{1.1.c}$$

$$\Delta w + (1-\mu)Bw = u \in L_2(\Sigma_T) \qquad \text{in } \Sigma_T \tag{1.1.d}$$

In (1.1.d), the boundary operator B takes the form

$$Bw = -\frac{\partial^2 w}{\partial \tau^2} - k\frac{\partial w}{\partial \eta} \tag{1.2}$$

where $\eta = (n_1, n_2)$ is the outward normal, $\tau = (-n_2, n_1)$, and k is the curvature of $\Gamma$. The constant $\mu$, $0 < \mu < ½$, represents Poisson's ratio.

We consider the problem of exact controllability of (1.1), i.e. given $(w_{T,1}, w_{T,2}) \in H_0^1(\Omega)\times H^{-1}(\Omega)$, we want to determine $T > 0$ and $u \in L_2(\Sigma_T)$ such that the solution, $(w, w_t)$, of (1.1) satisfies

$$w(T) = w_{T,1}; \qquad w_t(T) = w_{T,2}. \tag{1.3}$$

The problem of exact controllability for Euler-Bernoulli models with control on the boundary has attracted a lot of attention in recent years ([L-1], [L-2], [L-L], [L-T-1], [L-T-2],

---

[1]This material is based upon work supported under a National Science Foundation Graduate Fellowship.

[2]Research partially supported by National Science Foundation under Grant DMS-87-96320 and by Air Force Office of Scientific Research under Grant AFOSR-87-0321.

[L-T-3], [L-T-4]). However, in all these cases, the control action, $u$, is acting through different boundary conditions than (1.1.c)-(1.1.d). In [L-1], the Euler-Bernoulli model with control acting only through Neumann boundary conditions has been treated. The cases when only one control is active in Dirichlet boundary conditions or else control is active in both Dirichlet and Neumann boundary conditions are considered in [L-T-1] and [L-T-4]. Three papers which deal with the same boundary condition as in (1.1.c)-(1.1.d), but with $B \equiv 0$, are [L-1], [L-T-2], and [L-T-5]. But in [L-1] and [L-T-2], in order to obtain the exact controllability results on spaces of optimal regularity, it was necessary to assume that two controls are acting, one in each boundary condition. This limitation was due to a technical difficulty arising in certain controllability estimates, where, in order to dispense with a "higher order term", the simplest way was to add the second control. Since then, the question of whether the Euler- Bernoulli problem can be controlled with only one active control (e.g. bending) has become an open and physically appealing problem. A solution to exact controllability (and uniform stabilization) with control only in $\Delta w|_{\Sigma}$ is found in [L-T-5], but in spaces of finite energy, not of optimal regularity. At the same time, it has been recognized that in order to settle this question in spaces of optimal regularity, new controllability estimates are needed. The first progress in this direction is in [L-2] where the exact controllability for the model (1.1) with *one control* only was established. Although [L-2] settles the question of controllability with one control, $u \in L_2(\Sigma_T)$, the space of reachable states obtained is strictly smaller than $H_0^1(\Omega) \times H^{-1}(\Omega)$. However, regularity results for (1.1) give

$$u \in L_2(\Sigma_T) \Rightarrow (w, w_t) \in C[0,T; H_0^1(\Omega) \times H^{-1}(\Omega)] \tag{1.4}$$

(see [L-T-6]). Hence, $H_0^1(\Omega) \times H^{-1}(\Omega)$ is the natural, and in fact the optimal, space of exact controllability.

The first result where exact controllability in $\Delta w$ is established on the optimal space $H_0^1(\Omega) \times H^{-1}(\Omega)$ with one control only is a recent reprint, [L-3]. But [L-3] only considers the case when $B \equiv 0$. In fact, the assumption $B \equiv 0$ is critical for the analysis in [L-3] which is based on reduction of the plate problem to the Schrödinger problem. If $B \neq 0$, the problem loses a natural symmetry and this reduction is no longer valid.

The main contribution of this paper is that we treat a complete boundary operator, (1.1.d), as it arises in physical models and which includes moments of inertia realistically present in the system. For this model, we shall prove that, with one control acting as a bending moment, exact controllability holds on the maximal space $H_0^1(\Omega) \times H^{-1}(\Omega)$. In our case, because the boundary operator includes the term B, major new technical difficulties occur. The techniques of [L-3], which are based on microlocal estimates for the Schrödinger equation, are not applicable. Instead, we shall prove the necessary controllability estimates for the plate equation directly. The main technical contribution is the proof of new regularity estimates for the traces of the solutions to the perturbed plate equation (see Lemma 3.2).

Below we formulate our main result:

Theorem 1 : For any $T>0$ and any $(w_0, w_1)\in H_0^1(\Omega)\times H^{-1}(\Omega)$, $(w_{T,1}, w_{T,2})\in H_0^1(\Omega)\times H^{-1}(\Omega)$ there exists $u\in L_2(\Sigma_T)$ such that the solution of (1.1) satisfies

$$(w(T),\, w_t(T))=(w_{T,1}, w_{T,2}).$$

Remark 1.1 : If we consider the Kirchoff model with finite speed of propagation instead of the Euler-Bernoulli model, then the question of controllability with only one control is a simpler one. Indeed, the solutions to Kirchoff models display more regularity in the time variable. As a result, the controllability estimates are easier to obtain (see [L-T-3]). In view of this, our results assert that, among other things, the Kirchoff model in the limit, i.e., when the speed of propagation becomes infinite, displays the same controllability properties as the model with finite speed of propagation.

Remark 1.2 : The result of Theorem 1, together with regularity property (1.4), allow us to use abstract Riccati theory (see [F-L-T]). This, in turn, provides us with a solution to the stabilization problem where the feedback operator is based on the solution to the Algebraic Riccati Equation.

Remark 1.3 : One could consider a more general case when only a portion of the boundary is available to the control problem. The techniques of this paper can be easily adapted to provide a solution to this problem, assuming the boundary, $\Gamma$, satisfies some rather natural geometric conditions.

The paper is organized in the following way. In the second section, we provide some background material and we state the controllability inequality. The third and fourth sections are devoted to the proof of this inequality.

## II. Background Material and Controllability Inequality

It is enough to prove Theorem 1 for some $T>T_0>0$. Indeed, once we have the result for $T>T_0$, then an independent argument as in [L-1] allows us to deduce the same result for an arbitrary $T>0$.

We find it convenient to represent the solution to (1.1) in semigroup form. To accomplish this, we introduce the following operators:

Define $A: L_2(\Omega) \to L_2(\Omega)$ by:

$$Ay = \Delta^2 y \qquad \mathcal{D}(A) = \{y \in H^4(\Omega): y|_\Gamma = 0,\ \Delta y + (1-\mu)By|_\Gamma = 0\}. \tag{2.1}$$

Define $G: L_2(\Gamma) \to L_2(\Omega)$ by:

$$Gg = v \text{ iff } \quad \begin{cases} \Delta^2 v = 0 \\ v|_\Gamma = 0 \\ \Delta v + (1-\mu)Bv = g \end{cases}. \tag{2.2}$$

The sine and cosine operators corresponding to $A$ will be respectively denoted by:

$$S(t): L_2(\Omega) \to L_2(\Omega) \quad \text{and} \quad C(t): L_2(\Omega) \to L_2(\Omega). \tag{2.3}$$

Define $\mathcal{L}_T^i: L_2(\Sigma_T) \to L_2(\Omega)$, $i = 1, 2$, by:

$$\begin{aligned} \mathcal{L}_T^1 u &\equiv \int_0^T AS(T-\tau)\,Gu(\tau)d\tau \\ \mathcal{L}_T^2 u &\equiv \int_0^T AC(T-\tau)\,Gu(\tau)d\tau \end{aligned}. \tag{2.4}$$

By the same arguments as those in [L-T-6], we can show that the operator

$$\mathcal{L}_T \equiv \begin{bmatrix} \mathcal{L}_T^1 u \\ \mathcal{L}_T^2 u \end{bmatrix} \in \mathcal{L}(L_2(\Sigma_T) \to C[0,T; H_0^1(\Omega) \times H^{-1}(\Omega)]). \tag{2.5}$$

The solution to (1.1) can now be written as:

$$\begin{cases} w(t) = C(t)\,w_0 + S(t)\,w_1 + (\mathcal{L}_T^1 u)(t) \\ w_t(t) = -AS(t)\,w_0 + C(t)\,w_1 + (\mathcal{L}_T^2 u)(t) \end{cases}. \tag{2.6}$$

Thus, equation (1.1) is exactly controllable if and only if the operator $\mathcal{L}_T$ is from $L_2(\Sigma_T)$ onto $H_0^1(\Omega) \times H^{-1}(\Omega)$. The latter is equivilent to the statement: there exists a constant $C_T > 0$ such that

$$|\mathcal{L}_T^* v|_{L_2(\Sigma_T)} \geq C_T |v|_{H_0^1(\Omega) \times H^{-1}(\Omega)} \qquad \forall\, v \in H_0^1(\Omega) \times H^{-1}(\Omega). \tag{2.7}$$

Our next step is to compute $\mathcal{L}_T^*$.

<u>Proposition 2.1</u>: With $v = (v_0, v_1)$,

$$\mathcal{L}_T^* v = G^* A[S(T-t)A^{1/2} v_0 + C(T-t)A^{-1/2} v_1] \tag{2.8}$$

or, in a partial differential equation form,

$$\mathcal{L}_T^* v = \frac{\partial}{\partial \eta} \Psi \tag{2.9}$$

where $\Psi(t)$ is the solution to

$$\begin{cases} \Psi_{tt} = \Delta^2 \Psi \\ \Psi|_{\Gamma} = 0 \\ \Delta\Psi + (1-\mu) B \Psi = 0 \\ \Psi(T) = A^{1/2} v_1, \quad \Psi_t(T) = A^{-1/2} v_0 \end{cases} \tag{2.10}$$

Proof : From the interpolation result of [G-1], $\mathcal{D}(A^{1/4}) = H_0^1(\Omega)$. Therefore,

$$|y|_{H_0^1(\Omega)} = |A^{1/4} y|_{L_2(\Omega)} \quad \text{and} \quad |y|_{H^{-1}(\Omega)} = |A^{-1/4} y|_{L_2(\Omega)}.$$

By using the above fact and Fubini's theorem, it is easily seen that

$$(L_T^* u, v)_{H_0^1(\Omega)\times H^{-1}(\Omega)} = \langle u, G^* A[S(T-\tau) A^{1/2} v_0 + C(T-\tau) A^{-1/2} v_1] \rangle_{L_2[(0,T)\times\Gamma]}. \blacksquare$$

We find it convenient to introduce another change of variable. Let

$$z \equiv A_D^{-1} \Psi \tag{2.11}$$

where $\Psi$ satisfies (2.10) and

$$A_D \Psi \equiv \Delta\Psi \quad \forall \Psi \in H_0^1(\Omega) \cap H^2(\Omega).$$

Clearly,

$$z|_{\Gamma} = \Delta z|_{\Gamma} = 0. \tag{2.12}$$

We can easily show that $z(t)$ satisfies the equation

$$z_{tt} + \Delta^2 z = (1-\mu) D \left(k \frac{\partial}{\partial \eta} \Delta z\right)$$

$$\text{where } Dg = v \quad \text{iff} \quad \begin{cases} \Delta v = 0 \text{ in } \Omega \\ v|_{\Gamma} = g \end{cases}. \tag{2.13}$$

Moreover, since for all $t$

$$z(t) = A_D^{-1} \Psi(t) \quad \text{and} \quad z_t(t) = A_D^{-1} \Psi_t(t),$$

we can find, using interpolation theory ([G-1]), that

$$\begin{aligned} |A_D^{3/2} z(t)|_{L_2(\Omega)} &= |A_D^{3/2} A_D^{-1} \Psi(t)|_{L_2(\Omega)} = |A_D^{1/2} \Psi(t)|_{L_2(\Omega)} \sim |A^{1/4} \Psi(t)|_{L_2(\Omega)} \\ |A_D^{1/2} z_t(t)|_{L_2(\Omega)} &= |A_D^{-1/2} \Psi_t(t)|_{L_2(\Omega)} \sim |A^{-1/4} \Psi(t)|_{L_2(\Omega)}. \end{aligned} \tag{2.14}$$

Since

$$\frac{d}{dt}\left[ |A^{1/4} \Psi(t)|^2_{L_2(\Omega)} + |A^{-1/4} \Psi_t(t)|^2_{L_2(\Omega)} \right] = 0, \tag{2.15}$$

by (2.14) and (2.15), we obtain that for any $t_1, t_2$,

$$|A_D^{3/2} z(t_1)|^2_{L_2(\Omega)} + |A_D^{1/2} z_t(t_1)|^2_{L_2(\Omega)} \sim |A_D^{3/2} z(t_2)|^2_{L_2(\Omega)} + |A_D^{1/2} z_t(t_2)|^2_{L_2(\Omega)}. \tag{2.16}$$

Noting that

$$\frac{\partial}{\partial\eta}\Delta z = \frac{\partial}{\partial\eta}\Psi \tag{2.17}$$

and recalling Proposition 2.1, inequality (2.7) can be equivilently expressed as

Lemma 2.1 : (Controllability Inequality)

The result of Theorem 1 holds iff the following inequality is satisfied: there exists $T > 0$ and $C_T > 0$ such that for any $z(t)$ satisfying both equation (2.13) with boundary conditions (2.12) and the equivilence relation (2.16), we have

$$|\frac{\partial}{\partial\eta}\Delta z|_{L_2(\Sigma_T)} \geq C_T \left[|A_D^{3/2} z(T)|_{L_2(\Omega)} + |A_D^{1/2} z_t(T)|_{L_2(\Omega)}\right]. \tag{2.18}$$

Proof : Follows at once from (2.7), (2.9), and (2.11). Indeed, it is enough to note that from (2.10), (2.14), and (2.15) we have

$$|v_1|_{H^{-1}(\Omega)} \sim |A^{-1/4} A^{1/2} \Psi(T)|_{L_2(\Omega)} = |A^{1/4} \Psi(T)|_{L_2(\Omega)} \sim |A_D^{3/2} z(T)|_{L_2(\Omega)}.$$

$$|v_0|_{H_0^1(\Omega)} \sim |A^{1/4} A^{-1/2} \Psi_t(T)|_{L_2(\Omega)} \sim |A_D^{1/2} z_t(T)|_{L_2(\Omega)}. \blacksquare$$

## III. Proof of Controllability Inequality

Lemma 3.1 : Let $z$ be the solution to (2.12)-(2.13) with the property (2.16). Then

$$|\frac{\partial}{\partial\eta}\Delta z|^2_{L_2(\Sigma_T)} + |\frac{\partial}{\partial\eta} z_t|^2_{L_2(\Sigma_T)} \geq C_T \left[|A_D^{3/2} z(T)|^2_{L_2(\Omega)} + |A_D^{1/2} z_t(T)|^2_{L_2(\Omega)}\right]. \tag{3.1}$$

Proof : By multiplying both sides of equation (2.13) by $\vec{h}\cdot\nabla(\Delta z)$, where $\vec{h} = \vec{x} - \vec{x}_0$, $\vec{x}_0 \in R^2$, and using the boundary conditions (2.12), we can find

$$\begin{aligned}\int_{Q_T}\left\{|\nabla z_t|^2 + |\nabla(\Delta z)|^2\right\} d\Omega dt &\leq |(z_t, \vec{h}\cdot\nabla(\Delta z))_\Omega|_0^T + \frac{n}{2}|\int_{Q_T}\left\{|\nabla z_t|^2 - |\nabla(\Delta z)|^2\right\} d\Omega dt| \\ &+ \frac{3}{2} M_h \int_{\Sigma_T} |\frac{\partial z_t}{\partial\eta}|^2 d\Gamma dt + \frac{3}{2} M_h \int_{\Sigma_T} |\frac{\partial(\Delta z)}{\partial\eta}|^2 d\Gamma dt \\ &+ (1-\mu)|< k\frac{\partial(\Delta z)}{\partial\eta}, D^*(\vec{h}\cdot\nabla(\Delta z)) >_{L_2(\Sigma_T)}|.\end{aligned} \tag{3.2}$$

To bound the last term on the right-hand side of equation (3.2), note that

$$D^* \in \mathcal{L}(L_2(\Omega) \rightarrow L_2(\Gamma)).$$

Therefore,

$$(1-\mu)|<k\frac{\partial(\Delta z)}{\partial\eta}, D^*(\vec{h}\cdot\nabla(\Delta z))>_{L_2(\Sigma_T)}|$$
$$\leq(1-\mu)M_k\left\{\frac{1}{4\varepsilon}|\frac{\partial(\Delta z)}{\partial\eta}|^2_{L_2(\Sigma_T)}+\varepsilon M_h^2M_D^2|\nabla(\Delta z)|^2_{L_2(Q_T)}\right\}, \quad (3.3)$$

where $M_k$, $M_h$, and $M_D$ are constants depending respectively on $k$, $\vec{h}$, and the operator $D$. To bound the second term on the right-hand side of equation (3.2), we use the multiplier $\Delta z$ to get

$$|\int_{Q_T}\left\{|\nabla z_t|^2-|\nabla(\Delta z)|^2\right\}d\Omega dt|\leq(1-\mu)\frac{M_k}{4\varepsilon}|\frac{\partial(\Delta z)}{\partial\eta}|^2_{L_2(\Sigma_T)}+\varepsilon C|\nabla(\Delta z)|^2_{L_2(Q_T)}. \quad (3.4)$$

Considering the first term on the right-hand side of (3.2), we can show, using (2.16), that

$$|(z_t,\vec{h}\cdot\nabla(\Delta z))_\Omega|_0^2\leq\frac{M_h}{2\varepsilon}|z_t|^2_{C[0,T;L_2(\Omega)]}+\varepsilon\left[|A_D^{3/2}z(T)|^2_{L_2(\Omega)}+|A_D^{1/2}z_t(T)|^2_{L_2(\Omega)}\right]. \quad (3.5)$$

By substituting equations (3.3)-(3.5) into (3.2) and using the property (2.16), we arrive at

$$|\frac{\partial}{\partial\eta}\Delta z|^2_{L_2(\Sigma_T)}+|\frac{\partial}{\partial\eta}z_t|^2_{L_2(\Sigma_T)}+|z_t|^2_{C[0T;L_2(\Omega)]}$$
$$\leq C_T\left[|A_D^{3/2}z(T)|^2_{L_2(\Omega)}+|A_D^{1/2}z_t(T)|^2_{L_2(\Omega)}\right]. \quad (3.6)$$

By the well known compactness argument, we can show that

$$|z_t|^2_{C[0,T;L_2(\Omega)]}\leq C\left[|\frac{\partial}{\partial\eta}\Delta z|^2_{L_2(\Omega)}+|\frac{\partial}{\partial\eta}z_t|^2_{L_2(\Omega)}\right] \quad (3.7)$$

which, together with (3.6), gives us (3.1). ■

**Remark 3.1** : Inequality (3.1) together with the techniques of [L-T-3] imply the regularity result of (2.5).

Looking at inequality (3.1), it is clear that by eliminating the term $|\frac{\partial}{\partial\eta}z_t|^2_{L_2(\Sigma_T)}$ from (3.1), inequality (2.18) will follow. In fact, this is the main difficulty and novelty in this paper.

Let $\alpha>0$ be a given constant and define $\Sigma_{T_\alpha}\equiv\Gamma\times[-\alpha, T+\alpha]$.

Lemma 3.2 : For any $\varepsilon > 0$,

$$|\frac{\partial}{\partial\eta} z_t|^2_{L_2(\Sigma_T)} \le C_{T,\varepsilon}\left[|\frac{\partial}{\partial\eta}\Delta z|^2_{L_2(\Sigma_{T_\alpha})} + |z(T)|^2_{L_2(\Omega)} + |z_t(T)|^2_{L_2(\Omega)}\right]$$
$$+\varepsilon\left[|A_D^{3/2} z(T)|^2_{L_2(\Omega)} + |A_D^{1/2} z_t(T)|^2_{L_2(\Omega)}\right]. \tag{3.8}$$

Assuming that Lemma 3.2 is valid, we shall now prove the controllability inequality (2.18).

Proof of Lemma 2.1 : Combining the results of Lemma 3.1 and Lemma 3.2 we obtain, since $\varepsilon$ can be chosen to be arbitrarily small,

$$|A_D^{3/2} z(T)|^2_{L_2(\Omega)} + |A_D^{1/2} z_t(T)|^2_{L_2(\Omega)}$$
$$\le C_T\left[|\frac{\partial}{\partial\eta}\Delta z|^2_{L_2(\Sigma_{T_\alpha})} + |z(T)|^2_{L_2(\Omega)} + |z_t(T)|^2_{L_2(\Omega)}\right]. \tag{3.9}$$

Using our compactness argument, we obtain the estimate

$$|z(T)|^2_{L_2(\Omega)} + |z_t(T)|^2_{L_2(\Omega)} \le C_T |\frac{\partial}{\partial\eta}\Delta z|^2_{L_2(\Sigma_{T_\alpha})}. \tag{3.10}$$

Combining (3.10) with the equivilence relation in (2.16), we find

$$|A_D^{3/2} z(T+\alpha)|^2_{L_2(\Omega)} + |A_D^{1/2} z_t(T+\alpha)|^2_{L_2(\Omega)} \le \bar{C}_T \int_{-\alpha}^{T+\alpha} |\frac{\partial}{\partial\eta}\Delta z|^2_\Gamma\, dt. \tag{3.11}$$

Finally, introducing the new variable $\tilde{z}(t) \equiv z(t-\alpha)$ in (3.11) yields

$$|A_D^{3/2} \tilde{z}(T+2\alpha)|^2_{L_2(\Omega)} + |A_D^{1/2} \tilde{z}(T+2\alpha)|^2_{L_2(\Omega)} \le \bar{C}_T \int_0^{T+2\alpha} |\frac{\partial}{\partial\eta}\Delta\tilde{z}|^2_\Gamma dt. \tag{3.12}$$

Since both $\tilde{z}$ and $z$ are solutions to the same problem, (2.12)-(2.13) and (2.16), we obtain the statement of Lemma 2.1 with $T$ replaced by $T+2\alpha$. ■

Next we must prove Lemma 3.2.

Proof of Lemma 3.2 : Using the variation of parameter formula to write the solution of (2.12)-(2.13), we obtain

$$z(t) = e^{iA_D(T-t)}\tilde{z}_0 + e^{-iA_D(T-t)}\tilde{z}_1 - A_D^{-1}\int_t^T \frac{1}{2i}(e^{iA_D(t-\tau)} - e^{-iA_D(t-\tau)})Df(\tau)d\tau \tag{3.13}$$

$$\text{where } \begin{cases} f \equiv (1-\mu)\,k\frac{\partial}{\partial\eta}\Delta z \\ \tilde{z}_0 \equiv \tfrac{1}{2}z_0 + \tfrac{1}{2}A_D^{-1}z_1. \\ \tilde{z}_1 \equiv \tfrac{1}{2}z_0 - \tfrac{1}{2}A_D^{-1}z_1 \end{cases}$$

Define the following:

$$\begin{cases} A_1 \equiv \frac{\partial}{\partial\eta}A_D\, e^{iA_D(T-t)}\tilde{z}_0 \\ A_2 \equiv \frac{\partial}{\partial\eta}A_D\, e^{-iA_D(T-t)}\tilde{z}_1 \\ B_1 \equiv \tfrac{1}{2}\frac{\partial}{\partial\eta}\int_t^T e^{iA_D(t-\tau)}D\,f(\tau)d\tau \\ B_2 \equiv \tfrac{1}{2}\frac{\partial}{\partial\eta}\int_t^T e^{-iA_D(t-\tau)}D\,f(\tau)d\tau. \end{cases} \tag{3.14}$$

Then

$$\frac{\partial}{\partial\eta}z_t(t) = -i(A_1 - A_2) - (B_1 + B_2);$$

$$\frac{\partial}{\partial\eta}\Delta z(t) = (A_1 + A_2) + i(B_1 - B_2).$$

Hence

$$\left|\frac{\partial}{\partial\eta}z_t(t,x)\right|^2 - \left|\frac{\partial}{\partial\eta}\Delta z(t,x)\right|^2 = -4\,\mathrm{Re}B(t,x) \tag{3.15}$$

$$\begin{aligned} \text{where } B(t,x) &\equiv (A_1 + iB_1)\overline{(A_2 - iB_2)} \\ &= A_1\bar{A}_2 - iA_1\bar{B}_2 + iA_2\bar{B}_1 - B_1\bar{B}_2. \end{aligned} \tag{3.16}$$

Let $\Phi(t) \in C_0^\infty(R^2)$ be such that $\Phi(t) \equiv 1$ on $[0, T]$ and $\Phi(t) \equiv 0$ on $(-\infty, -\alpha) \cup (T+\alpha, \infty)$. Then by (3.15),

$$\begin{aligned} \left|\frac{\partial}{\partial\eta}z_t\right|^2_{L_2(\Sigma_T)} &\le \int_{-\infty}^{\infty}\Phi(t)\left|\frac{\partial}{\partial\eta}z_t\right|^2_{L_2(\Gamma)}dt = \int_{-\infty}^{\infty}\Phi(t)\left|\frac{\partial}{\partial\eta}\Delta z\right|^2_{L_2(\Gamma)}dt \\ &-4\,\mathrm{Re}\int_{-\infty}^{\infty}\Phi(t)\int_\Gamma B(t,x)\,dx\,dt \le \left|\frac{\partial}{\partial\eta}\Delta z\right|^2_{L_2(\Sigma_T)} + 4\left|\int_{-\infty}^{\infty}\Phi(t)\int_\Gamma B(t,x)\,dx\,dt\right|. \end{aligned} \tag{3.17}$$

By using the same arguments as in [L-3], we show that

$$|\int_{-\infty}^{\infty} \Phi(t) \int_{\Gamma} A_2 \bar{A}_1 \, dx \, dt| \le C_T \left[ |z_0|^2_{L_2(\Omega)} + |z_1|^2_{L_2(\Omega)} \right]. \tag{3.18}$$

In order to estimate the remaining three terms on the right hand side of (3.16), we need the following result.

<u>Proposition 3.1</u> : Let v be a solution of

$$\begin{cases} v_t = iA_D v - Dg \\ v(T) = v_0 \in \mathcal{D}(A_D^{1/2}) \end{cases}. \tag{3.19}$$

Then

$$|\frac{\partial}{\partial \eta} v|_{L_2(\Sigma_T)} \le C_T \left[ |A_D^{1/2} v_0|^2_{L_2(\Omega)} + |g|_{L_2(\Sigma_T)} \right]. \tag{3.20}$$

Proposition 3.1 will be proven in the next section.

<u>Remark 3.2</u> : Notice that inequality (3.20) does not follow from standard regularity theory for the Schrödinger equation. In fact, with $g \in L_2(\Sigma_T)$, one has $Dg \in L_2[0,T; H^{1/2-\varepsilon}(\Omega)]$ and standard regularity theory gives $v \in C[0,T; H^{1/2-\varepsilon}(\Omega)]$. This result would at most imply $\frac{\partial}{\partial \eta} v \in L_2[0,T; H^{-1/2-\varepsilon}(\Omega)]$. Instead, (3.20) allows us to gain one additional derivative on the boundary for $\frac{\partial}{\partial \eta} v$.

Assuming the validity of Proposition 3.1, we now continue with the proof of Lemma 3.2. Applying the result of Proposition 3.1 with $g \equiv 0$ we obtain

$$\begin{aligned} |A_1|^2_{L_2(\Sigma_{T\alpha})} + |A_2|^2_{L_2(\Sigma_{T\alpha})} &\le C_T \left[ |A_D^{3/2} \tilde{z}_0|^2_{L_2(\Omega)} + |A_D^{3/2} \tilde{z}_1|^2_{L_2(\Omega)} \right] \\ &\le C_T \left[ |A_D^{3/2} z_0|^2_{L_2(\Omega)} + |A_D^{1/2} z_1|^2_{L_2(\Omega)} \right]. \end{aligned} \tag{3.21}$$

Again applying the result of Propsition 3.1, but now with $v_0 \equiv 0$, we find

$$|B_1|^2_{L_2(\Sigma_{T\alpha})} + |B_2|^2_{L_2(\Sigma_{T\alpha})} \le C_T |f|^2_{L_2(\Sigma_{T\alpha})} = C_T |\frac{\partial}{\partial \eta} \Delta z|^2_{L_2(\Sigma_{T\alpha})}. \tag{3.22}$$

Hence

$$\begin{aligned} &\int_{-\infty}^{\infty} \Phi(t) \int_{\Gamma} \left[ |A_1 \bar{B}_2| + |A_2 \bar{B}_1| + |B_1 \bar{B}_2| \right] dx \, dt \\ &\le \int_{\Sigma_{T\alpha}} \left[ |A_1| |B_2| + |A_2| |B_1| + |B_1| |B_2| \right] \\ &\le C_T \left[ |A_D^{3/2} z_0|_{L_2(\Omega)} + |A_D^{1/2} z_1|_{L_2(\Omega)} + |\frac{\partial}{\partial \eta} \Delta z|_{L_2(\Sigma_{T\alpha})} \right] |\frac{\partial}{\partial \eta} \Delta z|_{L_2(\Sigma_{T\alpha})}. \end{aligned} \tag{3.23}$$

Combining (3.17), (3.18), and (3.23) gives us

$$\left|\frac{\partial}{\partial\eta}z_t\right|^2_{L_2(\Sigma_T)} \le C_T\left|\frac{\partial}{\partial\eta}\Delta z\right|^2_{L_2(\Sigma_{T_\alpha})} + C_T\left[|z_0|^2_{L_2(\Omega)} + |z_1|^2_{L_2(\Omega)}\right] + \varepsilon\left[|A_D^{3/2}z_0|^2_{L_2(\Omega)} + |A_D^{1/2}z_1|^2_{L_2(\Omega)}\right] \tag{3.24}$$

where ε can be taken to be arbitrarily small. Thus, the proof of Lemma 3.2 is complete. ∎

## IV. Proof of Proposition 3.1

Let

$$\mathrm{v}(t) = e^{iA_D(T-t)}\mathrm{v}_0 + \int_t^T e^{iA_D(T-t)}Dg(\tau)\,d\tau \equiv \mathrm{v}_1(t) + \mathrm{v}_2(t).$$

From the above equation, we can see that $\mathrm{v}(t)$ satisfies equation (3.15).

<u>Step 1</u>: We multiply equation (3.15) by $\vec{h}\cdot\nabla v$, where $\vec{h}|_\Gamma = \vec{\eta}$, and integrate by parts. This gives

$$\left|\frac{\partial}{\partial\eta}\mathrm{v}\right|_{L_2(\Sigma_T)} \le C_T\left[|Dg|_{L_1(0,T;\Omega)} + |A_D^{1/2}\mathrm{v}|_{L_\infty(0,T;\Omega)}\right]. \tag{4.1}$$

<u>Proof</u>: Using the multiplier $\vec{h}\cdot\nabla\bar{\mathrm{v}}$, we obtain

$$\mathrm{Im}\int_{Q_T}\mathrm{v}_t\vec{h}\cdot\nabla\bar{\mathrm{v}}\,d\Omega\,dt - \tfrac{1}{2}\int_{\Sigma_T}\left|\frac{\partial\mathrm{v}}{\partial\eta}\right|^2 d\Gamma\,dt + \int_{Q_T}H|\nabla\mathrm{v}|^2 d\Omega\,dt - \tfrac{1}{2}\int_{Q_T}|\nabla\mathrm{v}|^2\,\mathrm{div}\,\vec{h}\,d\Omega\,dt = -\mathrm{Im}\int_{Q_T}Dg\,\vec{h}\cdot\nabla\bar{\mathrm{v}}\,d\Omega\,dt. \tag{4.2}$$

Let

$$F = \vec{h}\,\overline{\Psi}\Psi_t.$$

Then

$$\mathrm{div}F = \overline{\Psi}\Psi_t\,\mathrm{div}\vec{h} + (\vec{h}\cdot\nabla\overline{\Psi})\Psi_t + (\vec{h}\cdot\nabla\Psi_t)\overline{\Psi}.$$

Therefore, by using the divergence theorem, we can find

$$\mathrm{Im}\int_{Q_T}\mathrm{v}_t\vec{h}\cdot\nabla\bar{\mathrm{v}}\,d\Omega\,dt = \tfrac{1}{2}\int_{\Sigma_T}\bar{\mathrm{v}}\mathrm{v}_t\,d\Gamma\,dt - \tfrac{1}{2}\int_{Q_T}\bar{\mathrm{v}}\mathrm{v}_t\,\mathrm{div}\vec{h} - \tfrac{1}{2}\int_\Omega[\bar{\mathrm{v}}(T)\vec{h}\cdot\nabla\mathrm{v}(T) - \bar{\mathrm{v}}(0)\vec{h}\cdot\nabla\mathrm{v}(0)]. \tag{4.3}$$

Combining equations (4.2) and (4.3), we get

$$\frac{1}{2}\int_{\Sigma_T}|\frac{\partial v}{\partial\eta}|^2\,d\Gamma\,dt=-\frac{1}{2}\int_{Q_T}\bar{v}v_t\,\mathrm{div}\vec{h}\,d\Omega\,dt$$

$$+\frac{1}{2}\int_{\Omega}[\bar{v}(0)\vec{h}\cdot\nabla v(0)-\bar{v}(T)\vec{h}\cdot\nabla v(T)]\,d\Omega+\int_{Q_T}H|\nabla v|^2\,d\Omega\,dt \tag{4.4}$$

$$-\frac{1}{2}\int_{Q_T}|\nabla v|^2\,\mathrm{div}\vec{h}\,d\Omega\,dt+\mathrm{Im}\int_{Q_T}D\,g\,\vec{h}\cdot\nabla\bar{v}\,d\Omega\,dt.$$

Next, using the multiplier $\bar{v}$ with equation (3.15), we obtain

$$|\int_{Q_T}v_t\bar{v}|d\Omega\,dt\le(1+C\varepsilon)|\nabla v|^2_{L_\infty[0,T;L_2(\Omega)]}+\frac{1}{4\varepsilon}|Dg|^2_{L_1[0,T;L_2(\Omega)]}. \tag{4.5}$$

By combining (4.5) with (4.4), we obtain our desired inequality, (4.1). ■

<u>Step 2</u> : Take $g\equiv 0$ in (3.15). Since $v_1$ is the solution to the resulting problem, we will use the multiplier $\bar{v}_{1,t}$. This yields

$$|A_D^{1/2}v_1(t)|^2_{L_2(\Omega)}=\text{constant}=|A_D^{1/2}v_0|^2_{L_2(\Omega)}. \tag{4.6}$$

Hence, for all $x\in H_0^1(\Omega)$,

$$\sup_t|A_D^{1/2}e^{iA_Dt}x|^2_{L_2(\Omega)}=|A_D^{1/2}x|^2_{L_2(\Omega)} \tag{4.7}$$

and therefore,

$$|A_D^{1/2}v_1|_{L_\infty[0,T;L_2(\Omega)]}=|A_D^{1/2}v_0|_{L_2(\Omega)}. \tag{4.8}$$

Thus, applying step 1 with $g\equiv 0$, we find

$$|\frac{\partial}{\partial\eta}v_1|^2_{L_2(\Sigma_T)}\le C_T|A_D^{1/2}v_0|^2_{L_2(\Omega)}. \tag{4.9}$$

<u>Step 3</u> : We shall prove

$$|v_2|_{L_\infty[0,T;H_0^1(\Omega)]}\le C_T|g|_{L_2(\Sigma_T)}. \tag{4.10}$$

We define the closed and densely defined operator $L: L_2(\Sigma_T)\to L_2(Q_T)$ by:

$$(Lf)(t)\equiv A_D\int_t^T e^{iA_D(t-\tau)}Df(\tau)d\tau.$$

Then we can easily show that

$$(L^*g)(t)\equiv D^*A_D\int_0^t e^{-iA_D(t-\tau)}g(\tau)d\tau.$$

If we let

$$\Psi(t) \equiv \int_0^t e^{-iA_D(t-\tau)} g(\tau) d\tau,$$

then $\Psi(t)$ satisfies

$$\begin{cases} \Psi_t = -iA_D\Psi + g \\ \Psi(0) = 0. \end{cases} \tag{4.11}$$

Similarly to the proof in step 1, we can find

$$|\frac{\partial}{\partial \eta}\Psi|_{L_2(\Sigma_T)} \le C_T \left[ |g|_{L_1[0,T;L_2(\Omega)]} + |A_D^{1/2}\Psi|_{L_\infty[0,T;L_2(\Omega)]} \right]. \tag{4.12}$$

But from (4.6), we have

$$|A_D^{1/2}\Psi(t)|_{L_2(\Omega)} \le \int_0^t |A_D^{1/2} g(\tau)|_{L_2(\Omega)} d\Omega \le |A_D^{1/2} g|_{L_1[0,T;L_2(\Omega)]}. \tag{4.13}$$

Combining (4.12) and (4.13) yields

$$|\frac{\partial}{\partial \eta}\Psi|_{L_2(\Sigma_T)} \le C_T \left[ |A_D^{1/2} g|_{L_1[0,T;L_2(\Omega)]} \right] = C_T |g|_{L_1[0,T;H_0^1(\Omega)]} \tag{4.14}$$

which means that

$$L^* \in \mathcal{L}(L_1[0,T;H_0^1(\Omega)] \to L_2(\Sigma_T)). \tag{4.15}$$

Hence,

$$L \in \mathcal{L}(L_2(\Sigma_T) \to L_\infty[0,T;H^{-1}(\Omega)]). \tag{4.16}$$

Let

$$Kf \equiv A_D^{-1} Lf.$$

Because of the regularity of $A_D^{-1}$, (4.16) is equivilent to the statement

$$K \in \mathcal{L}(L_2(\Sigma_T) \to L_\infty[0,T;H_0^1(\Omega)]). \tag{4.17}$$

In particular, this means that

$$|v_2|_{L_\infty[0,T;H_0^1(\Omega)]} \le C_T |g|_{L_2(\Sigma_T)} \tag{4.18}$$

as desired.

<u>Step 4</u> : Combining (4.7) and (4.10), we find

$$|A_D^{1/2} v|_{L_\infty[0,T;L_2(\Omega)]} \le C_T \left[ |A_D^{1/2} v_0|_{L_2(\Omega)} + |g|_{L_2(\Sigma_T)} \right]. \tag{4.19}$$

By substituting (4.19) into (4.1) and recalling that $D \in \mathcal{L}(L_2(\Gamma) \to L_2(\Omega))$, the desired result of Proposition 3.1 is found. ■

**References**


[G-1] P. Grisvard. "Caracterization de quelques espaces d'interpolation", *Arch. Rational Mech. Anal.*, **25** (1967), pp. 40-63.

[L-1] J. L. Lions. *Contrôlabilité exacte de systèmes distribués*, Masson, Paris, 1988.

[L-2] I. Lasiecka. "Exact controllability of a plate equation with one control acting as a bending moment", *Mercel Dekker*, (to appear).

[L-3] G. Lebeau. "Controle de l'equation de Schrödinger", *Report Universite de Paris-Sud Mathematiques*, 1989.

[F-L-T] F. Flandoli, I. Lasiecka and R. Triggiani. "Algebraic Riccati Equations with non-smooth observation arising in hyperbolic and Euler-Bernoulli equations", *Annali di Math. Pura Appl. (IV)* **25** (1988), pp. 307-382.

[L-L] J. Lagnese and J. L. Lions. *Modelling, Analysis and Control of Thin Plates*, Masson, Paris, 1988.

[L-T-1] I. Lasiecka and R. Triggiani. "Exact controllability of the Euler-Bernoulli equation with controls in the Dirichlet and Neumann boundary conditions: a non-conservative case", *SIAM J. Control Optim.* **27** (1989), pp. 330-373.

[L-T-2] I. Lasiecka and R. Triggiani. "Exact controllability of the Euler-Bernoulli equation with boundary controls for displacement and moment", *J. Math. Anal. Appl.*, **146**, No. 1 (1990), pp. 1-33.

[L-T-3] I. Lasiecka and R. Triggiani. "Exact controllability and uniform stabilization of Kirchoff plates with boundary control only on $\Delta w|_\Sigma$ and homogeneous boundary displacement", *J. Differential Equations*, (to appear).

[L-T-4] I. Lasiecka and R. Triggiani. Further results on exact controllability of the Euler-Bernoulli equation with controls on the Dirichlet and Neumann boundary conditions, *in* "Lecture Notes in Control and Information Sciences", Proceedings of Conference on Stabilization of Flexible Structures, Springer-Verlag, (to appear).

[L-T-5] I. Lasiecka and R. Triggiani. "Exact controllability and uniform stabilization of Euler-Bernoulli equations with boundary control only in $\Delta w|_\Sigma$", *Boll. Un. Mat. Ital.*, (to appear).

[L-T-6] I. Lasiecka and R. Triggiani. "Regularity theory for a class of nonhomogeneous Euler-Bernoulli equations: a cosine operator approach", *Boll. Un. Mat. Ital. (7)*, 3-B (1989), pp.199-228.

# A PROSPECTIVE LOOK AT SQP METHODS FOR SEMILINEAR PARABOLIC CONTROL PROBLEMS

F.-S. Kupfer and E. W. Sachs
Universität Trier, FB IV – Mathematik
Postfach 3825
D-5500 Trier
Federal Republic of Germany

## 1 Introduction

In this paper we present different optimization methods which can be used to solve optimal control problems with nonlinear parabolic differential equations. In particular, we want to show how more sophisticated methods can be implemented for these problems.

We consider the following optimization problem in infinite dimensions as the framework in which we present various algorithms:

$$\min f(x) \text{ s.t. } h(x) = 0 \tag{1.1}$$
$$f : X \to \mathbb{R}\,,\ h : X \to Z\,,$$

where $X$ is a real Hilbert space and $Z$ is a Banach space. Quite some progress has been made in the past few years in the analysis of quasi-Newton methods for infinite-dimensional problems. This includes recently also SQP methods and later we want to focus on some of these new results.

In Section 2 we introduce a decoupling of variables which is typical for control problems because the unknowns separate into state and control. We start our tour of optimization algorithms with brief comments on the gradient method, Newton's method and the BFGS method as a representative of the family of quasi-Newton methods [6]. In all the previous methods we view the state as a dependent variable so that the minimization is only with respect to the control. While this viewpoint is the correct one with regard to storage it has its disadvantages because each gradient evaluation for some control requires the computation of the corresponding state. In our application this means the solution of a nonlinear boundary value problem.

SQP (Sequential Quadratic Programming) methods, see e.g. [15], have turned out to be one of the most successful methods in nonlinear optimization. Here the nonlinear equality constraint is linearized at each iteration and a quadratic approximation of the Lagrangian is minimized. Since the calculation of second derivatives is very expensive quasi-Newton updates for the Hessian of the Lagrangian are commonly used. This, however, causes some difficulty since the positive definiteness of this Hessian can only be guaranteed on a subspace. Another drawback is the large dimension of the space of discretized variables which includes both states and controls. These two disadvantages of the full SQP method lead us to consider reduced SQP methods. They have the advantage that the requirement on positive definiteness is in line with the second order sufficiency condition and that the update is carried out in the control space.

In the third section we consider a semilinear parabolic differential equation

$$\begin{aligned}
y_t(t,x) &= (\Delta y)(t,x) + c(t,x,y(t,x)) && t \in (0,T), \quad x \in \Omega \\
y(0,x) &= y_0(x) && x \in \Omega \\
\frac{\partial y}{\partial n}(t,\xi) &= g(\xi)u(t) && t \in (0,T), \quad \xi \in \Gamma
\end{aligned} \tag{1.2}$$

and minimize

$$\int_0^T \int_\Omega l(t,x,y(t,x))\, dx\, dt + \int_0^T k(t,u(t))\, dt. \tag{1.3}$$

We are led by the objective to introduce the reader to modern methods in optimization and to show for an optimal control problem with a nonlinear partial differential equation how these methods can be implemented. It is not our goal in this paper to justify rigorously all the assumptions. This is by all means no trivial task and currently under research as well as the actual numerical computation. One can formulate the nonlinear partial differential equation as an integral equation in a Banach space, see e.g. [17], compute derivatives and adjoints which we interpret as classical solutions with data sufficiently smooth. It is conceivable that one remains during the iteration in spaces with smooth controls and states, if one starts in these spaces. The problem with this approach is that the convergence theory and conditions of second order sufficiency usually require a Hilbert space framework so that one cannot use any of the Banach spaces in [16].

The use of SQP methods in their reduced version with a quasi-Newton update presents a new approach to an efficient solution of (1.2) and (1.3). In this paper we formulate reduced SQP methods in a framework of separable variables which is well suited for optimal control problems. This setting suggests to use a representation of the nullspace of $h'$ which is nonstandard in the reduced SQP context. We interpret each step in the iteration of the reduced SQP method for the control problem and give the corresponding initial-boundary value problems which need to be solved.

The last section reviews some of the convergence results for the different algorithms. The various statements which describe local convergence properties give the reader some insight why one tends to prefer certain methods over others. This section includes recent results on the convergence of reduced SQP methods in infinite dimensions [11].

## 2 General Optimization Methods

In this section we discuss various optimization algorithms which include those that are commonly used for control problems and others which are not so well known in the control community.

We treat optimal control problems as infinite-dimensional optimization problems with equality constraints as defined in (1.1). The point $\hat{x}$ denotes the solution of the problem and capital letters with an asterisk are used for the adjoints of linear bounded operators and for dual spaces. Furthermore, Hilbert spaces will be identified with their duals. Let $f$ and $h$ be twice Fréchet-differentiable and let the Lagrangian of the problem be defined in the following way

$$L(x,l) = f(x) - l(h(x)) \quad , \quad x \in X \; , \; l \in Z^*.$$

Since some of the algorithms use linearized constraints we make a few comments about the linearization of $h$. Assume that each element of the nullspace of $h'(x)$ is the image of a linear bounded operator $T = T(x)$:

$$\begin{aligned}\mathcal{N}(h'(x)) &= \{p \in X : h'(x)p = 0\} = \{T(x)w : w \in W\} = \mathcal{R}(T(x)),\\ &\quad x \in X\,,\ T(x) \in \mathcal{L}(W,X)\,,\ W \text{ real Hilbert space.}\end{aligned}$$

In the finite-dimensional case $T(x)$ is any matrix providing a basis for the nullspace of the gradients of the constraints and also for our infinite-dimensional application this representation of the kernel is easily verified.

In control problems one usually distinguishes between the control variables and the state variables. Therefore, we assume that $X = Y \times U$ is a product space with Hilbert spaces $Y$ and $U$, $x = (y,u)$ and $Z = Y$. We have

$$h'(y,u)(\bar{y},\bar{u}) = h_1(y,u)\bar{y} + h_2(y,u)\bar{u}\,,$$

where $h_1(y,u) \in \mathcal{L}(Y)$ and $h_2(y,u) \in \mathcal{L}(U,Y)$ are the partial derivatives with respect to $y$ and $u$. Provided $h_1$ is bijective, each element of the nullspace can be expressed as follows:

$$(\bar{y},\bar{u}) \in \mathcal{N}(h'(y,u)) \quad \Leftrightarrow \quad (\bar{y},\bar{u}) = (-h_1(y,u)^{-1}h_2(y,u)\bar{u},\bar{u}).$$

This allows the following natural choice for the operator $T$ which is typical for problems with separable variables

$$T(y,u) = (-h_1(y,u)^{-1}h_2(y,u), I) \in \mathcal{L}(U, Y \times U) \text{ and } W = U. \tag{2.1}$$

The separability of variables can also be used directly to reduce the dimension of the problem. However, one has to assume that for each $u \in U$ there exists a unique $y = S(u) \in Y$ such that $h(y,u) = h(S(u),u) = 0$, where, in general, $S$ is a nonlinear operator. Then the original problem can be treated as an unconstrained problem in the variable $u$:

$$\min_{u \in U} f(S(u),u) = \varphi(u) \quad , \quad \varphi : U \to \mathbb{R}.$$

One iteration of the gradient method is defined in the following way

$$u^+ = u - \alpha\varphi'(u)\,,$$

where $\alpha \in \mathbb{R}$ is chosen from some stepsize rule. In the following the current and the next iterate are denoted by $u$ and $u^+$, respectively.

In order to compute $\varphi'$ which is just the reduced gradient of the original constrained problem note that by the implicit function theorem $S'(u)$ is given by

$$S'(u) = -h_1(S(u),u)^{-1}h_2(S(u),u)$$

and with (2.1) for $\nu \in U$

$$\begin{aligned}\varphi'(u)\nu &= f_1(S(u),u)S'(u)\nu + f_2(S(u),u)\nu\\ &= f'(S(u),u)(T(S(u),u)\nu).\end{aligned} \tag{2.2}$$

Therefore

$$\varphi'(u) = T(S(u),u)^* f'(S(u),u)\,. \tag{2.3}$$

Hence, one iteration of a gradient method is of the following form

**Algorithm 1 (Gradient Method)**

$$u^+ = u - \alpha T(S(u),u)^* f'(S(u),u).$$

All gradient type methods are known for a rather robust behavior initially but in the final phase they tend to converge at a very slow rate. In this situation one likes to switch to faster methods like Newton's method. An iteration is given by

**Algorithm 2 (Newton's Method)**

$$\begin{aligned} \varphi''(u)\Delta u &= -T(S(u),u)^* f'(S(u),u) \\ u^+ &= u + \Delta u. \end{aligned}$$

The major problem with Newton's method for optimal control problems is the computation of the step, which involves exact second order information of the objective function. Obviously, the differentiation of (2.3) is a very difficult task numerically and theoretically. In order to avoid an expensive calculation of $\varphi''$ one can use a quasi-Newton method instead. Here the second derivative of $\varphi$ is approximated by operators $B \in \mathcal{L}(U)$ which are updated after each calculation of the step. Hence an iteration has the following form in general

$$u^+ = u - B^{-1}\varphi'(u).$$

One iteration for the BFGS algorithm looks as follows, if we take into account the structure of separable variables

**Algorithm 3 (BFGS Method)**

$$\begin{aligned} B\Delta u &= -T(S(u),u)^* f'(S(u),u) \\ u^+ &= u + \Delta u \\ \eta &= T(S(u^+),u^+)^* f'(S(u^+),u^+) - T(S(u),u)^* f'(S(u),u) \\ B_+ &= B + \frac{\eta \otimes \eta}{< \eta, \Delta u >} - \frac{(B\Delta u) \otimes (B\Delta u)}{< \Delta u, B\Delta u >} \in \mathcal{L}(U), \end{aligned}$$

where $u \otimes v$ denotes the outer product in $U$.

Another approach to equality constrained optimization consists of a quadratic approximation of the Lagrangian and, at the same time, a linearization of the constraint. In this framework we are required to solve only a linear equation for the constraint per iteration as opposed to the solution of a nonlinear equation (computation of $S(u)$) for the methods considered previously. Methods of this type are called SQP (Sequential Quadratic Programming) methods . It can be shown under weak assumptions that the SQP-step can be decomposed into a restoration step and a minimization step which leads to the following representation of a full SQP-step

$$\begin{aligned} (y^+,u^+) - (y,u) &= -T(y,u)[T(y,u)^* M T(y,u)]^{-1} T(y,u)^* [f'(y,u) - MR(y,u)h(y,u)] \\ &\quad -R(y,u)h(y,u). \end{aligned}$$

Here $M \in \mathcal{L}(X)$ is an approximation to the Hessian of the Lagrangian and $R(y,u) \in \mathcal{L}(Y,X)$ is an arbitrary right-inverse of $h'(y,u)$ , i.e. $h'(y,u)R(y,u)v = v$ for all $v \in Y$. Obviously,

$$R(y,u) = (h_1(y,u)^{-1}, 0) \tag{2.4}$$

is a right-inverse of $h'$, where the second component is the null-operator in $\mathcal{L}(Y,U)$.

Using the definitions (2.4) and (2.1) for $R$ and $T$, the SQP-step can be written in the following way:

$$\begin{aligned}(y^+,u^+)-(y,u) &= T(y,u)\Delta u - R(y,u)h(y,u)\\ &= (-h_1(y,u)^{-1}h_2(y,u)\Delta u - h_1(y,u)^{-1}h(y,u), \Delta u)\\ &= (-h_1(y,u)^{-1}(h_2(y,u)\Delta u + h(y,u)), \Delta u)\ ,\end{aligned}$$

where $\Delta u$ is the solution of

$$T(y,u)^* M T(y,u)\Delta u = -T(y,u)^*[f'(y,u) - MR(y,u)h(y,u)].$$

This leads to the following iteration of a full SQP method

**Algorithm 4 (Full SQP Method)**

$$\begin{aligned}T(y,u)^* M T(y,u)\Delta u &= -T(y,u)^*[f'(y,u) - MR(y,u)h(y,u)]\\ h_1(y,u)\Delta y &= -h_2(y,u)\Delta u - h(y,u)\\ u^+ &= u + \Delta u\\ y^+ &= y + \Delta y\\ M_+ &= M + \mathit{Update} \in \mathcal{L}(X).\end{aligned}$$

Issues of full SQP methods for finite-dimensional problems are discussed in [15]. We have not specified an update for $M$ because we will see later that this method is not suitable for control problems with a 'large' state space. Also, standard secant updates are excluded unless one makes the strong assumption that the Hessian of the Lagrangian is positive definite on the entire space.

An alternative to the restrictive assumption of requiring $L''_{(y,u)}(\hat{y},\hat{u},\hat{l})$ positive definite is the application of a reduced SQP method. In this case the last part of the minimization step in the SQP-Newton-step is dropped and the remaining reduced Hessian $T(y,u)^* L''_{(y,u)} T(y,u)$ is approximated by an operator $B \in \mathcal{L}(U)$. This yields the reduced SQP-step

$$(y^+,u^+)-(y,u) = -T(y,u)B^{-1}T(y,u)^* f'(y,u) - R(y,u)h(y,u). \tag{2.5}$$

This approach can be motivated from the case of linear constraints where all iterates generated by an SQP method are feasible, except possibly the starting point. Hence $h(y,u)$ vanishes after the first iteration. Therefore the full SQP-step is identical with the reduced SQP-step, if $B = T(y,u)^* M T(y,u)$.

For finite-dimensional problems there is special emphasis on a particular choice for $R$ and $T$ in (2.5) namely $R$ is the Moore-Penrose pseudoinverse of $h'$ and $T$ is an orthonormal basis, taken from a smooth QR-decomposition , see [4], [12] and [3]. The general reduced SQP method (2.5) was investigated by [8] and with modifications by [9]. Moreover, it is well known for the finite-dimensional case that the cancellation of the term $T(y,u)^*MR(y,u)h(y,u)$ in the SQP-step results in a loss of the q-superlinear convergence. In general only a two-step q-superlinear rate can be achieved for reduced methods which is only a slight disadvantage.

Apart from the fact that it is reasonable to approximate only the positive definite portion of the Hessian, the resulting savings in storage is another characteristic advantage of reduced over full SQP methods. The importance of this property becomes even more evident in infinite-dimensional applications. In the full SQP method we have to update the operator $M$ which is defined on the whole space $X = Y \times U$. If the dimension of the discretized state space $Y$ is considerably higher than that of the control space $U$, the savings in storage can be enormous when a reduced method is used. In the next section this will be demonstrated for a parabolic boundary control problem.

On the other hand, we should mention that a reduced secant method in the formulation below needs two gradient evaluations at each iteration. For the semilinear parabolic control problem studied in the next section this means the solution of four linear initial-boundary value problems. We could avoid the second gradient evaluation by using $(y^+, u^+)$ in the definition of $\eta$, however, we are not able to retain superlinear convergence in this case. The additional gradient evaluation at the intermediate point $(y,u) + T(y,u)\Delta u$ can be seen in the following iteration of a reduced SQP method

**Algorithm 5 (Reduced SQP Method)**

$$
\begin{aligned}
B\Delta u &= -T(y,u)^* f'(y,u) \\
h_1(y,u)\Delta y &= -h_2(y,u)\Delta u - h(y,u) \\
u^+ &= u + \Delta u \\
y^+ &= y + \Delta y \\
\eta &= T((y,u) + T(y,u)\Delta u)^* f'((y,u) + T(y,u)\Delta u) - T(y,u)^* f'(y,u) \\
B_+ &= B + \frac{\eta \otimes \eta}{< \eta, \Delta u >} - \frac{(B\Delta u) \otimes (B\Delta u)}{< \Delta u, B\Delta u >}.
\end{aligned}
$$

At this point it should also be noted that with the separability approach we circumvent a problem which occurs often when reduced methods are studied in a finite-dimensional context. From the above definition of the right-inverse and the operator $T$ we can immediately deduce that $R$ and $T$ have the same smoothness properties as $h'$. Therefore, under standard smoothness assumptions for $f$ and $h$ we do not encounter the difficulty with the Lipschitz-continuity of $T(y,u)$ which appears when $T$ is chosen as orthonormal basis obtained from a QR-decomposition of the Jacobian of the constraints , see [5].

Let us summarize the arguments which give the motivation for the application of reduced SQP methods to parabolic boundary control problems. First they are reasonable since in general only the reduced Hessian is positive definite at the solution. This allows the use of secant update formulas maintaining positive definiteness. Furthermore, a fast convergence rate can be achieved and the reduction to the control space results in a significant decrease in storage. Moreover, the natural nullspace representation of $h'$ leads to Lipschitz-continuity of $T(y,u)$. Finally, the linearization of the constraint results in the solution of a linear equation as opposed to a nonlinear one in Algorithms 1–3.

## 3 Optimization Methods for Optimal Control Problems

In this section we want to consider the problem of controlling the semilinear parabolic differential equation through the boundary

$$\begin{array}{rcll} y_t(t,x) &=& (\Delta y)(t,x)+c(t,x,y(t,x)) & t\in(0,T),\quad x\in\Omega \\ y(0,x) &=& y_0(x) & x\in\Omega \\ \frac{\partial y}{\partial n}(t,\xi) &=& g(\xi)u(t) & t\in(0,T),\quad \xi\in\Gamma \end{array} \tag{3.1}$$

where $\Omega$ is a bounded domain in $\mathbb{R}^n$ with sufficiently smooth boundary $\Gamma$. The nonlinear function $c:[0,T]\times\bar{\Omega}\times\mathbb{R}\to\mathbb{R}$ is smooth and $g\in C(\Gamma)$ is fixed. The objective is to find $u$ such that

$$\int_0^T\int_\Omega l(t,x,y(t,x))\,dx\,dt+\int_0^T k(t,u(t))\,dt \tag{3.2}$$

is minimized. The nonlinear functions $l$ and $k$ are also assumed to be smooth. We denote by $k_u, l_y, c_y$ the derivatives with respect to the second and third variable, respectively.

In the preceding paragraph we presented various methods to solve optimal control problems. The goal of this section is to demonstrate the advantage of a reduced SQP method in particular for the parabolic boundary control problem (3.1), (3.2). Since this method requires a number of technical details we concentrate on the derivation of the method and not on the setup of the spaces, existence of optimal controls and solutions of the semilinear parabolic differential equation and Fréchet- differentiability. As mentioned in the introduction the latter is not a simple verification, e.g. by using a semigroup approach, but requires more work which is under investigation by the authors.

Let $U=U[0,T]$ denote the function space for the controls and $Y=Y([0,T]\times\Omega)$ the function space for the states. In the setup of the previous section the variables are $(y,u)\in Y\times U$. We consider the linearized constraint and compute its representation of the nullspace. The equation

$$h'(y,u)(\bar{y},\bar{u})=0$$

is satisfied, if $\bar{y}$ solves the linear partial differential equation for given $\bar{u}$

$$\begin{array}{rcll} \bar{y}_t(t,x) &=& (\Delta\bar{y})(t,x)+c_y(t,x,y(t,x))\bar{y}(t,x) & t\in(0,T),\quad x\in\Omega \\ \bar{y}(0,x) &=& 0 & x\in\Omega \\ \frac{\partial\bar{y}}{\partial n}(t,\xi) &=& g(\xi)\bar{u}(t) & t\in(0,T),\quad \xi\in\Gamma. \end{array} \tag{3.3}$$

Even in discretized form this does not suggest a parametrization of the nullspace through an orthonormal basis, obtained from a QR-decomposition, which is used in most reduced SQP methods for finite-dimensional problems. An obvious representation of all $(\bar{y},\bar{u})$ which satisfy (3.3) is through $\bar{u}$. Hence we set

$$T(y,u)\bar{u}=\begin{pmatrix}\bar{y}\\ \bar{u}\end{pmatrix}$$

where $\bar{y}\in Y$ solves (3.3). This parametrization is very natural and smooth dependence on the data yields Lipschitz-continuity or Fréchet-differentiability of $T$. We use this definition of $T$ to compute the gradient $T(y^+,u^+)^*f'(y^+,u^+)$, where we choose $(y^+,u^+)$ instead of $(y,u)$ for a more consistent formulation of the algorithms. This element is given by

$$\gamma^+(t) = (T(y^+,u^+)^* f'(y^+,u^+))(t) = \int_\Gamma g(\xi) d^+(t,\xi)\, d\xi + k_u(t,u^+(t)) \tag{3.4}$$

where $d^+$ is the solution of the adjoint equation (see e.g. [17])

$$\begin{array}{rcll} -d_t^+(t,x) & = & (\Delta d^+)(t,x) + c_y(t,x,y^+(t,x))d^+(t,x) & \\ & & + \, l_y(t,x,y^+(t,x)) & t \in (0,T), \;\; x \in \Omega \\ d^+(T,x) & = & 0 & x \in \Omega \\ \frac{\partial d^+}{\partial n}(t,\xi) & = & 0 & t \in (0,T), \;\; \xi \in \Gamma. \end{array} \tag{3.5}$$

In the gradient method one needs to compute the gradient (2.3) of the objective function $\varphi$ which can be achieved by (3.4).

**Gradient Method**

| | |
|---|---|
| Given | control $u$ and gradient $\gamma$ |
| Step 1 | Set $u^+ = u - \alpha\gamma$ |
| Step 2 | Compute the solution $y^+$ of the nonlinear b.v.p. (3.1) with control input $u^+$ |
| Step 3 | Compute the solution $d^+$ of the linear b.v.p. (3.5) |
| Step 4 | Compute the gradient $\gamma^+(\cdot)$ from (3.4) |

The step-size parameter $\alpha$ is selected according to some step size rule. We see that this method requires per iteration the solution of one linear and one nonlinear initial-boundary value problem. However, in general this method converges rather slowly so that we turn our attention to Newton's method. We consider again the computation of the gradient. It itself takes the solution of a linear initial-boundary value problem whose data depend on the solution of the nonlinear problem (3.1). Hence it is obvious that the analytical calculation of the derivative of the gradient is a rather tedious task. One should remember that this is already quite nasty for optimal control problems with ordinary differential equations and even more so for problems with partial differential equations.

In order to improve this situation we turn our attention to quasi-Newton methods. Here we approximate the Hessian of the objective function in a proper way.

**BFGS Method**

| | |
|---|---|
| Given | control $u$, gradient $\gamma$ and an operator $B \in \mathcal{L}(U)$ |
| Step 1 | Solve $B\Delta u = -\gamma$ |
| Step 2 | Set $u^+ = u + \Delta u$ |
| Step 3 | Compute the solution $y^+$ of the nonlinear b.v.p. (3.1) with control input $u^+$ |
| Step 4 | Compute the solution $d^+$ of the linear b.v.p. (3.5) |
| Step 5 | Compute the gradient $\gamma^+(\cdot)$ from (3.4) |
| Step 6 | Set $\eta = \gamma^+ - \gamma$ |
| Step 7 | Set $B_+ = B + \frac{\eta \otimes \eta}{<\eta, \Delta u>} - \frac{(B\Delta u)\otimes(B\Delta u)}{<\Delta u, B\Delta u>}$ |

If we compare the number of boundary value problems which we need to solve during one iteration of the BFGS method with those of the gradient method we see that they are the same. The big advantage of the BFGS method is that it exhibits locally a superlinear rate of convergence under certain assumptions. The only additional cost is the solution of the linear equation in Step 1 and the storage of $B$. Both aspects can be solved rather efficiently, see e.g. [6].

At this point we want to focus on SQP methods. These methods have the common feature that they linearize the nonlinear function which defines the equality constraint. Hence we expect for these methods to omit the solution of a nonlinear initial-boundary value problem. If we consider Algorithm 4 (full SQP method) from the previous section, then we notice that we have to store and update an operator $M \in \mathcal{L}(Y \times U)$. If one discretizes the time-dependent space $U$ by a $d_t$-dimensional space and the space- and time-dependent $Y = Y([0,T] \times \Omega)$ by a $d_t n d_x$-dimensional space where $n$ denotes the space dimension and $d_x$ the discretization in one space component, then the matrix to be stored for a full SQP update is of the size $\mathbb{R}^{d_t(1+nd_x)\times d_t(1+nd_x)}$. This shows that for parabolic differential equations of the type under consideration the full method is not recommendable. It is the main advantage of the reduced SQP method (Algorithm 5) that it allows to use the structure of the nullspace of $h'(y,u)$ by parametrizing the state.

The last missing ingredient for the reduced SQP-step (2.5) is the calculation of the restoration step. If we choose $R(y,u)$ according to (2.4), then

$$R(y,u)h(y,u) = \begin{pmatrix} y - w \\ 0 \end{pmatrix}$$

where $w \in Y$ solves

$$\begin{array}{rcll} w_t(t,x) & = & (\Delta w)(t,x) + c_y(t,x,y(t,x))w(t,x) & \\ & & + \, c(t,x,y(t,x)) - c_y(t,x,y(t,x))y(t,x) & t \in (0,T), \quad x \in \Omega \\ w(0,x) & = & y_0(x) & x \in \Omega \\ \frac{\partial w}{\partial n}(t,\xi) & = & g(\xi)u(t) & t \in (0,T), \quad \xi \in \Gamma. \end{array} \tag{3.6}$$

Note that the element $R(y,u)h(y,u)$ is not needed explicitly in Algorithm 5. However, it can be used to determine the new state $y^+$ in a convenient way. Therefore, recall that the new iterate in the reduced SQP method is given by (2.5):

$$\begin{pmatrix} y^+ \\ u^+ \end{pmatrix} = \begin{pmatrix} y \\ u \end{pmatrix} + T(y,u)\Delta u - R(y,u)h(y,u) = \begin{pmatrix} \bar{y} + w \\ u + \Delta u \end{pmatrix}$$

where $\bar{y}$ solves (3.3) with control input $\bar{u} = \Delta u$ and $w$ solves (3.6). A combination of these two formulas shows that $y^+$ then solves the following linear initial-boundary value problem

$$\begin{array}{rcll} y^+_t(t,x) & = & (\Delta y^+)(t,x) + c_y(t,x,y(t,x))y^+(t,x) & \\ & & + \, c(t,x,y(t,x)) - c_y(t,x,y(t,x))y(t,x) & t \in (0,T), \quad x \in \Omega \\ y^+(0,x) & = & y_0(x) & x \in \Omega \\ \frac{\partial y^+}{\partial n}(t,\xi) & = & g(\xi)u^+(t) & t \in (0,T), \quad \xi \in \Gamma. \end{array} \tag{3.7}$$

The step to compute a new correction for the update $B$ requires an additional effort because for reasons mentioned in the previous section we cannot use $y^+$ directly for this computation. The element

$$\begin{pmatrix} y \\ u \end{pmatrix} + T(y,u)\Delta u = \begin{pmatrix} y \\ u \end{pmatrix} + \begin{pmatrix} \bar{y} \\ \Delta u \end{pmatrix} = \begin{pmatrix} y+\bar{y} \\ u^+ \end{pmatrix}$$

can be computed by a solution of

$$\begin{array}{rcll} \bar{y}_t(t,x) &=& (\Delta\bar{y})(t,x) + c_y(t,x,y(t,x))\bar{y}(t,x) & t\in(0,T),\quad x\in\Omega \\ \bar{y}(0,x) &=& 0 & x\in\Omega \\ \frac{\partial\bar{y}}{\partial n}(t,\xi) &=& g(\xi)\Delta u(t) & t\in(0,T),\quad \xi\in\Gamma. \end{array} \tag{3.8}$$

Then we form

$$\tilde{y} = y + \bar{y}$$

to formulate

$$\begin{array}{rcll} -\tilde{d}_t(t,x) &=& (\Delta\tilde{d})(t,x) + c_y(t,x,\tilde{y}(t,x))\tilde{d}(t,x) & \\ && +\, l_y(t,x,\tilde{y}(t,x)) & t\in(0,T),\quad x\in\Omega \\ \tilde{d}(T,x) &=& 0 & x\in\Omega \\ \frac{\partial\tilde{d}}{\partial n}(t,\xi) &=& 0 & t\in(0,T),\quad \xi\in\Gamma. \end{array} \tag{3.9}$$

Then we compute the function $\tilde{\gamma}$ which is used in the calculation of the update

$$\tilde{\gamma}(t) = \int_\Gamma g(\xi)\tilde{d}(t,\xi)\,d\xi + k_u(t,u^+(t)) \tag{3.10}$$

(see also (3.4) and (3.5)).

In detail, the reduced SQP method for the semilinear boundary control problem (3.1),(3.2) requires the following steps

**Reduced SQP Method**

| | |
|---|---|
| Given | control $u$, state $y$, gradient $\gamma$ and an operator $B\in\mathcal{L}(U)$ |
| Step 1 | Solve $B\Delta u = -\gamma$ |
| Step 2 | Set $u^+ = u + \Delta u$ |
| Step 3 | Compute the solution $y^+$ of the linear b.v.p. (3.7) |
| Step 4 | Compute the solution $d^+$ of the linear b.v.p. (3.5) |
| Step 5 | Compute the gradient $\gamma^+(\cdot)$ from (3.4) |
| Step 6 | Compute the solution $\bar{y}$ of the linear b.v.p. (3.8) |
| Step 7 | Set $\tilde{y} = y + \bar{y}$ |
| Step 8 | Compute the solution $\tilde{d}$ of the linear b.v.p. (3.9) |
| Step 9 | Compute $\tilde{\gamma}(\cdot)$ from (3.10) |
| Step 10 | Set $\eta = \tilde{\gamma} - \gamma$ |
| Step 11 | Set $B_+ = B + \frac{\eta\otimes\eta}{<\eta,\Delta u>} - \frac{(B\Delta u)\otimes(B\Delta u)}{<\Delta u, B\Delta u>}$ |

Note that the solution for (3.8) and (3.7) involves the same linear parabolic equation with different inhomogeneous terms. The adjoint equations (3.5) and (3.9) are not the same, because the coefficients in the differential equation differ.

## 4 Convergence Results

Here we give a review of the convergence behavior of the methods discussed in the previous paragraphs. Throughout this section we make use of the following definition: Let $V_1, V_2$ normed linear spaces, $D \subset V_1$ an open set and $\hat{v} \in D$. A mapping $F : D \to V_2$ satisfies a Lipschitz-condition at $\hat{v}$ in $D$, $F \in Lip(D)$, if there exists a constant $M > 0$ such that

$$\|F(v) - F(\hat{v})\|_{V_2} \leq M\|v - \hat{v}\|_{V_1} \quad \text{for all } v \in D.$$

Recall that $X = Y \times U$ where $Y$ and $U$ are real Hilbert spaces. Since the convergence results for Algorithms 1–3 are well known we state them only very briefly.

**Gradient Method**

Let $\varphi$ be twice continuously Fréchet-differentiable and assume a convexity condition. Then the standard result yields that the gradient method with an appropriate stepsize converges to the solution, cf. [13] . Moreover, if $u_k \neq \hat{u}$ for all k, the convergence is q-linear:

$$\|u_{k+1} - \hat{u}\| \leq q\|u_k - \hat{u}\| \text{ for some } q \in (0, 1).$$

This convergence can be fairly slow, in particular, when $u_k$ is close to $\hat{u}$.

**Newton's Method**

The reward for the high price paid in computing the second derivative of $\varphi$ at each step, is the q-quadratic convergence rate of Newton's method. The following result can be found in [13], for example.

**Theorem 4.1** *Let $\varphi$ be twice Fréchet-differentiable and $\varphi' \in Lip(D)$ for a neighborhood $D$ of $\hat{u}$. Furthermore, assume that $\varphi''(\hat{u})$ is bijective.*
*If $\|u_0 - \hat{u}\|$ is sufficiently small, then the sequence $\{u_k\}$ from Newton's method converges to $\hat{u}$ at a q-quadratic rate.*

**BFGS Method**

At least in the finite-dimensional case the standard BFGS-method for unconstrained minimization (see e.g. [6]) and the full SQP-BFGS method are known for their q-superlinear convergence (see e.g. [7]) under appropriate assumptions. However, as mentioned before the SQP method suffers from the drawback that in general the Hessian of the Lagrangian is not positive definite on the entire space. Unfortunately, the use of the reduced version to overcome this difficulty is at the expense of the one-step superlinear convergence rate: the best we can expect from a reduced SQP method is superlinear convergence in two steps, see e.g. [1], [18].

Griewank [10] shows local q-linear convergence of the sequence $\{u_k\}$ from Algorithm 3 . Moreover, he proves q-superlinear convergence, if in addition the initial discrepancy $B_0 - \varphi''(\hat{u})$ is compact. The following theorem is a special case of the results in [10].

**Theorem 4.2** *Let the assumptions of Theorem 4.1 hold and assume in addition that $\varphi''(\hat{u})$ is positive definite. Let the sequence $\{u_k\}$ be generated by Algorithm 3 with $B_0 \in \mathcal{L}(U)$ selfadjoint and positive definite.*

*Then there exist positive scalars $\delta$ and $\epsilon$ such that if $\|u_0 - \hat{u}\| < \epsilon$ and $\|B_0 - \varphi''(\hat{u})\| < \delta$, then $\{u_k\}$ converges q-linearly to $\hat{u}$. Moreover, if $B_0 - \varphi''(\hat{u})$ is compact, then the rate of convergence is q-superlinear.*

**Reduced SQP Method**

Since we are primarily interested in the application of reduced SQP methods to optimal control problems, we present the convergence behavior of these methods in a more detailed way. Therefore, we apply the general theory from [11] to the separability approach. For sake of simplicity we often use $x$ for the variable instead of $(y, u)$. Recall that $f$ and $h$ denote the objective and the constraint of the original problem (1.1).

Suppose there is a neighborhood $D := \{x : \|x - \hat{x}\| < \rho\}$ of the solution $\hat{x} = (\hat{y}, \hat{u})$ such that the following assumption holds which is standard in the context of constrained minimization:

(A): $f$ and $h$ are twice Fréchet-differentiable on $D$ and $f'', h'' \in Lip(D)$:
$\|f''(x) - f''(\hat{x})\|$ , $\|h''(x) - h''(\hat{x})\| \leq M\|x - \hat{x}\|$, $x \in D$, $M > 0$.

The definitions (2.4) and (2.1) of $R$ and $T$ and the assumptions in Section 2 lead with (A) to the Fréchet- differentiability of $T : D \to \mathcal{L}(U, Y \times U)$ and $R : D \to \mathcal{L}(Y, Y \times U)$ with $T', R' \in Lip(D)$.

For abbreviation we set

$$\hat{H} = T(\hat{x})^* L''_x(\hat{x}, \hat{l}) T(\hat{x}).$$

We consider for a moment a reduced SQP method with a general BFGS-update. Therefore, we replace the updating procedure in Algorithm 5 by the formula

$$B_{k+1} = B_k + \frac{\eta_k \otimes \eta_k}{< \eta_k, w_k >} - \frac{(B_k w_k) \otimes (B_k w_k)}{< w_k, B_k w_k >}, \tag{4.1}$$

where $\eta_k, w_k \in U, B_k \in \mathcal{L}(U)$ and $\eta_k, w_k$ are not specified any further. In this way it is possible to establish convergence results for reduced SQP-BFGS methods where the sequences $\{\eta_k\}$ and $\{w_k\}$ are chosen from the variety of possible choices proposed in the literature [12]. We will see that the following condition for these sequences is essential to prove convergence:

$$< \eta_k, w_k > > 0 \quad \text{and} \quad \|\eta_k - \hat{H} w_k\| \leq K \max\{\|x_k - \hat{x}\|, \|x_{k+1} - \hat{x}\|\} \|w_k\| \tag{4.2}$$

for a constant $K > 0$ and all $k \geq 0$.

It is shown by [11] that locally a two-step q-linear rate can be achieved for a reduced SQP-BFGS method, if the approximation error $\|B_k - \hat{H}\|$ is sufficiently small at each iteration. If (4.2) is satisfied, the latter condition can be guaranteed from a bounded deterioration property for the sequence $\{B_k\}$. In particular, if (A) and the classical second order sufficient optimality condition for problem (1.1) hold, i.e. there exists $m > 0$ such that

$$< v, L''_x(\hat{x}, \hat{l}) v > \geq m\|v\|^2 \text{ for all } v \in \mathcal{N}(h'(\hat{x})), \tag{4.3}$$

then the property (4.2) can easily be verified for

$$\begin{aligned} w_k &= -B_k^{-1}T(x_k)^*f'(x_k) \qquad \text{and} \\ \eta_k &= T(x_k + T(x_k)w_k)^*f'(x_k + T(x_k)w_k) - T(x_k)^*f'(x_k). \end{aligned} \tag{4.4}$$

These are the choices taken in Algorithm 5 and in Steps 9–11 of the reduced SQP algorithm for the semilinear parabolic boundary control problem from Section 3. Consequently, if (A) and (4.3) are valid, we obtain local q-linear convergence in two steps for the sequence $\{(y_k, u_k)\}$ from Algorithm 5, provided the update is skipped in the case $\Delta u = 0$. We point out again that the same convergence rate can be achieved for any other choice of the input of the update formula, as long as (4.2) is satisfied. This can be important, if a standard choice of $\eta_k$ or $w_k$ cannot be calculated exactly and some approximation has to be used, a situation that is likely to occur if the computation of $\eta_k$ involves the solution of boundary value problems.

We now address the superlinear convergence behavior. It is shown in [11] that the Powell-condition [14] is sufficient for two-step q-superlinear convergence of reduced SQP methods also in Hilbert space, i.e. the limit

$$\lim_{k\to\infty} \frac{\|(B_k - \hat{H})B_k^{-1}T(x_k)^*f'(x_k)\|}{\|x_{k+1} - x_k\|} = 0 \tag{4.5}$$

implies that

$$\lim_{k\to\infty} \frac{\|x_{k+1} - \hat{x}\|}{\|x_{k-1} - \hat{x}\|} = 0 \,,$$

if the sequence $\{x_k\}$ converges to $\hat{x}$.

To prove the consistency condition (4.5) we make use of the following general result on the BFGS-update which is valid in an arbitrary real Hilbert space $U$. Hence, it is applicable to constrained and unconstrained optimization and independent of any particular method. It is an immediate consequence of a more general theorem proven by Griewank [10] in his study of secant methods to solve nonlinear operator equations in Hilbert space.

**Theorem 4.3** *Let $\{B_k\}$ be generated by the BFGS-update formula (4.1) where $B_0 \in \mathcal{L}(U)$ is bijective, selfadjoint, positive definite and $\lambda_*(B_0) > 0$. Furthermore, assume that $\{\eta_k\}$ and $\{w_k\}$ satisfy*

$$< \eta_k, w_k > \; > 0 \quad \textit{and} \quad \|\eta_k - \hat{D}w_k\| \le \alpha_k \|w_k\| \,, \; k \ge 0 \,,$$

*for some positive definite, selfadjoint and bijective operator $\hat{D} \in \mathcal{L}(U)$ and a sequence $\{\alpha_k\}$ with the property $\sum_{k=0}^{\infty} \alpha_k < \infty$.*
*If $B_0 - \hat{D}$ is compact, then*

$$\lim_{k\to\infty} \frac{\|(B_k - \hat{D})w_k\|}{\|w_k\|} = 0.$$

Here $\lambda_*(B) = \inf\{\|B - C\| : \mathcal{L}(U) \ni C \text{ compact}\}$ denotes the essential norm of $B$. The restrictive assumption $\lambda_*(B_0) > 0$ can be replaced by requiring that $U$ is infinite-dimensional and for the finite-dimensional case the above result was proven in [2].

Theorem 4.3 indicates the crucial role of compactness and the importance of property (4.2) in the infinite-dimensional convergence analysis of quasi-Newton methods. To our knowledge a convergence result for the full SQP method (Algorithm 4) has not yet been established in infinite-dimensional spaces. Theorem 4.3 can serve as a tool to prove superlinear convergence also for this method, if the BFGS-formula is used in an appropriate way.

To finish the discussion on convergence rates, we apply Theorem 4.3 to show local two-step q-superlinear convergence of the reduced SQP method from Algorithm 5. The proof, see [11], uses the linear convergence result just mentioned and the extended Powell-condition (4.5).

**Theorem 4.4** *Assume (A) and (4.3).*
*Let the sequence* $\{(y_k, u_k)\}, (y_k, u_k) \neq (\hat{y}, \hat{u})$*, be generated by Algorithm 5 with* $B_0 \in \mathcal{L}(U)$ *selfadjoint and positive definite.*
*Then there exist positive scalars* $\delta$ *and* $\epsilon$ *such that if* $\|(y_0, u_0) - (\hat{y}, \hat{u})\| < \epsilon$ *and* $\|B_0 - \hat{H}\| < \delta$*, then* $\{(y_k, u_k)\}$ *converges to* $(\hat{y}, \hat{u})$ *at a two-step q-linear rate. Moreover, if* $B_0 - \hat{H}$ *is compact, then the rate of convergence is two-step q-superlinear:*

$$\lim_{k \to \infty} \frac{\|(y_{k+1}, u_{k+1}) - (\hat{y}, \hat{u})\|}{\|(y_{k-1}, u_{k-1}) - (\hat{y}, \hat{u})\|} = 0.$$

Theorem 4.4 shows that the compactness of the starting discrepancy is significant for fast convergence of reduced SQP methods. This condition always appears when a quasi-Newton method is applied to an infinite-dimensional problem. The same can be said of the smoothness properties of $f$ and $h$ required in (A): a comparison of the assumptions in the convergence theorems of this section shows that they are very similar for all the methods discussed in this paper. The verification of these assumptions in the application of a reduced SQP method to semilinear parabolic boundary control problems will be the subject of our future work. Another point in this context will be the setup of the spaces. In view of the approach taken to prove second order sufficiency conditions it seems likely that an adaption of the convergence theory to the specific needs of control problems is necessary for a successful treatment of the open questions.

## References


[1] R. H. Byrd. An example of irregular convergence in some constrained optimization methods that use the projected Hessian. *Math. Programming*, 32:232–237, 1985.

[2] R. H. Byrd and J. Nocedal. A tool for the analysis of quasi- Newton methods with application to unconstrained minimization. *SIAM J. Numer. Anal.*, 26:727–739, 1989.

[3] R. H. Byrd and J. Nocedal. An analysis of reduced Hessian methods for constrained optimization. *Math. Programming*, to appear.

[4] T. F. Coleman and A. R. Conn. On the local convergence of a quasi–Newton method for the nonlinear programming problem. *SIAM J. Numer. Anal.*, 21:755–769, 1984.

[5] T. F. Coleman and D. C. Sorensen. A note on the computation of an orthonormal basis for the null space of a matrix. *Math. Programming*, 29:234–242, 1984.

[6] J. E. Dennis and R. B. Schnabel. *Numerical Methods for Unconstrained Optimization and Nonlinear Equations.* Prentice-Hall, Englewood Cliffs, N.J, 1983.

[7] R. Fontecilla, T. Steihaug, and R. A. Tapia. A convergence theory for a class of quasi-Newton methods for constrained optimization. *SIAM J. Numer. Anal.*, 24:1133–1151, 1987.

[8] D. Gabay. Reduced quasi–Newton methods with feasibility improvement for nonlinearly constrained optimization. *Mathematical Programming Study*, 16:18–44, 1982.

[9] J. Ch. Gilbert. On the local and global convergence of a reduced quasi-Newton method. *Optimization*, 20:421–450, 1989.

[10] A. Griewank. Rates of convergence for secant methods on nonlinear problems in Hilbert space. In J. P. Hennart, editor, *Numerical Analysis, Proceedings Guanajuato, Mexico 1984*, pages 138–157. Springer, 1986.

[11] F.-S. Kupfer. An infinite-dimensional convergence theory for reduced SQP methods in Hilbert space. Technical report, Universität Trier, Fachbereich IV - Mathematik, 1990.

[12] J. Nocedal and M. L. Overton. Projected Hessian updating algorithms for nonlinearly constrained optimization. *SIAM J. Numer. Anal.*, 22:821–850, 1985.

[13] J.M. Ortega and W.C. Rheinboldt. *Iterative Solution of Nonlinear Equations in Several Variables.* Academic Press, New York, 1970.

[14] M. J. D. Powell. The convergence of variable metric methods for nonlinearly constrained optimization calculations. In O. L. Mangasarian, R. R. Meyer, and S. M. Robinson, editors, *Nonlinear Programming 3*, pages 27–63. Academic Press, 1978.

[15] R. A. Tapia. On secant updates for use in general constrained optimization. *Math. Comp.*, 51:181–203, 1988.

[16] F. Tröltzsch. *Optimality Conditions for Parabolic Control Problems and Applications.* Teubner, Leipzig, 1984.

[17] F. Tröltzsch. On the semigroup approach for the optimal control of semilinear parabolic equations including distributed and boundary control. *Zeitschr. f. Analysis und ihre Anwendungen*, 8:431–443, 1989.

[18] Y. Yuan. An only 2-step q-superlinear convergence example for some algorithms that use reduced Hessian approximations. *Math. Programming*, 32:224–231, 1985.

# The Hilbert Uniqueness Method: A Retrospective *

John E. Lagnese
Department of Mathematics
Georgetown University
Washington, DC 20057 USA

## 1 Introduction

The purpose of this paper is to give a brief overview of certain aspects of recent developments in the area of exact controllability of distributed parameter systems. Our starting point is a 1986 paper of J.-L. Lions in which is described a systematic, general method for attacking exact controllability problems for linear distributed parameter systems [14]. This method, called the *Hilbert Uniqueness Method* (HUM) by its author, provides a powerful, constructive means for solving a wide variety of exact controllability problems for partial differential equations. The reader is referred to [15], where HUM is systematically applied to a large and diverse collection of distributed parameter control problems.

It was subsequently pointed out in [11],[12],[13],[16],[17] and by others that HUM (and its first cousin, the *Reverse Hilbert Uniqueness Method* –RHUM) may be understood, at the abstract level, as a version of a well-known duality theory of exact controllability of linear evolutionary systems. (see e.g., [4, Theorem 2.1]). This observation cannot, however, account for the substantial progress made in exact controllability of distributed parameter systems since the introduction of HUM. Indeed, this success is precisely due to the *ad hoc*, distributed parameter systems approach to exact controllability adopted by Lions, based on new types of *a priori* estimates for solutions of various classes of partial differential equations that were originally developed outside of the immediate context of exact controllability theory.

Roughly speaking, the theoretical basis of HUM is the observation that if one has uniqueness of solutions of a linear evolutionary system in a Hilbert space it is possible to introduce a Hilbert space norm $\|\cdot\|_F$ based on the uniqueness property in such a way that the dual system is exactly controllable to the dual space $F'$. The exact controllability problem is thereby transfered to the problem of identifying or otherwise characterizing the couple $F, F'$. The latter is essentially a problem in partial differential equations when the original evolutionary system is a distributed parameter system: can *a priori* estimates of $\|\cdot\|_F$ be obtained in terms of norms in spaces which are both intrinsic to the given problem and which are readily identifiable? Fortunately, techniques developed in the early 1980's

*Research supported by the Air Force Office of Scientific Research through grant AFOSR 88–0337.

for deriving *a priori* estimates in the context of *uniform stabilization* of partial differential equations (e.g. [3],[8]) were available to provide a framework to attack the latter problem, at least for a number of distributed parameter control problems of interest. Indeed, one might speculate that it was an "observability" estimate for solutions of the wave equation with boundary observation (see [6]), derived by essentially the same multiplier methods as were originally employed in [3], that provided the catalyst for the introduction of HUM.

In fact, in practice it is common to first derive an *a priori* estimate leading to a uniqueness result and then to use that estimate as the starting point for the application of HUM. Each such estimate leads to *some* exact controllability theorem. However, the apparent emphasis of many authors on the derivation of *a priori* estimates has tended to obscure the simple duality principle underlying the method as well as the fact that the estimates themselves are not really part of the basic principle but rather are the means by which one identifies the space $F$ or, more commonly, some other space $G$ that is dense in $F$. (Of course, at the practical level identification of $F$ is *the crucial point* since, otherwise, the exact controllability problem cannot be considered solved in any real sense.) Moreover, while the various estimates are obtained by similar methods (such as the use of multipliers), they appear to have a different structure from one problem to the next, and it is often difficult to discern any common threads running through them. Further, the control and state spaces that one is led to consider on the basis of the estimates often have extremely weak topologies and are certainly nonstandard in the context of classical distributed parameter systems. For example, certain estimates lead to control spaces that are not even spaces of distributions, and some components of the corresponding solutions may not be continuous functions of time into any space. One then must ask in what sense the exact controllability problem has been solved. To the uninitiated, each problem may appear to require a separate treatment.

In this paper the basic principle underlying the Hilbert Uniqueness Method will be described in an abstract framework general enough to be applicable to many distributed parameter control problems of interest. Our goal is to present HUM in a general manner that both retains the distributed parameter systems flavor of the method and parallels the way the method is actually employed in applications to specific control problems. In terms of our general description of HUM, we do not claim any particular novelty; what is done here is equivalent to what can already be found in the work of Lasiecka and Triggiani (see, e.g., [11],[12],[13],[17] and Remark 2.3 below), and at certain points we have exactly adopted their framework (as in the proof of Proposition 2.5 below), although at others we have taken a somewhat different point of view. In fact, the main motivation for this paper is Russell's review [16] of [15], and what we have attempted to do is extend the basic duality structure outlined briefly in [16] to a setting sufficiently general to cover a variety of interesting distributed parameter control problems, particularly boundary control problems.

The principle of HUM will be described in the next section in the context of the reachability problem for the first order linear control system $\dot{y} = \mathcal{A}y + \mathcal{B}u$. In Section 3 we consider the situation in which the first order system arises from a second order control system $\ddot{w} = Aw + Bu$, a common occurrence in practice. Naturally, stronger results obtain in this special case than hold in the general case and, in addition, it is possible to identify a "generic" space of reachable states in terms of spaces intrinsic to the second order system; that is to say, we can identify a particular lower bound for the controllability operator, in terms of a such spaces, that holds for many second order distributed parameter control systems. This estimate in some sense ties together the diverse collection of *a priori* estimates obtained in the process of applying HUM to specific distributed parameter control systems. Examples related to boundary control of elastic plates and of Maxwell's system are presented in Section 4 to illustrate how specific control problems can be framed within the general theory.

In preparing this paper, I have benefited greatly from discussions about HUM that I have had from time to time with G. Leugering, I. Lasiecka, D. L. Russell and, particularly, R. Triggiani. It is a pleasure to acknowledge their contributions. I also wish to thank A. Bensoussan for making available preprints of his related works [1],[2].

# 2 First Order Control Systems

Let $\mathcal{H}$ be a Hilbert space with dual space $\mathcal{H}'$. The scalar product between two elements $h_1$ and $h_2$ in $\mathcal{H}$ is denoted by $(h_1, h_2)_{\mathcal{H}}$, and the duality pairing between elements $h' \in \mathcal{H}'$ and $h \in \mathcal{H}$ is denoted by $\langle h', h\rangle_{\mathcal{H}}$. We denote by $\Lambda_{\mathcal{H}}$ the Riesz isomorphism of $\mathcal{H}$ onto $\mathcal{H}'$. $\mathcal{H}'$ is itself a Hilbert space under the scalar product

$$(h_1', h_2')_{\mathcal{H}'} = (\Lambda_{\mathcal{H}}^{-1} h_1', \Lambda_{\mathcal{H}}^{-1} h_2')_{\mathcal{H}} = \langle h_1', \Lambda_{\mathcal{H}}^{-1} h_2'\rangle_{\mathcal{H}}.$$

Let $T > 0$ be fixed and $H^1(0,T;\mathcal{H}')$ be the Hilbert space consisting of functions $f : (0,T) \to \mathcal{H}'$ such that $f$ and its strong derivative $\dot{f}$ ($\dot{f} = df/dt$) belong to $L^2(0,T;\mathcal{H}')$, topologized by

$$\left(\int_0^T [\|f(t)\|_{\mathcal{H}'}^2 + \|\dot{f}(t)\|_{\mathcal{H}'}^2]dt\right)^{1/2}.$$

We may identify $L^2(0,T;\mathcal{H})$ with the dual of $L^2(0,T;\mathcal{H}')$ and with this identification we have the dense and continuous embedding

$$L^2(0,T;\mathcal{H}) \subset (H^1(0,T;\mathcal{H}'))'.$$

We will usually write $L^2(\mathcal{H})$, $H^1(\mathcal{H}')$, etc., in place of $L^2(0,T;\mathcal{H})$, $H^1(0,T;\mathcal{H}')$, etc., when the value of $T$ is clear from context.

Let $\mathcal{U}$ be another Hilbert space, $\mathcal{A}$ be a linear operator in $\mathcal{H}$ with domain $D_{\mathcal{A}}$, and $\mathcal{B} \in \mathcal{L}(\mathcal{U}', (H^1(\mathcal{H}'))')$. ($\mathcal{L}(X,Y)$ denotes the space of bounded linear operators from $X$ to $Y$.) We assume that $\mathcal{A}$ is the generator of a $C_0$-semigroup of bounded linear operators on $\mathcal{H}$. Consider the following control system:

$$\dot{y} = \mathcal{A}y + \mathcal{B}u, \quad y(0) = 0, \quad u \in \mathcal{U}'. \tag{2.1}$$

Our purpose is to identify or otherwise characterize the *reachable set*

$$\mathcal{R}_T = \{y(T) \,|\, u \in \mathcal{U}', \ y \text{ satisfies (2.1)}\}. \tag{2.2}$$

**Remark 2.1.** The choice of $(H^1(\mathcal{H}'))'$ as the space of control outputs is dictated primarily by applications to boundary control problems for partial differential equations that will be discussed in Section 4. This space is sufficiently general for many applications. However, one may treat more general classes of control outputs such as $(H^k(\mathcal{H}'))'$ or $(H^k(D_{\mathcal{A}'}))'$, $k \geq 0$, with only minor modifications of the theory presented below. Here $\mathcal{A}'$ denotes the dual operator of $\mathcal{A}$; $D_{\mathcal{A}'}$ is the domain of $\mathcal{A}'$ endowed with the graph norm of $\mathcal{A}'$.

Since the range of $\mathcal{B}$ is in a very weak space, the sense in which equation (2.1) is to be understood needs to be clarified. If $\mathcal{B}u$ is in the stronger space $L^2(\mathcal{H})$, the solution of (2.1) is unambiguously defined by the variation of constants formula

$$y(t) = \int_0^t S(t-s)(\mathcal{B}u)(s)\,ds, \quad 0 \leq t \leq T, \tag{2.3}$$

where $S(t), t \geq 0$, is the semigroup on $\mathcal{H}$ generated by $\mathcal{A}$. If $\phi^0 \in \mathcal{H}'$, from (2.3) we have

$$\begin{aligned} \langle \phi^0, y(T)\rangle_{\mathcal{H}} &= \int_0^T \langle \phi^0, S(T-s)(\mathcal{B}u)(s)\rangle_{\mathcal{H}}\,ds \\ &= \int_0^T \langle S'(T-s)\phi^0, (\mathcal{B}u)(s)\rangle_{\mathcal{H}}\,ds \\ &= \int_0^T \langle \phi(s), (\mathcal{B}u)(s)\rangle_{\mathcal{H}}\,ds = \langle \phi, \mathcal{B}u\rangle_{L^2(\mathcal{H})}, \end{aligned} \tag{2.4}$$

where $S'(t)$ is the dual semigroup of $S(t)$ and $\phi(s) = S'(T-s)\phi^0$. The dual semigroup acts in $\mathcal{H}'$ and is generated by the dual operator $\mathcal{A}'$ of $\mathcal{A}$. Therefore $\phi$ is a mild solution of

$$\dot{\phi}(t) = -\mathcal{A}'\phi(t), \ (t < T), \ \ \phi(T) = \phi^0. \tag{2.5}$$

The variational equation

$$\langle \phi^0, y(T)\rangle_{\mathcal{H}} = \langle \phi, \mathcal{B}u\rangle_{L^2(\mathcal{H})}, \quad \forall \phi^0 \in \mathcal{H}', \tag{2.6}$$

characterizes those states $y(T)$ that may be reached through the action of controls $u \in \mathcal{U}'$ such that $\mathcal{B}u \in L^2(\mathcal{H})$. A similar characterization will be given for the full set $\mathcal{R}_T$ and, simultaneously, the meaning of the solution of (2.1) when $\mathcal{B}u \in (H^1(\mathcal{H}'))'$ will be elucidated. This is done by the transposition method.

To motivate things, let $y$ be a strong solution of (2.1), $\phi$ be the solution of

$$\dot{\phi}(t) = -\mathcal{A}'\phi(t) + g, \ (t < T), \ \ \phi(T) = \phi^0, \tag{2.7}$$

where $\phi^0 \in D_{\mathcal{A}'}$ and $g \in L^\infty(\mathcal{H}')$, $\dot{g} \in L^1(\mathcal{H}')$. Then $\phi$ is a strong solution of (2.7) and we have

$$\begin{aligned} \langle \phi, \mathcal{B}u\rangle_{L^2(\mathcal{H})} &= \int_0^T \langle \phi, \dot{y} - \mathcal{A}y\rangle_{\mathcal{H}} dt \\ &= \langle \phi^0, y(T)\rangle_{\mathcal{H}} - \langle g, y\rangle_{L^2(\mathcal{H})}. \end{aligned} \tag{2.8}$$

Equation (2.8) is essentially the definition of the solution of (2.1), provided we interpret the various brackets $\langle \cdot, \cdot \rangle$ as duality pairings in spaces different from those indicated in (2.8). For example, if $\mathcal{B}u \in (H^1(\mathcal{H}'))'$, the left bracket is to be interpreted in the $(H^1(\mathcal{H}'))' - H^1(\mathcal{H}')$ duality pairing

$$\langle \mathcal{B}u, \phi \rangle_{H^1(\mathcal{H}')},$$

provided $\phi \in H^1(\mathcal{H}')$. The duality pairings to be chosen on the right side of (2.8) depend on what must be assumed about $\phi^0$ and $g$ to assure that $\phi \in H^1(\mathcal{H}')$.

**Lemma 2.1** *Assume that $\phi^0 \in D_{\mathcal{A}'}$ and $g \in L^1(D_{\mathcal{A}'}) \cap L^2(\mathcal{H}')$. Then the solution of (2.7) satisfies $\phi \in H^1(\mathcal{H}')$. Moreover,*

$$\|\phi\|_{H^1(\mathcal{H}')} \leq C \left( \|\phi^0\|_{D(\mathcal{A}')} + \|g\|_{L^1(D_{\mathcal{A}'})} + \|g\|_{L^2(\mathcal{H}')} \right). \tag{2.9}$$

**Proof.** The assumptions $\phi^0 \in D_{\mathcal{A}'}$, $g \in L^1(D_{\mathcal{A}'})$, imply that the solution of (2.7) is strongly differentiable and satisfies the differential equation almost everywhere, $\dot{\phi} \in L^1(\mathcal{H}')$, and

$$\|\phi\|_{L^\infty(D_{\mathcal{A}'})} \leq C \left( \|\phi^0\|_{D(\mathcal{A}')} + \|g\|_{L^1(D_{\mathcal{A}'})} \right).$$

If also $g \in L^2(\mathcal{H}')$ then

$$\dot{\phi} = -\mathcal{A}'\phi + g \in L^2(\mathcal{H}')$$

and

$$\|\dot{\phi}\|_{L^2(\mathcal{H}')} \leq C \left( \|\phi^0\|_{D(\mathcal{A}')} + \|g\|_{L^1(D_{\mathcal{A}'})} + \|g\|_{L^2(\mathcal{H}')} \right). \quad \square$$

We introduce the Banach space

$$\mathcal{X} = L^1(D_{\mathcal{A}'}) \cap L^2(\mathcal{H}')$$

with

$$\|g\|_{\mathcal{X}} = \|g\|_{L^1(D_{\mathcal{A}'})} + \|g\|_{L^2(\mathcal{H}')}.$$

One has the dense and continuous embeddings

$$L^2(D_{\mathcal{A}'}) \subset \mathcal{X}, \quad \mathcal{X}' \subset L^2((D_{\mathcal{A}'})').$$

We now rewrite (2.8) as

$$\langle y(T), \phi^0 \rangle_{D_{\mathcal{A}'}} - \langle y, g \rangle_{\mathcal{X}} = \langle \mathcal{B}u, \phi \rangle_{H^1(\mathcal{H}')}, \quad \forall \phi^0 \in D_{\mathcal{A}'}, \ \forall g \in \mathcal{X},$$

where $\phi$ satisfies (2.7).

**Proposition 2.2** *If $\mathcal{B}u \in (H^1(\mathcal{H}'))'$, there is a unique pair*

$$(y^0, y) \in (D_{\mathcal{A}'})' \times \mathcal{X}'$$

*such that*

$$\langle y^0, \phi^0 \rangle_{D_{\mathcal{A}'}} - \langle y, g \rangle_{\mathcal{X}} = \langle \mathcal{B}u, \phi \rangle_{H^1(\mathcal{H}')}, \quad \forall \phi^0 \in D_{\mathcal{A}'}, \ \forall g \in \mathcal{X}. \tag{2.10}$$

**Proof.** It is simply a matter of observing that the mapping $(\phi^0, g) \mapsto \langle \mathcal{B}u, \phi\rangle_{H^1(\mathcal{H}')}$ is linear and, according to Lemma 2.1, continuous from $D_{\mathcal{A}'} \times \mathcal{X}$ into $\Re$. □

*By definition*, the element $y$ provided by Proposition 2.2 is the solution of (2.1) and $y^0$ is its value at $T$. This convention is justified by the following result, which also describes the sense in which $y$ satisfies (2.1)

**Proposition 2.3** *Let $(y^0, y) \in (D_{\mathcal{A}'})' \times L^2((D_{\mathcal{A}'})')$ satisfy*

$$\langle y^0, \phi^0\rangle_{D_{\mathcal{A}'}} - \int_0^T \langle y(t), g(t)\rangle_{D_{\mathcal{A}'}}\, dt = \langle \mathcal{B}u, \phi\rangle_{H^1(\mathcal{H}')}, \qquad \forall \phi^0 \in D_{\mathcal{A}'},\ \forall g \in L^2(D_{\mathcal{A}'}), \tag{2.11}$$

*where $\phi$ is the solution of* (2.7) *corresponding to $\phi^0$ and $g$. Then $y$ satisfies, in the sense of distributions on* $(0,T)$,

$$\frac{d}{dt}\langle y(t), \phi^0\rangle_{D_{\mathcal{A}'}} = \langle y(t), \mathcal{A}'\phi^0\rangle_{D_{\mathcal{A}'}} + ((\Lambda^{-1}\mathcal{B}u)(t), \phi^0)_{\mathcal{H}'} - \frac{d}{dt}\Big(\frac{d}{dt}(\Lambda^{-1}\mathcal{B}u)(t), \phi^0\Big)_{\mathcal{H}'}, \quad \forall \phi^0 \in D_{(\mathcal{A}')^2}, \tag{2.12}$$

*where $\Lambda$ is the Riesz isomorphism of $H^1(\mathcal{H}')$ onto its dual. If, moreover, the map $t \to y(t) : [0,T] \to (D_{\mathcal{A}'})'$ is continuous at $t = T$ (resp., at $t = 0$), then*

$$y(T) = y^0, \quad (\textit{resp.},\ y(0) = 0.)$$

**Proof.** Let $\phi^0 \in D_{(\mathcal{A}')^2}$ and set

$$\phi(t) = \alpha(t)\phi^0, \quad g(t) = \dot\alpha(t)\phi^0 + \alpha(t)\mathcal{A}'\phi^0, \tag{2.13}$$

where

$$\alpha \in C^1([0,T]), \quad \alpha(0) = \alpha^0, \quad \alpha(T) = \alpha^1, \tag{2.14}$$

with $\alpha^0$ and $\alpha^1$ fixed, but arbitrary, constants. Then $\phi$ is the solution of

$$\dot\phi = -\mathcal{A}'\phi + g, \quad \phi(T) = \alpha^1\phi^0.$$

Substitution of (2.13) into (2.11) yields

$$\begin{aligned} \alpha^1\langle y^0, \phi^0\rangle_{D_{\mathcal{A}'}} &- \int_0^T \left[\dot\alpha(t)\langle y(t), \phi^0\rangle_{D_{\mathcal{A}'}} + \alpha(t)\langle y(t), \mathcal{A}'\phi^0\rangle_{D_{\mathcal{A}'}}\right] dt \\ &= \langle \mathcal{B}u, \alpha(\cdot)\phi^0\rangle_{H^1(\mathcal{H}')}. \end{aligned} \tag{2.15}$$

We have

$$\begin{aligned} \langle \mathcal{B}u, \alpha(\cdot)\phi^0\rangle_{H^1(\mathcal{H}')} &= (\Lambda^{-1}\mathcal{B}u, \alpha(\cdot)\phi^0)_{H^1(\mathcal{H}')} \\ &= \int_0^T \left[\alpha(t)((\Lambda^{-1}\mathcal{B}u)(t), \phi^0)_{\mathcal{H}'} + \dot\alpha(t)\Big(\frac{d}{dt}(\Lambda^{-1}\mathcal{B}u)(t), \phi^0\Big)_{\mathcal{H}'}\right] \\ &= \int_0^T [\xi(t)\alpha(t) + \dot\xi(t)\dot\alpha(t)]dt, \end{aligned} \tag{2.16}$$

where

$$\xi(\cdot) = ((\Lambda^{-1}\mathcal{B}u)(\cdot), \phi^0)_{\mathcal{H}'} \in H^1(0,T). \tag{2.17}$$

With (2.16), (2.17), identity (2.15) takes the form

$$\begin{aligned}\alpha^1\langle y^0, \phi^0\rangle_{D_{\mathcal{A}'}} &- \int_0^T \left[\dot{\alpha}(t)\langle y(t), \phi^0\rangle_{D_{\mathcal{A}'}} + \alpha(t)\langle y(t), \mathcal{A}'\phi^0\rangle_{D_{\mathcal{A}'}}\right] dt \\ &= \int_0^T [\xi(t)\alpha(t) + \dot{\xi}(t)\dot{\alpha}(t)]dt.\end{aligned} \tag{2.18}$$

This identity holds for every $\phi^0 \in D_{(\mathcal{A}')^2}$ and every $\alpha$ which satisfies (2.14). Since $\alpha^0$ and $\alpha^1$ are arbitrary, it follows from (2.18) that $y$ satisfies, in the sense of distributions on $(0,T)$, the variational equation (2.12) and, in a weak sense made precise by (2.18), the end conditions

$$\langle y^0 - y(T), \phi^0\rangle_{D_{\mathcal{A}'}} = 0, \quad \langle y(0), \phi^0\rangle_{D_{\mathcal{A}'}} = 0, \quad \forall \phi^0 \in D_{\mathcal{A}'}. \tag{2.19}$$

If $y$ is continuous from $[0,T]$ into $(D_{\mathcal{A}'})'$ then, in particular, (2.19) holds in the strict sense, so that

$$y(T) = y^0, \quad y(0) = 0. \quad \square$$

If we set $g = 0$ in (2.10), it follows that *elements* $y(T) \in \mathcal{R}_T$ *are characterized by the variational equation*

$$\langle y(T), \phi^0\rangle_{D_{\mathcal{A}'}} = \langle \mathcal{B}u, \phi\rangle_{H^1(\mathcal{H}')} = \langle u, \mathcal{B}'\phi\rangle_{\mathcal{U}} = (u, \Lambda_{\mathcal{U}}\mathcal{B}'\phi)_{\mathcal{U}'}, \quad \forall \phi^0 \in D_{\mathcal{A}'}, \tag{2.20}$$

where $\phi$ satisfies (2.5). Let us introduce the *linear space of observations*

$$\mathcal{O} = \{u \,|\, u = \Lambda_{\mathcal{U}}\mathcal{B}'\phi, \;\; \phi \text{ satisfies (2.5) with } \phi^0 \in D_{\mathcal{A}'}\} \subset \mathcal{U}'.$$

It follows from (2.20) that controls $u \in \overline{\mathcal{O}}^{\perp}$, the orthogonal complement in $\mathcal{U}'$ of the closure of $\mathcal{O}$, simply steer the zero state to itself, therefore

$$\mathcal{R}_T = \{y(T) \,|\, u \in \overline{\mathcal{O}}, \;\; y \text{ satisfies (2.1)}\}.$$

**Proposition 2.4** *Controls* $u \in \overline{\mathcal{O}}$ *driving* 0 *to a state* $y^0 \in \mathcal{R}_T$ *are unique. If* $u \in \overline{\mathcal{O}}$ *drives* 0 *to* $y^0$*, then* $u$ *is the control of minimum norm among all controls in* $\mathcal{U}'$ *that drive* 0 *to* $y^0$.

**Proof.** If $u \in \overline{\mathcal{O}}$ and $v \in \overline{\mathcal{O}}$ drive 0 to the same state $y^0$, then

$$(u - v, \Lambda_{\mathcal{U}}\mathcal{B}'\phi)_{\mathcal{U}'} = 0, \quad \forall \phi^0 \in D_{\mathcal{A}'},$$

hence $u - v \in \overline{\mathcal{O}}^{\perp}$, so that $u - v = 0$.

Let $u$ be a control in $\mathcal{U}'$ that drives 0 to a state $y^0$. We may write

$$u = u_0 + u_1, \quad u_0 \in \overline{\mathcal{O}}, \quad u_1 \in \overline{\mathcal{O}}^{\perp}.$$

But $u_1$ drives 0 to itself, hence $u_0$ drives 0 to $y^0$ and $\|u\|^2_{\mathcal{U}'} \geq \|u_0\|^2_{\mathcal{U}'}$. $\square$

In order to proceed further, we need to impose the following hypothesis:

*Observability Assumption:*

$$\phi^0 \in D_{\mathcal{A}'}, \quad \|\mathcal{B}'\phi\|_{\mathcal{U}} = 0 \Longleftrightarrow \phi^0 = 0. \tag{2.21}$$

If (2.21) holds, we may introduce a Hilbert norm

$$\|\phi^0\|_F = \|\mathcal{B}'\phi\|_{\mathcal{U}},$$

and a Hilbert space

$$F = \text{completion of } D_{\mathcal{A}'} \text{ in } \|\cdot\|_F.$$

The space $F$ will, in general, depend on $T$. Since

$$\|\mathcal{B}'\phi\|^2_{\mathcal{U}} \leq C\|\phi\|^2_{H^1(\mathcal{H}')} \leq C_T\|\phi^0\|^2_{D_{\mathcal{A}'}},$$

we have the dense and continuous embeddings

$$D_{\mathcal{A}'} \subset F, \quad F' \subset (D_{\mathcal{A}'})'.$$

For $\psi^0 \in D_{\mathcal{A}'}$ we may define $\Lambda_F\psi^0 \in (D_{\mathcal{A}'})'$ by $y(T) = \Lambda_F\psi^0$, where $y(T)$ is the solution of (2.20) corresponding to $u = \Lambda_{\mathcal{U}}\mathcal{B}'\psi$, $\psi$ denoting the solution of (2.5) with $\psi(T) = \psi^0$. ($\Lambda_F$ corresponds to the controllability Grammian in the finite dimensional case.) From (2.20) we have

$$\langle \Lambda_F\psi^0, \phi^0\rangle_{D_{\mathcal{A}'}} = (\psi^0, \phi^0)_F, \quad \forall \psi^0, \phi^0 \in D_{\mathcal{A}'}. \tag{2.22}$$

Therefore $\Lambda_F$ extends to an operator in $\mathcal{L}(F, F')$, this extension being precisely the Riesz isomorphism of $F$ onto $F'$. As a consequence we have

$$\mathcal{R}_T = \text{ range of } \Lambda_F = F'.$$

**Remark 2.2.** The uniqueness property (2.21), together with construction of the corresponding Hilbert space $F$ whose dual characterizes the reachable states of (2.1), is the reason for the terminology *Hilbert Uniqueness Method.*

To find the control $u_0$ in $\overline{\mathcal{O}}$ that steers 0 to a given element $y^0 \in F'$, define $\psi^0 = \Lambda_F^{-1}y^0 \in F$, let $\psi$ be the "solution" of

$$\dot{\psi} = -\mathcal{A}'\psi, \quad (t < T), \quad \psi(T) = \psi^0, \tag{2.23}$$

and define $u_0 = \Lambda_{\mathcal{U}}\mathcal{B}'\psi$. Then the above construction formally gives $y(T) = y^0$. However, when $\psi^0 \in F$, it is not clear how the solution of (2.23) is defined and, consequently, what is the meaning of $u_0$. However, $u_0$ can be made precise as follows. Take a sequence $\{\psi_n^0\} \in D_{\mathcal{A}'}$ such that $\psi_n^0 \to \psi^0$ in $F$. Then $u_n =: \Lambda_{\mathcal{U}}\mathcal{B}'\psi_n$ converges in $\mathcal{U}'$ (from the definition of the $F$-norm) and therefore

$$u_0 =: \lim_n \Lambda_{\mathcal{U}}\mathcal{B}'\psi_n$$

is a well-defined element in $\overline{\mathcal{O}}$. Furthermore, if $y_n^0 =: \Lambda_F\psi_n^0$ we have $y_n^0 \to y^0$ in $F'$. Let $y_n$ and $y$ be the solutions of

$$\dot{y}_n = \mathcal{A}y_n + \mathcal{B}u_n, \quad y_n(0) = 0,$$

$$\dot{y} = \mathcal{A}y + \mathcal{B}u_0, \quad y(0) = 0.$$

By the definition of these solutions,

$$\langle y_n^0 - y(T), \phi^0 \rangle_{D_{\mathcal{A}'}} - \langle y_n - y, g \rangle_{\mathcal{X}} = \langle \mathcal{B}u_n - \mathcal{B}u_0, \phi \rangle_{H^1(\mathcal{H}')}, \quad \forall \phi^0 \in D_{\mathcal{A}'}, \forall g \in \mathcal{X}.$$

It follows that

$$y_n \to y \text{ weak* in } \mathcal{X}', \quad y_n^0 \to y(T) \text{ weak* in } (D_{\mathcal{A}'})'.$$

Since $y_n^0 \to y^0$ strongly in $F'$ and, *a fortiori*, in $(D_{\mathcal{A}'})'$, we have $y(T) = y^0$. That is, $u_0$ is the unique control in $\overline{\mathcal{O}}$ that drives 0 to $y^0$.

**Remark 2.3.** An alternative formulation of the foregoing description of the reachable set is given in [11],[12],[17] and may be described in terms of the control-to-state map $u \mapsto y(T) : \mathcal{U}' \mapsto (D_{\mathcal{A}'})'$. If fact, denoting this map by $L_T$, from (2.20) we deduce that $L_T \in \mathcal{L}(\mathcal{U}', (D_{\mathcal{A}'})')$. The kernal of $L_T$ is $\overline{\mathcal{O}}^{\perp}$, and its dual operator satisfies

$$L_T'\phi^0 = B'\phi, \quad \forall \phi^0 \in D_{\mathcal{A}'},$$

where $\phi^0, \phi$ satisfy (2.5). The observability assumption is equivalent to the hypothesis that $L_T'$ is one-to-one which is in turn equivalent to $\overline{\mathrm{Rg}(L_T)} = (D_{\mathcal{A}'})'$, i.e., the system (2.1) is approximately controllable. Thus observability (sometimes called *distinguishability*; c.f. [4]) and approximate controllability are "dual" properties. The operator $\Lambda_F$ is given in terms of $L_T$ by

$$\Lambda_F = L_T \Lambda_{\mathcal{U}} L_T' \quad \text{on } D_{\mathcal{A}'}.$$

The minimum norm control that steers 0 to an element $y^0 \in \Lambda_F(D_{\mathcal{A}'})$ is

$$u_0 = \Lambda_{\mathcal{U}} B'\psi = \Lambda_{\mathcal{U}} L_T'\psi^0 = \Lambda_{\mathcal{U}} L_T'(L_T \Lambda_{\mathcal{U}} L_T')^{-1} y^0. \quad \Box$$

Although the reachable set at time $T$ is identified as the space $F'$ (when the observability assumption is satisfied), the reachability problem is still not solved in any practical sense unless $F'$, or a dense subspace of it, can be identified in terms of spaces intrinsic to the problem. In order to do this, it is necessary and sufficient to establish *a priori* estimates of the form

$$\|\mathcal{B}'\phi\|_{\mathcal{U}}^2 \geq c^2 \|\phi^0\|_{\mathcal{H}_1}^2, \quad \forall \phi^0 \in D_{\mathcal{A}'}. \tag{2.24}$$

More precisely, one has

**Proposition 2.5** *Let $\mathcal{H}_1$ be a Hilbert space such that $D_{\mathcal{A}'}$ is continuously and densely embedded in $\mathcal{H}_1$. If* (2.24) *holds, then $\mathcal{H}_1'$ is in the reachable set of* (2.1). *Conversely, if $\mathcal{H}_1'$ is in the reachable set of* (2.1) *and if*

$$D = \{u \in \mathcal{U}' \,|\, y(T) = L_T u \in \mathcal{H}_1'\} \quad \textit{is dense in } \mathcal{U}',$$

*then* (2.24) *must hold.*

**Proof.** If (2.24) holds then

$$F \subset \mathcal{H}_1, \quad \mathcal{H}_1' \subset F',$$

hence $\mathcal{H}_1'$ is in the reachable set of (2.1). For the converse, define $\mathcal{L}_T = L_T|_D$. It is a densely defined, closed linear operator from $D \subset \mathcal{U}'$ into $\mathcal{H}_1'$. Its dual $\mathcal{L}_T'$: $\mathrm{Dom}(\mathcal{L}_T') \subset \mathcal{H}_1 \mapsto \mathcal{U}$ satisfies $D_{\mathcal{A}'} \subset \mathrm{Dom}(\mathcal{L}_T')$ and $\mathcal{L}_T'|_{D'_{\mathcal{A}}} = L_T'$. It is standard theory that $\mathrm{Rg}(\mathcal{L}_T) = \mathcal{H}_1'$ if, and only if, $\mathcal{L}_T'$ is continuously invertible. The latter property is equivalent to (2.24). □

Property (2.24) is sometimes referred to as *continuous $\mathcal{H}_1$ observability*. Thus exact controllability to $\mathcal{H}_1$ and continuous $\mathcal{H}_1$ observability are equivalent properties, a duality relation that has frequently been pointed out in the control literature (e.g., [4],[11],[12],[17] in infinite dimensional contexts). Therefore, deriving *a priori* estimates of the form (2.24) is the crucial issue in exact controllability problems. An application of the last proposition to Maxwell's system of will be given in Section 4.

In cases where $\mathcal{A}$ is a skew-adjoint operator arising from a linear, second order system (a situation considered in the next section), there is an intrinsic space $\mathcal{H}_1$ for which the estimate (2.24) often can be established. In fact, it is sometimes possible to obtain two-sided estimates of $\|\cdot\|_F$ in terms of $\|\cdot\|_{\mathcal{H}_1}$, so that $F = \mathcal{H}_1$ in such cases.

# 3 Second Order Control Systems

Let $V$ and $H$ be Hilbert spaces with $V$ dense and continuously embedded in $H$. We identify $H$ with its dual space so that we have as usual $V \subset H \subset V'$. Let $A$ be the Riesz isomorphism of $V$ onto $V'$, and set

$$D_A = \{v \in V \,|\, Av \in H\}, \quad \Delta_A = \{v \in V \,|\, Av \in V\},$$

$$\|v\|_{D_A} = \|Av\|_H, \quad \|v\|_{\Delta_A} = \|Av\|_V.$$

$D_A$ and $\Delta_A$ are Hilbert spaces and $A$ is an isomorphism of $D_A$ onto $H$ and of $\Delta_A$ onto $V$. We have the algebraic and topological inclusions

$$\Delta_A \subset D_A \subset V \subset H \subset V' \subset D_A' \subset \Delta_A'.$$

Furthermore, $A$ extends to an isomorphism of $H$ onto $D_A'$ and of $V'$ onto $\Delta_A'$. The spaces $V'$, $D_A'$ and $\Delta_A'$ are Hilbert spaces under the scalar product

$$(v, w)_X = (A^{-1}v, A^{-1}w)_Y,$$

where $(X, Y)$ stands for any one of the pairs $(V', V)$, $(D_A', H)$ or $(\Delta_A', V')$.

Let $\mathcal{U}$ be a Hilbert space whose dual $\mathcal{U}'$ will be the space of controls, and let $T > 0$ be fixed. In keeping with the notation of Section 2, we shall write $L^2(H)$, $H^1(V)$, etc., in place of $L^2(0,T;H)$, $H^1(0,T;V)$, etc. We identify $L^2(H)$ with its dual space, so that

$$H^1(V) \subset L^2(H) \subset (H^1(V))'.$$

Let $B \in \mathcal{L}(\mathcal{U}', (H^1(V))')$. We consider the following control problem:

$$\ddot{w} + Aw = Bu, \quad w(0) = \dot{w}(0) = 0, \quad u \in \mathcal{U}'. \tag{3.1}$$

We wish to identify the reachable set

$$\mathcal{R}_T = \{(w(T), \dot{w}(T)) | u \in \mathcal{U}', \ w \text{ satisfies } (3.1)\}. \tag{3.2}$$

The problem (3.1) may be transformed to a first order system of the type considered in Section 2. Set

$$y_1 = w, \ y_2 = \dot{w}, \ y = \begin{pmatrix} y_1 \\ y_2 \end{pmatrix}, \ \mathcal{A} = \begin{pmatrix} 0 & I \\ -A & 0 \end{pmatrix}, \ \mathcal{B} = \begin{pmatrix} 0 \\ B \end{pmatrix}.$$

Define

$$\mathcal{H} = H \times V', \ (\text{so that } \mathcal{H}' = H \times V).$$

Then $\mathcal{B} \in \mathcal{L}(\mathcal{U}', (H^1(\mathcal{H}'))')$ and (3.1) becomes

$$\dot{y} = \mathcal{A}y + \mathcal{B}u, \ \ y(0) = 0. \tag{3.3}$$

**Lemma 3.1** *The operator $\mathcal{A}$, as an unbounded operator in $H \times V'$ with domain $V \times H$, is skew-adjoint and, therefore, generates a group of unitary operators on $H \times V'$.*

**Proof.** It is obvious that $\mathcal{A}$ is an isomorphism of $V \times H$ onto $H \times V'$ and, therefore, $\mathcal{A}$ is closed. So we have only to check that $-\mathcal{A} \subset \mathcal{A}^*$, the adjoint of $\mathcal{A}$. If $y = (y_1, y_2)$ and $z = (z_1, z_2)$ are in $V \times H$ we have

$$\begin{aligned} (\mathcal{A}y, z)_{\mathcal{H}} &= (y_2, z_1)_H - (Ay_1, z_2)_{V'} \\ &= (y_2, z_1)_H - \langle z_2, y_1 \rangle_V \\ &= (y_2, z_1)_H - (z_2, y_1)_H = -(y, \mathcal{A}z)_{\mathcal{H}}. \ \square \end{aligned}$$

The theory of Section 2 may therefore be applied to (3.3). The problem (3.3) has a unique solution in the sense of Proposition 2.2 and, if the observability assumption (2.20) is satisfied, the reachable set may be identified with the space $F'$. However, because of the special structure of the operator $\mathcal{A}$, one can obtain stronger results than those provided by Section 2.

First, we need to identify the dual operator $\mathcal{A}'$ of $\mathcal{A}$.

**Lemma 3.2** *The dual of $\mathcal{A}$ is the operator in $H \times V$ defined by*

$$\mathcal{A}' = \begin{pmatrix} 0 & -A \\ I & 0 \end{pmatrix}, \ \ D_{\mathcal{A}'} = V \times D_A.$$

**Proof.** the operator $\mathcal{A}'$ is related to the adjoint operator $\mathcal{A}^* = -\mathcal{A}$ by

$$\mathcal{A}' = \Lambda_{\mathcal{H}} \mathcal{A}^* \Lambda_{\mathcal{H}'}^{-1},$$

where $\Lambda_{\mathcal{H}}$ is the Riesz isomorphism of $\mathcal{H} = H \times V'$ onto $\mathcal{H}' = H \times V$. Thus

$$\Lambda_{\mathcal{H}} = \begin{pmatrix} I & 0 \\ 0 & A^{-1} \end{pmatrix}, \ \ \Lambda_{\mathcal{H}}^{-1} = \begin{pmatrix} I & 0 \\ 0 & A \end{pmatrix},$$

so that $\mathcal{A}'$ is the indicated matrix. The domain of $\mathcal{A}'$ consists of elements $\phi = (\phi_1, \phi_2)$ such that $\Lambda_{\mathcal{H}}^{-1}\phi = (\phi_1, A\phi_2) \in D(\mathcal{A}) = V \times H$; thus $D(\mathcal{A}') = V \times D_A$. $\square$

**Proposition 3.3** *The solution $y = (y_1, y_2)$ of (3.3) (guaranteed by Proposition 2.2) satisfies*

$$y_1 \in C(V'), \;\; y_2 = \dot{y}_1 \in Y' \subset L^2(D'_A),$$

*where* $Y = L^1(D_A) \cap L^2(V)$,

$$\|f\|_Y = \|f\|_{L^1(D_A)} + \|f\|_{L^2(V)}.$$

*In addition,* $w = y_1$ *satisfies, in the sense of distributions on* $(0,T)$,

$$\frac{d}{dt}\langle w(t), \phi_1^0\rangle_V - \langle \dot{w}(t), \phi_1^0\rangle_{D_A} = 0, \;\; \forall \phi_1^0 \in D_A, \tag{3.4}$$

$$\begin{aligned}\frac{d}{dt}\langle \dot{w}(t), \phi_2^0\rangle_{D_A} + \langle w(t), A\phi_2^0\rangle_V &= ((\Lambda^{-1}Bu)(t), \phi_2^0)_V \\ &\quad -\frac{d}{dt}(\frac{d}{dt}(\Lambda^{-1}Bu)(t), \phi_2^0)_V, \;\; \forall \phi_2^0 \in \Delta_A,\end{aligned} \tag{3.5}$$

*where* $\Lambda$ *denotes the Riesz isomorphism of* $H^1(V)$ *onto* $(H^1(V))'$.

**Proof.** The solution of (3.3) guaranteed by Proposition 2.2 satisfies

$$(y_1^0, y_2^0) \in (D_{A'})' = V' \times D'_A, \;\; (y_1, y_2) \in \mathcal{Y}',$$

where

$$\mathcal{Y} = (L^1(V) \cap L^2(H)) \times (L^1(D_A) \cap L^2(V)).$$

Let $\phi = (\phi_1, \phi_2)$ be the solution of the system (corresponding to (2.7))

$$\begin{cases} \dot{\phi}_1 = A\phi_2 + g_1, \;\; \dot{\phi}_2 = -\phi_1 + g_2, \\ \phi_1(T) = \phi_1^0, \;\; \phi_2(T) = \phi_2^0. \end{cases} \tag{3.6}$$

The solution of (3.3) is defined by (2.10), i.e.,

$$\begin{aligned}\langle (y_1^0, y_2^0), (\phi_1^0, \phi_2^0)\rangle_{D_{A'}} &- \langle (y_1, y_2), (g_1, g_2)\rangle \\ &= \langle \mathcal{B}u, \phi\rangle_{H^1(\mathcal{H}')} = \langle Bu, \phi_2\rangle_{H^1(V)}.\end{aligned} \tag{3.7}$$

We have purposely omitted for the time being reference to the specific spaces in the second duality pairing of the left side of (3.7). The regularity possessed by the solution of (3.7) is determined by the spaces that $\phi^0 = (\phi_1^0, \phi_2^0)$ and $g = (g_1, g_2)$ must belong to in order to assure that the linear functional $(\phi^0, g) \mapsto \langle Bu, \phi_2\rangle_{H^1(V)}$ is continuous. But it is standard that if $(\phi_1^0, \phi_2^0) \in D_{A'} = V \times D_A$ and if $g_1 \in L^1(V)$, $g_2 \in L^1(D_A) \cap L^2(V)$, then

$$\|\phi_1\|_{L^\infty(V)} + \|\phi_2\|_{L^\infty(D_A)} + \|\dot{\phi}_2\|_{L^2(V)} \le C\left(\|(\phi_1^0, \phi_2^0)\|_{V \times D_A} + \|g_1\|_{L^1(V)} + \|g_2\|_Y\right).$$

Thus $(\phi^0, g) \mapsto \langle Bu, \phi_2\rangle_{H^1(V)}$ is linear and continuous from $(V \times D_A) \times (L^1(V) \times Y)$ into $\Re$. It follows that the unique solution of (3.7) satisfies, in particular, $y_1 \in L^\infty(V')$. The duality pairing in the second term of (3.7) is between $L^\infty(V') \times Y'$ and $L^1(V) \times Y$. One may pass from $L^\infty(V')$ to $C(V')$ by a standard approximation technique.

One has $y_2 = \dot{y}_1$ since

$$\int_0^T [\langle y_1(t), \dot{g}_2(t)\rangle_V + \langle y_2(t), g_2(t)\rangle_{D_A}]\, dt = 0, \quad \forall g_2 \in C_0^\infty(0,T; D_A). \tag{3.8}$$

In fact, set $\phi_1^0 = \phi_2^0 = 0$, $g_1 = \dot{g}_2$ in (3.6). Then

$$\ddot{\phi}_2 = -\dot{\phi}_1 + \dot{g}_2 = -A\phi_2, \ (t < T), \ \ \phi_2(T) = \dot{\phi}_2(T) = 0.$$

Therefore $\phi_2 \equiv 0$ and we obtain (3.8) from (3.7).

Equations (3.4) and (3.5) may be derived by choosing in (3.7)

$$g_1(t) = -\alpha_2(t)A\phi_2^0 + \dot{\alpha}_1(t)\phi_1^0, \ \ g_2(t) = \alpha_1(t)\phi_1^0 + \dot{\alpha}_2(t)\phi_2^0,$$

where $\alpha_i$ is an arbitrary $C^1([0,T])$ function and $(\phi_1^0, \phi_2^0) \in D_A \times \Delta_A$. Then $\phi_1(t) = \alpha_1(t)\phi_1^0$ and $\phi_2(t) = \alpha_2(t)\phi_2^0$ satisfy the adjoint system (3.6) and have end values $\phi_1(T) = \alpha_1(T)\phi_1^0$, $\phi_2(T) = \alpha_2(T)\phi_2^0$, respectively. If these quantities are then substituted into (3.7), equations (3.4) and (3.5) result. In addition, since $w$ is continuous into $V'$, the traces of $w$ at at $t = 0$ and at $t = T$ equal 0 and $y_1^0$, respectively. If $\dot{w}$ is continuous into $D'_A$ (this is not guaranteed by Proposition 3.3), then the traces of $\dot{w}$ agree with 0 and $y_2^0$ at $t = 0$ and $t = T$, respectively. □

The observability assumption (2.21), in the present context, takes the form

$$(\phi_1^0, \phi_2^0) \in V \times D_A, \ \ \|B'\phi_2\|_{\mathcal{U}} = 0 \Longleftrightarrow \phi_1^0 = \phi_2^0 = 0, \tag{3.9}$$

where $\phi = (\phi_1, \phi_2)$ is the solution of

$$\begin{cases} \dot{\phi}_1 = A\phi_2, \ \ \dot{\phi}_2 = -\phi_1, \ \ (t < T), \\ \phi_1(T) = \phi_1^0, \ \ \phi_2(T) = \phi_2^0. \end{cases} \tag{3.10}$$

Under assumption (3.9), we may introduce a norm

$$\|(\phi_1^0, \phi_2^0)\|_F = \|B'\phi_2\|_{\mathcal{U}}$$

and a space

$$F = \text{completion of } V \times D_A \text{ in } \|\cdot\|_F.$$

We have

$$V \times D_A \subset F, \ \ F' \subset V' \times D'_A,$$

and $F'$ coincides with the set of reachable states of (3.1) at time $T$. Let $\Lambda_F$ be the Riesz isomorphism of $F$ onto $F'$. If $(w^0, w^1) \in F'$, define

$$(\phi_1^0, \phi_2^0) = \Lambda_F^{-1}(w^0, w^1), \ \ u_0 = \Lambda_{\mathcal{U}} B'\phi_2,$$

where $(\phi_1, \phi_2)$ solves (3.10). Then $u_0$ is the control of minimum $\mathcal{U}'$ norm among all controls driving $(0,0)$ to $(w^0, w^1)$, where the dynamics are described by (3.1). Since, in general, it will not be the case that $\phi_2 \in H^1(V)$ when $(\phi_1^0, \phi_2^0) \in F$, the precise meaning of $u_0$ is

$$u_0 = \lim_n \Lambda_{\mathcal{U}} B'\phi_{2,n} \ \text{ strongly in } \mathcal{U}',$$

where $(\phi_{1,n}, \phi_{2,n})$ is the solution of (3.10) corresponding to $(\phi_{1,n}^0, \phi_{2,n}^0)$, where $(\phi_{1,n}^0, \phi_{2,n}^0) \subset V \times D_A$ converges to $(\phi_1^0, \phi_2^0)$ in $F$.

**Corollary 3.4** *If*

$$\|B'\phi_2\|_{\mathcal{U}}^2 \geq c^2\|(\phi_1^0,\phi_2^0)\|_{V'\times H}^2, \quad \forall(\phi_1^0,\phi_2^0) \in V \times D_A, \tag{3.11}$$

*where* $(\phi_1, \phi_2)$ *solves* (3.10), *then the reachable set of* (3.1) *at time* $T$ *contains* $V \times H$ . *Conversely, if* $V \times H$ *is in the reachable set at time* $T$ *and if*

$$\{u \in \mathcal{U}' \mid (w(T), \dot{w}(T)) \in V \times H\} \text{ is dense in } \mathcal{U}',$$

*then* (3.11) *must hold.*

The *a priori* estimate (3.11) is one that is known to hold for many boundary control problems for hyperbolic partial differential equations. An illustration of this is provided in the first example of the next section.

# 4 Examples

In this section two examples will be presented to illustrate how the above theory applies to specific distributed parameter control systems.

## 4.1 Kirchhoff plate equation with boundary controls acting in shear force and in bending and twisting moments

Let $\Omega$ be a bounded region in $\Re^2$ with smooth boundary $\Gamma$. Let $X_0 = (x_0, y_0)$ be a fixed but otherwise arbitrary point of $\Re^2$, and set

$$\Gamma_+ = \{X \in \Gamma \mid (X - X_0) \cdot \nu > 0\}, \quad \Gamma_- = \Gamma - \Gamma_+,$$

where $\nu$ denotes the unit normal to $\Gamma$ pointing towards the exterior of $\Omega$. Note that $\Gamma_\pm$ depend on the choice of $X_0$. We consider the plate equation

$$\ddot{w} - \gamma^2\Delta\ddot{w} + \gamma^2\Delta^2 w = 0 \text{ in } Q = \Omega \times (0,T) \tag{4.1}$$

with boundary conditions

$$w = \frac{\partial w}{\partial \nu} = 0 \text{ on } \Sigma_- = \Gamma_- \times (0,T), \tag{4.2}$$

$$\begin{cases} \gamma^2[\Delta w + (1-\mu)P_1 w] = u_0, \\ \gamma^2\left[\dfrac{\partial \Delta w}{\partial \nu} + (1-\mu)P_2 w - \dfrac{\partial \ddot{w}}{\partial \nu}\right] = -u_2 + \dfrac{\partial u_1}{\partial \tau} \text{ on } \Sigma_+ = \Gamma_+ \times (0,T), \end{cases} \tag{4.3}$$

and initial conditions

$$w(\cdot,0) = \frac{\partial w}{\partial t}(\cdot,0) = 0 \text{ in } \Omega. \tag{4.4}$$

In the above, $\Delta$ is the ordinary Laplacian in $\Re^2$, $\gamma^2$ is a constant of order $O(h^2)$, $h$ denoting the uniform thickness of the plate, and $\mu \in (0,1)$ is another constant (Poisson's ratio).

We specifically assume that $\Gamma_{\pm} \neq \emptyset$. $\tau$ is the positively oriented unit tangent vector to $\Gamma_+$, and $P_1$ and $P_2$ are boundary operators which satisfy the Green's formula

$$(\Delta^2 u, v)_{L^2(\Omega)} = a(u,v) + \int_\Gamma \left[ v\left(\frac{\partial \Delta u}{\partial \nu} + (1-\mu)P_2 u\right) - (\Delta u + (1-\mu)P_1 u)\frac{\partial v}{\partial \nu}\right] d\Gamma$$

where

$$a(u,v) = \int_\Omega \left[\left(\frac{\partial^2 v}{\partial x^2}\right)^2 + \left(\frac{\partial^2 v}{\partial y^2}\right)^2 + 2\mu \frac{\partial^2 v}{\partial x^2}\frac{\partial^2 v}{\partial y^2} + 2(1-\mu)\left(\frac{\partial^2 v}{\partial x \partial y}\right)^2\right] dxdy.$$

The specific forms of these operators may be found in [5] or in [10]. The quantities $u_0$, $u_1$ and $u_2$ are the controls. They correspond, respectively, to a bending moment about the tangent vector to $\Gamma$, a twisting moment about the normal to $\Gamma$ and to an edge shear force acting perpendicularly to the faces of the plate.

The spaces $L^2(\Omega)$, $H^k(\Omega)$, $L^2(\Gamma)$ denote the standard real $L^2$ and Sobolev spaces over $\Omega$ or $\Gamma$ as notation implies. We set

$$H = \{v \mid v \in H^1(\Omega),\ v = 0 \text{ on } \Gamma_-\},$$

$$V = \{v \mid v \in H^2(\Omega),\ v = \partial v/\partial \nu = 0 \text{ on } \Gamma_-\}.$$

The norms in these spaces are taken to be

$$\|v\|_H = \left(\int_\Omega (v^2 + \gamma^2 |\nabla v|^2) dxdy\right)^{1/2}, \quad \|v\|_V = [\gamma^2 a(v,v)]^{1/2}.$$

That $\|\cdot\|_V$ is a norm equivalent to the standard $H^2(\Omega)$ norm is a version of Korn's Lemma. We further choose

$$\mathcal{U} = (L^2(\Sigma_+))^3$$

and we identify $\mathcal{U}$ with its dual space $\mathcal{U}'$.

If one forms the $L^2(Q)$ scalar product of (4.1) with a test function $\phi \in L^2(0,T;V) =: L^2(V)$ and uses the above Green's formula one obtains

$$(\ddot{w}, \phi)_{L^2(H)} + (w, \phi)_{L^2(V)} = \int_0^T \int_{\Gamma_+} \left(u_2\phi + u_1 \frac{\partial \phi}{\partial \tau} + u_0 \frac{\partial \phi}{\partial \nu}\right) d\Gamma dt.$$

This variational equation is the same as

$$\ddot{w} + Aw = Bu \quad \text{in } L^2(V'), \quad w(0) = \dot{w}(0) = 0. \tag{4.5}$$

where $u = (u_0, u_1, u_2) \in \mathcal{U}$ and where $B \in \mathcal{L}(\mathcal{U}, L^2(V'))$ is defined by

$$\langle Bu, \phi \rangle_{L^2(V)} = \int_0^T \int_{\Gamma_+} \left(u_2\phi + u_1 \frac{\partial \phi}{\partial \tau} + u_0 \frac{\partial \phi}{\partial \nu}\right) d\Gamma dt, \quad \forall \phi \in L^2(V).$$

Problem (4.5) has a unique solution with $(w,\dot{w}) \in C([0,T]; H \times V')$.

Let us write down the inequality (3.11). The dual operator $B'$ is defined by

$$B'\phi = \left(\frac{\partial \phi}{\partial \nu}, \frac{\partial \phi}{\partial \tau}, \phi\right)\Big|_{\Sigma_+}, \quad \forall \phi \in L^2(V),$$

so that (3.11) is

$$\|B'\phi_2\|_{\mathcal{U}}^2 = \int_0^T \int_{\Gamma_+} (\phi_2^2 + |\nabla \phi_2|^2) d\Gamma dt \geq c^2 \|(\phi_1^0, \phi_2^0)\|_{V' \times H}^2, \qquad (4.6)$$

$$\forall (\phi_1^0, \phi_2^0) \in V \times D_A.$$

Set $\eta = \phi_2 \in C([0,T]; D_A)$. Then $\dot{\eta} = -\phi_1 \in C([0,T]; V)$ and

$$\ddot{\eta} + A\eta = 0 \ \text{ in } C([0,T]; H), \ \ \eta(T) = \phi_2^0 \in D_A, \ \ \dot{\eta}(T) = -\phi_1^0 \in V, \qquad (4.7)$$

and (4.6) is equivalent to

$$\int_0^T \int_{\Gamma_+} (\eta^2 + |\nabla \eta|^2) d\Gamma dt \geq c^2 \|(\eta^0, \eta^1)\|_{H \times V'}^2, \ \ \forall (\eta^0, \eta^1) \in D_A \times V. \qquad (4.8)$$

However, (4.7) signifies that $\eta$ is a solution of

$$\ddot{\eta} - \gamma^2 \Delta \ddot{\eta} + \gamma^2 \Delta^2 \eta = 0 \ \text{ in } Q,$$

$$\eta = \frac{\partial \eta}{\partial \nu} = 0 \ \text{ on } \Sigma_-,$$

$$\begin{cases} \gamma^2[\Delta \eta + (1-\mu) P_1 \eta] = 0, \\ \gamma^2 \left[\dfrac{\partial \Delta \eta}{\partial \nu} + (1-\mu) P_2 \eta - \dfrac{\partial \ddot{\eta}}{\partial \nu}\right] = 0 \ \text{ on } \Sigma_+, \end{cases}$$

with final data in the space $D_A \times V$. It follows from the estimate in [10, Lemma V.5.1] by the trick of "weakening the norm" that (4.8) is satisfied for all sufficiently large $T$ whenever $\eta$ is a solution of the last system with data having the indicated regularity.

It follows from the general theory that the reachable reachable set $F'$ of the system (4.1)–(4.4) contains $V \times H$, where $F$ is completion of $V \times D_A$ in the norm

$$\|(\phi_1^0, \phi_2^0)\|_F = \left(\int_0^T \int_{\Gamma_+} (\phi_2^2 + |\nabla \phi_2|^2) d\Gamma dt\right)^{1/2}, \quad T > T_0.$$

Given $(w^0, w^1) \in V \times H$, the minimum norm control in $\mathcal{U}$ that drives $(0,0)$ to $(w^0, w^1)$ at time $T$ is defined by

$$u_0 = \frac{\partial \eta}{\partial \nu}\Big|_{\Sigma_+}, \quad u_1 = \frac{\partial \eta}{\partial \tau}\Big|_{\Sigma_+}, \quad u_2 = \eta|_{\Sigma_+},$$

$\eta$ given by (4.7) with $(\phi_1^0, \phi_2^0) = \Lambda_F^{-1}(w^0, w^1)$.

## 4.2 Maxwell's equations with control acting through a tangentially flowing current in the boundary

Let $\Omega \subset \Re^3$ be a bounded region with smooth boundary $\Gamma$. We consider Maxwell's system

$$\begin{cases} \dot{E} - \text{curl } H = 0, \quad \dot{H} + \text{curl } E = 0, \\ \text{div } E = \text{div } H = 0 \quad \text{in } Q, \end{cases} \tag{4.9}$$

$$\nu \times H = u \quad \text{on } \Sigma, \tag{4.10}$$

$$H(0) = E(0) = 0, \tag{4.11}$$

where $Q = \Omega \times (0,T)$ and $\Sigma = \Gamma \times (0,T)$. The control $u$ represents a density of current flowing tangentially in $\Gamma$ at each instant $t$.

We set

$$\mathcal{L}^2(\Omega) = (L^2(\Omega))^3, \quad \mathcal{H}^k(\Omega) = (H^k(\Omega))^3, \quad \mathcal{L}^2(\Gamma) = (L^2(\Gamma))^3, \quad \mathcal{H}^k(\Gamma) = (H^k(\Gamma))^3.$$

The norm and scalar product in $\mathcal{L}^2(\Omega)$ are denoted by $\|\cdot\|$ and $(\cdot,\cdot)$, respectively. We also introduce the spaces (the notation is adopted from [7])

$$J = \text{ closure in } \mathcal{L}^2(\Omega) \text{ of } \{\chi \,|\, \chi \in C^\infty(\overline{\Omega}), \ \text{div } \chi = 0\},$$

$$\hat{J} = \text{ closure in } \mathcal{L}^2(\Omega) \text{ of } \{\chi \,|\, \chi \in C_0^\infty(\Omega), \ \text{div } \chi = 0\}.$$

For $k \geq 1$ we set

$$J_\nu^k = \{\chi \,|\, \chi \in J \cap \mathcal{H}^k(\Omega), \ \nu \cdot \chi = 0 \ \text{ on } \Gamma\},$$

$$J_\tau^k = \{\chi \,|\, \chi \in J \cap \mathcal{H}^k(\Omega), \ \nu \times \chi = 0 \ \text{ on } \Gamma\},$$

with the topology in each case that inherited from $\mathcal{H}^1(\Omega)$. In addition, define

$$J_\nu^* = \{\chi \,|\, \chi \in J_\nu^2, \ \nu \times \text{curl } \chi = 0 \text{ on } \Gamma\},$$

$$J_\tau^* = \{\chi \,|\, \chi \in J_\tau^2, \ \nu \cdot \text{curl } \chi = 0 \text{ on } \Gamma\},$$

each with the $\mathcal{H}^2(\Omega)$ topology. We have the dense and continuous embeddings

$$J_\tau^* \subset J_\tau^1 \subset J, \quad J_\nu^* \subset J_\nu^1 \subset \hat{J}.$$

If $J$ (resp., $\hat{J}$) is identified with its dual space, we therefore also have

$$J \subset (J_\tau^1)' \subset (J_\tau^*)', \quad \hat{J} \subset (J_\nu^1)' \subset (J_\nu^*)'.$$

The mapping $\phi \mapsto \text{curl } \phi$ is an isomorphism from $X$ onto $Y$, where $(X,Y)$ stands for any one of the pairs

$$(J_\tau^k, J_\nu^{k-1}), \quad (J_\nu^k, J_\tau^{k-1}), \quad (J_\tau^*, J_\nu^1), \quad (J_\nu^*, J_\tau^1), \quad (k \geq 1),$$

and where $J_\nu^0 = \hat{J}$, $J_\tau^0 = J$ (see [7]). Therefore we may renorm $J_\nu^1$, $J_\tau^1$, $J_\nu^*$ and $J_\tau^*$ by setting

$$\|\phi\|_{J_\nu^1} = \|\text{curl } \phi\|, \quad \|\phi\|_{J_\tau^1} = \|\text{curl } \phi\|, \quad \|\phi\|_{J_\nu^*} = \|\text{curl curl } \phi\|, \quad \|\phi\|_{J_\tau^*} = \|\text{curl curl } \phi\|.$$

These norms are equivalent to the corresponding Sobolev norms. Since

$$(\text{curl } \phi, \psi) = (\phi, \text{curl } \psi), \quad \forall \phi \in J^1_\tau, \psi \in J^1_\nu,$$

the map curl extends to a isomorphism of $J$ onto $(J^1_\nu)'$ and of $\hat{J}$ onto $(J^1_\tau)'$. We have

$$\begin{cases} \langle \text{curl } \phi, \psi \rangle_{J^1_\nu} = (\phi, \text{curl } \psi), & \forall \phi \in J, \psi \in J^1_\nu, \\ \langle \text{curl } \phi, \psi \rangle_{J^1_\tau} = (\phi, \text{curl } \psi), & \forall \phi \in \hat{J}, \psi \in J^1_\tau. \end{cases} \tag{4.12}$$

In addition, curl extends to an element in $\mathcal{L}((J^1_\nu)', (J^*_\tau)')$ and in $\mathcal{L}((J^1_\tau)', (J^*_\nu)')$, both isomorphisms, through the formulas

$$\begin{cases} \langle \text{curl } \phi, \psi \rangle_{J^*_\tau} = \langle \phi, \text{curl } \psi \rangle_{J^1_\nu}, & \forall \phi \in (J^1_\nu)', \psi \in J^*_\tau, \\ \langle \text{curl } \phi, \psi \rangle_{J^*_\nu} = \langle \phi, \text{curl } \psi \rangle_{J^1_\tau}, & \forall \phi \in (J^1_\tau)', \psi \in J^*_\nu. \end{cases} \tag{4.13}$$

To obtain the abstract formulation of (4.9), (4.10), let $(\phi, \psi) \in L^2(0,T; J^1_\tau \times J^1_\nu) := L^2(J^1_\tau \times J^1_\nu)$, and form

$$\begin{aligned} 0 &= \int_0^T [(\dot{H} + \text{curl } E, \phi) + (\dot{E} - \text{curl } H, \psi)] dt \qquad (4.14) \\ &= \int_0^T [((\dot{H}, \dot{E}), (\phi, \psi))_{J \times J} - ((H, E), (\text{curl } \psi, -\text{curl } \phi))_{J \times \hat{J}}] dt \\ &\quad - \int_0^T \int_\Gamma \psi \cdot u \, d\Gamma dt. \end{aligned}$$

We now consider various choices of control and state spaces.

### 4.2.1 Exact controllability to $J \times \hat{J}$ with $L^2(\mathcal{L}^2(\Gamma))$ controls, under a geometric condition on $\Gamma$.

We choose as the control space

$$\mathcal{U}' = L^2(\mathcal{L}^2(\Gamma)) = \mathcal{U},$$

and the state space

$$\mathcal{H} = (J^1_\tau)' \times (J^1_\nu)'.$$

We identify $J \times \hat{J}$ with its dual space, so that $\mathcal{H}' = J^1_\tau \times J^1_\nu$. Define an operator $\mathcal{A}$ in $\mathcal{H}$ by $D_\mathcal{A} = J \times \hat{J}$,

$$\mathcal{A} = \begin{pmatrix} 0 & -\text{curl} \\ \text{curl} & 0 \end{pmatrix}. \tag{4.15}$$

By using (4.12) and the properties of curl enunciated above, one sees that $\mathcal{A}$ is a skew-adjoint operator in $\mathcal{H}$. The dual of $\mathcal{A}$ is therefore given by

$$\mathcal{A}' = -\Lambda_\mathcal{H} \mathcal{A} \Lambda_\mathcal{H}^{-1} = \begin{pmatrix} 0 & \text{curl} \\ -\text{curl} & 0 \end{pmatrix}, \tag{4.16}$$

$$D_{\mathcal{A}'} = J^*_\tau \times J^*_\nu,$$

where $\Lambda_{\mathcal{H}}$ is the canonical isomorphism of $\mathcal{H}$ onto $\mathcal{H}'$ and satisfies

$$\Lambda_{\mathcal{H}}^{-1} = \begin{pmatrix} \text{curl curl} & 0 \\ 0 & \text{curl curl} \end{pmatrix}.$$

We define the control operator $\mathcal{B}$ by

$$\langle \mathcal{B}u, (\phi,\psi)\rangle = \int_0^T \int_\Gamma \psi \cdot u \, d\Gamma dt, \quad \forall(\phi,\psi) \in L^2(J_\tau^1 \times J_\nu^1). \tag{4.17}$$

We have

$$|\langle \mathcal{B}u, (\phi,\psi)\rangle| \le C\|u\|_{\mathcal{U}'}\|\psi\|_{L^2(J_\nu^1)} \le C\|u\|_{\mathcal{U}'}\|(\phi,\psi)\|_{L^2(\mathcal{H}')},$$

so that $\mathcal{B} \in \mathcal{L}(\mathcal{U}', L^2(\mathcal{H}))$. The dual operator $\mathcal{B}' \in \mathcal{L}(L^2(\mathcal{H}'), \mathcal{U})$ is given by

$$\mathcal{B}'(\phi,\psi) = \psi|_\Sigma, \tag{4.18}$$

In view of (4.17), (4.14) may be written

$$((\dot H, \dot E), (\phi,\psi))_{L^2(J\times\hat J)} = \langle (H,E), \mathcal{A}'(\phi,\psi)\rangle_{L^2(\mathcal{H}')} + \langle \mathcal{B}u, (\phi,\psi)\rangle_{L^2(\mathcal{H}')},$$

that is to say,

$$\dot y = \mathcal{A}y + \mathcal{B}u \quad \text{in } L^2(\mathcal{H}), \tag{4.19}$$

where $y = (H,E)$. With the initial condition $y(0) = 0$, (4.19) has a unique solution in $C([0,T];\mathcal{H})$.

The observability condition (2.21) is:

$$(\phi^0,\psi^0) \in J_\tau^* \times J_\nu^*, \quad \int_0^T \int_\Gamma |\psi|^2 d\Gamma\, dt = 0 \Leftrightarrow \phi^0 = \psi^0 = 0,$$

where

$$\begin{cases} \dot\phi + \text{curl}\,\psi = 0, \quad \dot\psi - \text{curl}\,\phi = 0, \\ \text{div}\,\phi = \text{div}\,\psi = 0 \quad \text{in } Q, \end{cases} \tag{4.20}$$

$$\phi(T) = \phi^0, \quad \psi(T) = \psi^0 \quad \text{in } \Omega. \tag{4.21}$$

The following result is proved in [9, Lemma 3.3].

**Theorem 4.1** *Assume that $\Gamma$ is star-shaped with respect to some point in $\Re^3$. Then there exists $T_0 > 0$ such that for all $T > T_0$*

$$\int_0^T \int_\Gamma |\psi|^2 d\Gamma\, dt \ge c^2(T - T_0)\|(\phi^0,\psi^0)\|^2_{J\times\hat J}, \quad \forall(\phi^0,\psi^0) \in J_\tau^* \times J_\nu^*.$$

With the space $F$ defined as the completion of $J_\tau^* \times J_\nu^*$ in the norm

$$\|(\phi^0,\psi^0)\|_F = \left(\int_0^T \int_\Gamma |\psi|^2 d\Gamma\, dt\right)^{1/2},$$

we therefore have

$$F \subset J \times \hat J \subset F',$$

and $J \times \hat J$ is in the reachable set of (4.9)–(4.11). Given $(H^0, E^0) \in J \times \hat J$, the minimum norm control in $L^2(\mathcal{L}^2(\Gamma))$ that drives $(0,0)$ to $(H^0,E^0)$ is given by

$$u = \psi|_\Gamma$$

with $(\phi,\psi)$ given by (4.20), (4.21) and with $(\phi^0,\psi^0) = \Lambda_F^{-1}(H^0,E^0)$.

### 4.2.2 Exact controllability to $(J_\tau^1)' \times (J_\nu^1)'$ with $(H^1(\mathcal{L}^2(\Gamma)))'$ controls, under a geometric condition on $\Gamma$.

Here we choose

$$\mathcal{U} = H^1(\mathcal{L}^2(\Gamma))$$

so that, identifying $L^2(\mathcal{L}^2(\Gamma))$ with its dual space, the control space is

$$\mathcal{U}' = (H^1(\mathcal{L}^2(\Gamma)))'. \tag{4.22}$$

The choices of $\mathcal{H}$, $\mathcal{A}$ and $D_\mathcal{A}$ are the same as in the last subsection. The only difference is the operator $\mathcal{B}$, now defined by the duality pairing

$$\langle \mathcal{B}u, (\phi, \psi)\rangle = \langle u, \psi\rangle_{H^1(\mathcal{L}^2(\Gamma))}.$$

We have

$$\langle \mathcal{B}u, (\phi, \psi)\rangle \leq \|u\|_{(H^1(\mathcal{L}^2(\Gamma)))'} \|\psi\|_{H^1(\mathcal{L}^2(\Gamma))} \leq C\|u\|_{(H^1(\mathcal{L}^2(\Gamma)))'} \|(\phi, \psi)\|_{H^1(\mathcal{H}')}.$$

Therefore $\mathcal{B} \in \mathcal{L}(\mathcal{U}', (H^1(\mathcal{H}'))')$, and the abstract formulation of (4.9), (4.10) is

$$\dot{y} = \mathcal{A}y + \mathcal{B}u \ \text{ in } (H^1(\mathcal{H}'))'.$$

With $y(0) = 0$, the last equation has a unique solution whose properties are delineated in Propositions 2.2 and 2.3 above.

The dual operator $\mathcal{B}' \in \mathcal{L}((H^1(\mathcal{H}'), \mathcal{U})$ is again given by (4.18). To identify the reachable set of the system we have to consider

$$\|(\phi^0, \psi^0)\|_F^2 = \|\mathcal{B}'(\phi, \psi)\|_\mathcal{U}^2 = \int_0^T \int_\Gamma (|\psi|^2 + |\dot{\psi}|^2) d\Gamma\, dt,$$

where $(\phi, \psi)$ satisfies (4.20), (4.21). According to [9, Lemma 3.1], we have

**Theorem 4.2** *Assume that $\Gamma$ is star-shaped with respect to some point in $\Re^3$. Then there exists $T_0 > 0$ such that for all $T > T_0$,*

$$\int_0^T \int_\Gamma (|\psi|^2 + |\dot{\psi}|^2) d\Gamma\, dt \geq c^2(T - T_0)\|(\phi^0, \psi^0)\|^2_{J_\tau^1 \times J_\nu^1}, \ \ \forall (\phi^0, \psi^0) \in J_\tau^* \times J_\nu^*.$$

It follows that the reachable set of (4.9)–((4.11), with controls satisfying (4.22), contains $(J_\tau^1 \times J_\tau^1)'$. Given $(H^0, E^0)$ in this space, the minimum norm control is given by

$$u_{\min} = \Lambda_\mathcal{U} \psi|_\Sigma,$$

where $(\phi, \psi)$ is given by (4.20), (4.21) with $(\phi^0, \psi^0) = \Lambda_F^{-1}(H^0, E^0)$, and where $\Lambda_\mathcal{U}$ the canonical isomorphism of $\mathcal{U}$ onto $\mathcal{U}'$. For $\xi \in \mathcal{U}$ one may write

$$\Lambda_\mathcal{U}\xi = \xi - \frac{d^2\xi}{dt^2},$$

where $\frac{d^2\xi}{dt^2} \in \mathcal{L}(\mathcal{U}, \mathcal{U}')$ is *defined* by

$$\langle \frac{d^2\xi}{dt^2}, \eta\rangle = -\int_0^T \int_\Gamma \dot{\xi} \cdot \dot{\eta}\, d\Gamma\, dt, \ \ \forall \xi, \eta \in \mathcal{U}.$$

Therefore

$$u_{\min} = \psi|_\Sigma - \left.\frac{d^2\psi}{dt^2}\right|_\Sigma.$$

### 4.2.3 Exact controllability to $(J_\tau^1)' \times (J_\nu^1)'$ without geometric restrictions.

When the star-shapedness requirement on $\Gamma$ is removed, it is necessary to work with control and state spaces with weaker topologies than before. Accordingly, we choose

$$\mathcal{H} = (J_\tau^* \times J_\nu^*)'.$$

The operator $\mathcal{A}$ is still given by the matrix (4.15), but with

$$D_\mathcal{A} = (J_\tau^1 \times J_\nu^1)'.$$

By using (4.13) it is seen that $\mathcal{A}$ is a skew-adjoint operator in $\mathcal{H}$. Its dual $\mathcal{A}'$ is an operator in $\mathcal{H}' = J_\tau^* \times J_\nu^*$ given by the matrix in (4.16) with

$$D_{\mathcal{A}'} = \{(\phi^0, \psi^0) | (\phi^0, \psi^0) \in J_\tau^* \times J_\nu^*, \text{ curl } \phi^0 \in J_\nu^*, \text{ curl } \psi^0 \in J_\tau^*\}.$$

The control space will be the dual of

$$\mathcal{U} = H^1(\mathcal{L}^2(\Gamma)) \oplus L^2(U),$$

where $U$ is a certain Hilbert space which satisfies $\mathcal{H}^1(\Gamma) \subset U \subset \mathcal{L}^2(\Gamma)$. Thus

$$\mathcal{U}' = (H^1(\mathcal{L}^2(\Gamma)))' \oplus L^2(U') \subset (H^1(\mathcal{L}^2(\Gamma)))' \oplus L^2(\mathcal{H}^{-1}(\Gamma)).$$

To define $U$, we introduce the closed subspace of $\mathcal{L}^2(\Gamma)$

$$\mathcal{L}_\tau^2 = \{\chi \in \mathcal{L}^2(\Gamma) | \nu \times \chi = 0 \text{ on } \Gamma\}.$$

For smooth functions $\chi$ defined in $\overline{\Omega}$ we have (see [9, Section 4.3])

$$\text{curl } \chi|_\Gamma = \nu \times \frac{\partial \chi}{\partial \nu} + \sigma \times \chi$$

where $\sigma = (\sigma_1, \sigma_2 . \sigma_3)$ is a formally self-adjoint tangential operator of order one on $\Gamma$:

$$\int_\Gamma \tilde{\chi} \cdot (\sigma \times \chi) d\Gamma = \int_\Gamma (\sigma \times \tilde{\chi}) \cdot \chi \, d\Gamma, \quad \forall \chi, \tilde{\chi} \in \mathcal{H}^1(\Gamma).$$

If, therefore, $\chi \in \mathcal{L}_\tau^2$ then

$$\int_\Gamma \tilde{\chi} \cdot \text{curl } \chi|_\Gamma d\Gamma = \int_\Gamma \tilde{\chi} \cdot (\sigma \times \chi) d\Gamma = \int_\Gamma (\sigma \times \tilde{\chi}) \cdot \chi \, d\Gamma. \tag{4.23}$$

Consequently, if $\chi \in \mathcal{H}^1(\Gamma)$, we may *define* curl $\chi \in \mathcal{L}_\tau^2$ by

$$(\text{curl } \chi, \tilde{\chi})_{\mathcal{L}_\tau^2} = \langle \sigma \times \tilde{\chi}, \chi \rangle_{\mathcal{H}^1(\Gamma)}, \quad \forall \tilde{\chi} \in \mathcal{L}_\tau^2.$$

Then curl$\in \mathcal{L}(\mathcal{H}^1(\Gamma), \mathcal{L}_\tau^2)$. In particular, we have $\nu \times \text{curl } \chi = 0$ for all $\chi \in \mathcal{H}^1(\Gamma)$.

We introduce on $\mathcal{H}^1(\Gamma)$ the norm

$$\|\chi\|_U = \left(\|\text{curl } \chi\|_{\mathcal{L}^2(\Gamma)}^2 + \|\chi\|_{\mathcal{L}^2(\Gamma)}^2\right)^{1/2} \tag{4.24}$$

and define $U$ as the completion of $\mathcal{H}^1(\Gamma)$ in $\|\cdot\|_U$. For $u = u_0 + u_1 \in \mathcal{U}'$ we define the control operator by

$$\langle \mathcal{B}u, (\phi, \psi)\rangle = \langle u_0, \psi\rangle_{H^1(\mathcal{L}^2(\Gamma))} + \langle u_1, \psi\rangle_{\mathcal{L}^2(U)}, \quad \forall(\phi, \psi) \in L^2(J_\tau^* \times J_\nu^*).$$

We have

$$\begin{aligned} |\langle \mathcal{B}u, (\phi, \psi)\rangle| &\leq \|u_0\|_{(H^1(\mathcal{L}^2(\Gamma)))'} \|\psi\|_{H^1(\mathcal{L}^2(\Gamma))} + \|u_1\|_{L^2(U')}\|\psi\|_{L^2(U)} \\ &\leq C\|u\|_{\mathcal{U}'} \left[\|(\phi, \psi)\|_{H^1(J_\tau^1 \times J_\nu^1)} + \|(\phi, \psi)\|_{L^2(J_\tau^* \times J_\nu^*)}\right]. \end{aligned}$$

Therefore we have, in particular, $\mathcal{B} \in \mathcal{L}(\mathcal{U}', (H^1(J_\tau^* \times J_\nu^*))') = \mathcal{L}(\mathcal{U}', (H^1(\mathcal{H}'))')$, so that the theory of Section 2 may be applied. To do so, we have to consider

$$\begin{aligned} \|(\phi^0, \psi^0)\|_F^2 &= \|\mathcal{B}'(\phi, \psi)\|_{\mathcal{U}} = \|\phi\|^2_{H^1(\mathcal{L}^2(\Gamma))} + \|\psi\|^2_{L^2(U)} \\ &= \int_0^T \int_\Gamma (2|\psi|^2 + |\dot{\psi}|^2 + |\text{curl } \psi|^2) d\Gamma\, dt, \end{aligned}$$

where $(\phi, \psi)$ satisfy (4.20), (4.21) with $(\phi^0, \psi^0) \in D_{\mathcal{A}'}$. According to [9, Lemma 3.2] we have

$$\int_0^T \int_\Gamma (|\psi|^2 + |\dot{\psi}|^2 + |\text{curl } \psi|^2) d\Gamma\, dt \geq c^2(T - T_0)\|(\phi^0, \psi^0)\|^2_{J_\tau^1 \times J_\nu^1}, \quad \forall(\phi^0, \psi^0) \in D_{\mathcal{A}'},$$

provided $T > T_0$ with a suitable $T_0$. The reachable set of our problem therefore contains $(J_\tau^1)' \times (J_\nu^1)'$. If $(H^0, E^0)$ is in this space then, with $(\phi^0, \psi^0) = \Lambda_F^{-1}(H^0, E^0)$, the control of minimum norm in $\mathcal{U}'$ steering $(0,0)$ to $(H^0, E^0)$ is given by

$$u_{\min} = u_0 + u_1$$

where, as in the last subsection,

$$u_0 = \psi|_\Sigma - \left.\frac{d^2\psi}{dt^2}\right|_\Sigma,$$

and where

$$u_1 = \Lambda_{\mathcal{U}_1} \mathcal{B}'(\phi, \psi) = \Lambda_{\mathcal{U}_1} \psi|_\Sigma,$$

$\Lambda_{\mathcal{U}_1}$ denoting the canonical isomorphism from $\mathcal{U}_1 := L^2(U)$ onto $\mathcal{U}_1' = L^2(U')$. From (4.23) and the definition (4.24) of the norm on $U$ it is seen that

$$\Lambda_{\mathcal{U}_1} \chi = \chi + \sigma \times \text{curl } \chi, \quad \forall \chi \in \mathcal{U}_1.$$

Therefore

$$u_1 = \psi|_\Sigma + \sigma \times \text{curl } \psi|_\Sigma = \psi|_\Sigma + \sigma \times \dot{\phi}|_\Sigma.$$

# References


[1] A. Bensoussan, *On the general theory of exact controllability for skew-symmetric operators*, preprint.

[2] A. Bensoussan, *Some remarks on the exact controllability of Maxwell's equations*, preprint.

[3] G. Chen, *Energy decay estimates and exact boundary value controllability for the wave equation in a bounded domain*, J. Math. Pures Appl., 58 (1979), 249–274.

[4] S. Dolecki and D. L. Russell, *A general theory of observation and control*, SIAM J. Control and Opt., 15, (1977), 185–220.

[5] G. Duvaut and J.-L. Lions, *Les Inéquations en Mécanique et en Physique*, Dunod, Paris, 1972.

[6] L. F. Ho, *Observabilité frontière de l'equation des ondes*, C.R. Acad. Sci. Paris Sér. I, 302 (1986), 443–446.

[7] O. A. Ladyzhenskaya and V. A. Solonikov, *The linearization principle and invariant manifolds for problems of magnetohydrodynamics*, J. Soviet Nath., 8, (1977), 384–422.

[8] J. E. Lagnese, *Decay of solutions of wave equations in a bounded region with boundary dissipation*, J. Diff. Eqs. 50, (1983), 163–182.

[9] J. E. Lagnese, *Exact boundary controllability of Maxwell's equations in a general region*, SIAM J. Control and Opt., 27, (1989), 374–388.

[10] J. E. Lagnese and J.-L. Lions, *Modelling, Analysis and Control of Thin Plates*, Recherches en Mathématiques Appliquées, Vol. 6, Masson, Paris, 1988.

[11] I. Lasiecka, *Controllability of a viscoelastic Kirchhoff plate*, Internat. Ser. in Numerical Math., 91 (1989), 237–247.

[12] I. Lasiecka and R. Triggiani, *Exact controllability of the wave equation with Neumann boundary control*, Appl. Math. and Opt., 19 (1989), 243–290.

[13] I. Lasiecka and R. Triggiani, *Exact controllability of the Euler-Bernoulli equation with controls in the Dirichlet and Neumann boundary conditions: a nonconservative case*, SIAM J. Control and Opt., 27 (1989), 330–372.

[14] J.-L. Lions, *Exact controllability, stabilization and perturbations for distributed parameter systems*, SIAM Review, 30 (1988), 1–68.

[15] J.-L. Lions, *Contrôlabilité Exacte, Perturbations et Stabilisation de Systèmes Distribués. Tome 1, Contrôlabilité Exacte; Tome 2, Perturbations*, Recherches en Mathématiques Appliquées, Vols. 8 and 9, Masson, Paris, 1988.

[16] D. L. Russell, Review of *Contrôlabilité Exacte, Perturbations et Stabilisation de Systèmes Distribués*, **Bull. Amer. Math. Soc.**, **22** (1990), 353–356.

[17] R. Triggiani, *Exact boundary controllability on $L^2(\Omega) \times H^{-1}(\Omega)$ of the wave equation with Dirichlet boundary control action of a portion of the boundary, and related problems*, **Appl. Math. and Opt.**, **18** (1988), 241–277.

# On control and stabilization of a rotating beam by applying moments at the base only*

G.Leugering
Department of Mathematics
Georgetown University
Washington, D.C. 20057

June, 1990

**Abstract**

A decomposition–method introduced in [Le1] is used to decompose the rigid–body motion from the elastic vibration of a slowly rotating beam. On the base of the transformed system, a controller is constructed that steers all oscillations of the beam to rest in finite time. In addition, the beam is thereby driven to zero angular velocity. A second result is concerned with strong feed–back stabilizability.

## 1 The control problem

In this note I want to give another example of how a simple idea of decomposing complex systems, introduced in [Le1], can be effectively used for the design of controls. The example which I will discuss here is that of a slowly rotating beam. This mechanical substructure has been the subject of many research articles in control theory. It is of some importance in the area of control of flexible space–structures. The typical model is given by a long thin flexible beam which is attached to a cylinder -called the base - at one end. The cylinder has its axis perpendicular to the plane of bending of the beam (plane motion is assumed), and this is also the axis of rotation. The control objective is to counteract disturbations of a slewing maneuver at a given angular velocity, which may be introduced to the system from various sources. In particular, it is of great importance to extinguish all vibrations of the structure due to bending. In addition, the controlled – non oscillatory – structure should approach a desired angular velocity (here set to be equal to zero ) as fast as possible. For the sake of brevity, I dispense with any mechanical derivation of such systems, and refer to [DK], [BL]. Let $w$ represent the vertical displacement of the centerline of the beam in the rotated frame. Let $\varphi$ denote the angular velocity at which the frame is rotated. The equlibrium towards which the system should ultimately be

*This work has been supported by the Deutsche Forschungsgemeinschaft (DFG), Heisenbergreferat, Le 595-3-1

controlled is $w = 0$, $\varphi = 0$. ($\varphi \neq 0$ could be handled equally well. ) The only control that can be applied is a torque, i.e. the moment applied to the base–cylinder. The system is then given by

$$(CP)\quad \begin{cases} \ddot{w} + \gamma^2 w'''' + \alpha x \ddot{\varphi} = & 0, & (x,t) \in (0,\ell) \times (0,\infty) \\ \dot{\varphi} + \alpha \int_0^\ell x \ddot{w} ds + \nu\varphi = & M, & t \in (0,\infty) \end{cases}$$

together with the boundary conditions

$$(bc)\quad w(0) = w'(0) = w''(\ell) = w'''(\ell) = 0, \quad t \in (0,\infty),$$

and initial conditions

$$(ic)\quad w(\cdot,0) = w_0, \quad \dot{w}(\cdot,0) = w_1, \quad x \in (0,\ell), \quad \varphi(0) = \varphi_0.$$

For simplicity of notation, the variables $x, t$ are supressed wherever it is felt unmistakable. A dot indicates a time–derivative, a prime denotes a spatial derivation. It should be emphazised that the term $\nu\varphi$ is viewed as a previously implemented feed–back control which, in the absence of the active control $M$, guarantees $w = 0$, $\varphi = 0$ as the eqilibrium. In Fact, it already gives asymptotic stability. One might, therefore, think of just cancelling this term by $M$ and insert the second equation into the first, to end up with a Sobolev–type equation in terms of $w$. However, this procedure would obscure the perturbation–viewpoint, i.e. that active - open-loop - controllers should be used to perform control actions related to disturbations, rather than to change the whole built–in control set– up. *In addition, asymptotic stability of* $\varphi = 0$ *could not be achieved that way!* The control process (CP) is also assumed to be already properly rescaled to a non–dimensional form. A system of this sort can be classified under hybrid control systems, as it consists of a partial differential equation together with an ordinary differential equation. As for the mathematical analysis of (CP) I refer to [DM],[Le2] and the bibliographies therein. The purpose of this note is to decompose the rigid body motion – the second equation in (CP) – from the purely oscillatory motion – the first equation in (CP) – by means of a similarity transform. The procedure is as in [Le1], eventhough the theorem there does not apply to this situation directly.

# 2 Decomposition

It is obvious that, upon integration and use of (bc), the system (CP) is equivalent to

$$(CP')\quad \begin{cases} \ddot{w} + \gamma^2 w'''' + \alpha x \ddot{\varphi} & = & 0 \\ \dot{\varphi} - \alpha\beta w''(0) + \mu\varphi & = & m \end{cases}$$

Here, with $(1 - \frac{\alpha}{3}\ell^3) =: \zeta$ and $\alpha$ small enough, one has $\beta = \gamma^2/\zeta > 1$, $\mu = \nu/\zeta$, $m = M/\zeta$. The paramter $\alpha$ is supposed to model the strenght of the coupling between rotation and bending. This parameter is small in real systems. The second equation in (CP') can be integrated to

$$\varphi(t) = \alpha\beta \int_0^\infty e^{-\mu s} w''(0, t-s) ds + \int_{-\infty}^t e^{-\mu(t-s)} m(s) ds$$

and differentiated again to give

$$\begin{aligned}\dot\varphi &= \alpha\beta\int_0^\infty e^{-\mu s}\dot w''(0,t-s)ds \\ &\quad + m(t) - \mu\int_{-\infty}^t e^{-\mu(t-s)}m(s)ds \\ &= \alpha\beta\int_{-\infty}^t e^{-\mu(t-s)}(\dot w''(0,s) - \frac{\mu}{\alpha\beta}m(s))ds + m(t) \\ &=: \beta y(t) + m(t)\,.\end{aligned}$$

This results in the equivalent system

$$(CP'')\quad \left\{\begin{array}{l} \ddot w + \gamma^2 w'''' + \alpha\beta x\cdot y + \alpha x\cdot m \;=\; 0 \\ \dot y = -\mu y + \alpha(\dot w''(0) - \frac{\mu}{\alpha\beta}m) \end{array}\right. .$$

Upon defining $w =: x_1$, $\dot w = x_2$, as usual, on obtains the matrix formulation

$$(M)\quad \dot{\begin{pmatrix} x_1 \\ x_2 \\ y \end{pmatrix}} = \begin{pmatrix} O & I & O \\ -A & O & -\alpha\beta P \\ O & \alpha T & -\mu I \end{pmatrix}\begin{pmatrix} x_1 \\ x_2 \\ y \end{pmatrix} + \begin{pmatrix} 0 \\ \alpha p \\ -\frac{\mu}{\beta} \end{pmatrix}\cdot m$$

Here $A$ is the self-adjoint positive definite operator in $L^2(0,\ell)$

$$Aw = w'''',\quad D(A) = \{w\in H^4(0,\ell)|\; w(0) = w'(0) = w''(\ell) = w'''(\ell) = 0\},$$

with

$$\begin{aligned}&\sigma(A) = \sigma_d(A) = \{0 < \lambda_1 < \lambda_2 < \ldots < \lambda_n < \ldots \lim_{n\to\infty}\lambda_n = \infty\}, \\ &\{\phi_n\}_1^\infty:\quad A\phi_n = \lambda_n\phi_n,\quad <\phi_n,\phi_m> = \delta_{n,m},\quad w = \sum_n <w,\phi_n>\phi_n \\ &A^s w := \sum \lambda_n^s <w,\phi_n>\phi_n, \\ &D(A^s) = \{w\in H := L^2(0,\ell)|\sum_n \lambda^{2s}|<w,\phi_n>|^2 < \infty\},\end{aligned}$$

and $T$, $P$ and $p$ are defined by

$$\begin{aligned}Tw &= w''(0),\quad D(T) = D(A^{\frac{3}{4}}), \\ Pw &= p\cdot w,\quad p = x.\end{aligned}$$

$T$ is relatively $A$–compact and has, therefore, a small $A$–bound. In addition to that it satisifies the inequality

$$|Tw| \le \ell^2\|Aw\|,\quad \forall w\in D(A),$$

and even more,

$$|Tw| \le \ell^2\|A^{\frac{3}{4}}w\|,\quad \forall w\in D(A^{\frac{3}{4}}) \Rightarrow T\cong \delta_0''\in D(A^{\frac{3}{4}})^*$$

In order for the third component in (M) to make any sense, it seems appropriate to introduce a somewhat stronger energy space than the usual one, i.e.

$$E = D(A^{\frac{3}{4}}) \times D(A^{\frac{1}{4}}) \times \mathbf{R}. \tag{1}$$

This can be called a shifted energy space. It has the property that the wave–type operator $\mathcal{A}$ associated with the beam equation, namely

$$\mathcal{A} = \begin{pmatrix} O & I \\ -A & O \end{pmatrix}, \tag{2}$$

has the dense domain $D(\mathcal{A}) := D(A^{\frac{5}{4}}) \times D(A^{\frac{3}{4}})$ in E, and is maximally dissipative there. In this space set–up one certainly has well–posedness in the strong sense for the system (M).

As in [Le1] one may introduce a (similarity) transformation $\Gamma$ acting within the "energy-space" as follows

$$\mathbf{x} := (x_1, x_2)^T, \quad \mathcal{Q} := (Q_1, Q_2) : D(A^{\frac{3}{4}}) \times D(A^{\frac{1}{4}}) \longrightarrow \mathbf{R},$$

$$\begin{pmatrix} \mathbf{x} \\ y \end{pmatrix} = \begin{pmatrix} I & O \\ \mathcal{Q} & I \end{pmatrix} \begin{pmatrix} \xi \\ \eta \end{pmatrix} =: \Gamma \begin{pmatrix} \xi \\ \eta \end{pmatrix}.$$

This defines a boundedly invertible operator, with the inverse obtained from $\Gamma$ by just changing the sign in front of $\mathcal{Q}$. In addition, $\Gamma D(\mathcal{A}) \subseteq D(\mathcal{A})$. Multiplying from the left by $\Gamma^{-1}$ one obtains

$$\dot{\begin{pmatrix} \xi_1 \\ \xi_2 \\ \eta \end{pmatrix}}$$

$$= \begin{pmatrix} O & I & O \\ -A - \alpha\beta P Q_1 & -\alpha\beta P Q_2 & -\alpha P \\ \left\{ \begin{matrix} -\mu Q_1 + \\ Q_2(A + \alpha P Q_1) \end{matrix} \right\} & \left\{ \begin{matrix} -\mu Q_2 - Q_1 + \\ \alpha\beta Q_2 P Q_2 + \alpha T \end{matrix} \right\} & \left\{ \begin{matrix} -\mu I + \\ \alpha\beta Q_2 P \end{matrix} \right\} \end{pmatrix} \begin{pmatrix} \xi_1 \\ \xi_2 \\ \eta \end{pmatrix}$$
$$+ \begin{pmatrix} 0 \\ \alpha p \\ -Q_2 p - \frac{\mu}{\beta} \end{pmatrix} \cdot m$$

As in [Le1], one tries to eliminate the first two entries in the third row of the matrix above. This ammounts to solving the operator equations

$$\begin{aligned} -\mu Q_1 + Q_2(A + \alpha P Q_1) &= O \\ -\mu Q_2 - Q_1 + \alpha\beta Q_2 P Q_2 + \alpha T &= O \end{aligned} \tag{3}$$

The meaning of the system (3) is

$$-\mu Q_1(f) + A^* Q_2(f) + \alpha\beta Q_2(x) Q_1(f) = 0$$
$$-\mu Q_2(g) - Q_1(g) + \alpha\beta Q_2(x) Q_2(g) + \alpha \delta_0''(g) = 0$$

for all $(f,g) \in D(\mathcal{A})$. The functionals $Q_1$, $Q_2$ are in fact elements $q_1 \in D(A^{\frac{3}{4}})^*$, $q_2 \in D(A^{\frac{1}{4}})^*$. Upon eliminating $q_1$ with the aid of the first equation, one obtains the following inhomogenous nonlinear elliptic boundary value problem.

$$\mu^2 q_2 + (1 - \frac{\alpha\beta}{\mu} q_2(x))^{-1} A q_2 + \alpha\beta\mu q_2(x) q_2 = -\alpha\mu\delta_0'' \tag{4}$$

This semi-linear problem admits a unique small solution in $D(A^{\frac{1}{4}})$ , if $\alpha$ is small. This is because, after the application of the resolvent $R(-\mu^2, A) := (\mu^2 + A)^{-1}$ of $A$ to (4) from the left, one obtains

$$q_2 + C(q_2)R(-\mu^2, A)Aq_2 + \alpha\beta\mu q_2(x)R(-\mu^2, A)q_2 = -\mu\alpha R(-\mu^2, A)\delta_0'',$$

where $C(q_2)$ is the expression in front of $Aq_2$ in (4). This, in turn, may be solved by a fixed-point argument, for a small enough $\alpha$. The fixed-point-map is

$$\begin{aligned} &\Lambda(r) = q, \quad \|r\|_{D(A^{\frac{1}{4}})} \leq \varepsilon, \quad \text{where } q \text{ is the solution of} \\ &q + C(r)AR(-\mu^2, A)q + \alpha\beta\mu r(x)R(-\mu^2, A)q = -R(-\mu^2, A)\delta_0''. \end{aligned} \tag{5}$$

This equation, obtained by freezing the occurence of $q_2$ in the nonlinear parts, is a linear inhomogenous problem, which admits a unique small solution $q$ for small $\alpha$. In addition, it is easy to verify that $\Lambda$ is in fact a contraction, mapping the $\varepsilon$-ball of $r$'s into itself. The resulting fixed-point is $q_2$ and, a forteriori, $q_1$ is determined by $q_2$, and has the right regularity, namely $q_1 \in D(A^{\frac{3}{4}})^*$. Having solved the equations (3), the matrix-system takes on the following block-triangular form

$$\begin{aligned} \begin{pmatrix} \xi_1 \\ \xi_2 \\ y \end{pmatrix}^{\cdot} &= \begin{pmatrix} O & I & O \\ -A - \alpha\beta PQ_1 & -\alpha\beta PQ_2 & -\alpha P \\ O & O & -\mu I + \alpha\beta Q_2 P \end{pmatrix} \begin{pmatrix} \xi_1 \\ \xi_2 \\ \eta \end{pmatrix} \\ &+ \begin{pmatrix} 0 \\ \alpha p \\ -Q_2 p - \frac{\mu}{\beta} \end{pmatrix} \cdot m \end{aligned} \tag{6}$$

Using the fact that $w = \xi_1, \dot{w} = \xi_2$, this can now be rewritten as follows

$$(\Gamma CP)\begin{cases} \ddot{w} + Aw + \alpha x \cdot q_1(w) + \alpha x \cdot q_2(\dot{w}) + \alpha x \cdot \eta = \alpha x \cdot m \\ \dot{\eta} = -\mu\eta + \alpha\beta q_2(x)\eta - (q_2(x) + \frac{\mu}{\beta}) \cdot m \end{cases}$$

# 3 Exact controllability

Upon integration, the second equation of $(\Gamma CP)$ gives

$$\eta = -\frac{1}{\beta}(\rho + \mu) \int_0^t e^{-\sigma(t-s)} m(s) ds \tag{7}$$

where $\sigma := \mu - \alpha\beta q_2(x)$, $\rho := \beta q_2(x)$. In fact, we have assumed zero initial-histories for the variable $y$ and the control $m$, which is in accordance with the applications, since the process is considered at rest before the loading impacts occur. Once again inserted into the first equation, this ammounts to

$$\ddot{w} + Aw + \alpha x \cdot q_1(w) + \alpha x \cdot q_2(\dot{w}) = \alpha x \cdot (m + \frac{1}{\beta}(\rho + \mu) \int_0^t e^{-\sigma(t-s)} m(s) ds \tag{8}$$

One considers the control problem for (8), neglecting the perturbations in terms of $q_1, q_2$ first. Set $\frac{1}{\beta}(\rho + \mu) := c$ and multiply the simplified equation (8) with $\phi_j$. Note that the original equation is not diagonal. This leads to the infinite sequence of initial value problems

$$\begin{aligned} \ddot{w}_j + \lambda_j w_j &= \alpha < x, \phi_j > (m(t) + c \int_0^t e^{-\sigma(t-s)} m(s) ds), \\ w_j(0) = w_{0,j}, \quad & \dot{w}_j(0) = w_{1,j} \ \forall j. \end{aligned}$$

With

$$\begin{aligned} h_j &:= \alpha < x, \phi_j > \\ w_j^0 &:= \cos \lambda_j^{\frac{1}{2}}(T) w_{0,j} + \frac{1}{\lambda_j^{\frac{1}{2}}} \sin \lambda_j^{\frac{1}{2}}(T) w_{1,j} \\ w_j^1 &:= -\lambda_j^{\frac{1}{2}} \sin \lambda_j^{\frac{1}{2}}(T) w_{0,j} + \cos \lambda_j^{\frac{1}{2}}(T) w_{1,j} \end{aligned}$$

and the requirement that

$$w_j(T) = 0, \ \ \dot{w}_j(T) = 0, \ \ \forall j,$$

one obtains

$$\begin{aligned} -w_j^0 \frac{\lambda_j^{\frac{1}{2}}}{h_j} &= \Im\{(1 + \frac{c}{i\lambda_j^{\frac{1}{2}} + \sigma}) \int_0^T e^{i\lambda_j^{\frac{1}{2}}(T-s)} m(s) ds \\ &\quad - \frac{c}{i\lambda_j^{\frac{1}{2}} + \sigma} \int_0^T e^{-\sigma(T-s)} m(s) ds\} \\ -w_j^1 \frac{1}{h_j} &= \Re(1 + \frac{c}{i\lambda_j^{\frac{1}{2}} + \sigma}) \int_0^T e^{i\lambda_j^{\frac{1}{2}}(T-s)} m(s) ds \\ &\quad - \frac{c\sigma}{\lambda_j^{\frac{1}{2}}} \Im \frac{c}{i\lambda_j^{\frac{1}{2}} + \sigma} \int_0^T e^{-\sigma(T-s)} m(s) ds. \end{aligned} \tag{9}$$

Here it is first assumed that $m \in H_0^1(0,T)$, and then, by density, the equations are extended to $L^2(0,T)$-controls . Now, by (7)

$$0 = \eta(T) = -c \int_0^T e^{-\sigma(T-s)} m(s) ds. \tag{10}$$

(Nonzero initial history for $\eta$ would not cause any difficulty as well.) Hence, (9) constitutes what is known as the moment problem associated with a beam under the distributed control $x \cdot m$. It is plain that $x \in D(A^{\frac{1}{4}})$. The condition (10) is just an additional moment condition. But the system of exponentials $e^{\lambda_j^{\frac{1}{2}} t} \cup e^{-\sigma t}$ still admits a unique, in fact, minimal norm– solution of (9),(10) in an arbitrarily small time interval $(0, T)$, for all initial data satisfying $w_0 \in D(A^{\frac{3}{4}})$, $w_1 \in D(A^{\frac{1}{4}})$. See Krabs[Kra], for a very nice treatment of moment problems related to beam (and other) equations. By standard perturbation arguments one may now include the neglected terms $\alpha x \cdot (q_1(w) + q_2(\dot{w}))$. On the state space $E$, these are bounded and small perturbations. The resulting norm–minimal control $m \in L^2(0, T)$ is then extended by $m(T + t) \equiv 0$, $\forall t > 0$, so that the whole state remains at equilibrium. The variable $\eta$ being controlled to zero, implies $y(T + t) = 0$ and in turn $\dot{\varphi}(T + t) = 0$. But from $m(T + t) \equiv 0$, one infers that, because of $w''(0, T + t) \equiv 0$, $\varphi(T + t) \equiv 0$. Hence, exact null–controllability of the original system (CP) obtains.

**Theorem:** (Exact controllability)
*Let $w_0 \in D(A^{\frac{3}{4}})$, $w_1 \in D(A^{\frac{1}{4}})$, $\varphi_0 \in \mathbf{R}$ and $T > 0$ be given, and let $\alpha$ be sufficiently small. Then there exists a norm–minimal control $M \in L^2(0, \infty)$ such that the solution $w, \varphi$ of (CP) has the following properties*

$$w_M(T + t) = \dot{w}_M(T + t) = 0, \varphi_M(T + t) = 0 \quad \forall t > 0$$

Concerning the well-posedness of the control–process, as stated in the theorem, some remarks are in order.

**Remark:**
*Formally, the transform $\Gamma$ leads to the desired decomposition (6), only if applied to strong data. Therfore, in order to obtain from (6) well–posedness in the "mild" sense of the theorem, one first of all starts with strong initial data, obtains strong solutions of (6) and, hence, strong solutions of (CP). The mild initial–data of (6) are then – by density – approximated by strong data. One, naturally, defines the resulting limits of solutions to (CP) as mild solutions thereof.*

# 4 Stabilizability

It is interesting to observe from ($\Gamma$CP) – and this is another nice feature of the decomposition method – that the frequency spectrum of (CP), in the absence of controls $m(t)$, consists of two parts. An "oscillatory" part, and a "creep" part. The oscillatory spectrum consists of infinetely many eigenfrequencies in the stable half–plane *approaching the imaginary axis at infinity.* The creep part consists of a negative real eigenvalue, namely $\sigma = \mu - \alpha\beta q_2(x)$. *As a result, there is no uniform decay without further control !* This turns out to be very important for stabilization devices based on $M$. Again, one starts with (8), where $\alpha$ is – for the time being – set equal to zero. The spatial part $x$ of the control satifies the following property

$$< x, \phi_j > = \frac{1}{\lambda_j} < x, A\phi_j > = \frac{1}{\lambda_j}\phi_j''(0).$$

Hence, the inner product $< A^{\frac{1}{4}}\dot{w}, A^{\frac{1}{4}}x >$, which naturally appears in any energy estimate based on the topology of $E_0 := D(A^{\frac{3}{4}}) \times D(A^{\frac{1}{4}})$, satisfies

$$< A^{\frac{1}{4}}\dot{w}, A^{\frac{1}{4}}x > = \sum_j \frac{1}{\lambda_j^{\frac{1}{2}}} < \dot{w}, \phi_j > \phi_j''(0) = (A^{-\frac{1}{2}}\dot{w})''(0),$$

which is well defined for $\dot{w} \in D(A^{\frac{1}{4}})$. It, therefore, makes perfect ( mathematical ) sense to define a feed-back control as follows

$$h(t) := -(A^{-\frac{1}{2}}\dot{w})''(0).$$

As a matter of fact, $(A^{-\frac{1}{2}}\cdot)''$ is a bounded operator which is given explicitly. This definition gives dissipation of energy.

$$\frac{d}{dt}E_0(t) = -\alpha((A^{-\frac{1}{2}}\dot{w}(t))''(0))^2. \tag{11}$$

One wonders, whether this dissipation is enough to ensure uniform exponential decay of $w, \dot{w}$. The answer is in the negative. As the operator

$$\mathcal{A} := \begin{pmatrix} O & I \\ -A & O \end{pmatrix}$$

generates a strongly conitinous (unitary) group $S(t)$ of bounded operators in the "energy-space" $D(A^{\frac{3}{4}}) \times D(A^{\frac{1}{4}})$, and the input-map

$$\mathcal{B}\dot{w}(t) := -\begin{pmatrix} 0 \\ \alpha x \end{pmatrix} \cdot (A^{-\frac{1}{2}}\dot{w})''(0)$$

is bounded and has a finite dimensional range, we conclude by Proposition 2.1 of Lasiecka and Triggiani [LT] that the semigroup associated with the feedback law above, i.e. the semigroup $S_{\mathcal{B}}(t)$ generated by the operator $\mathcal{A} + \mathcal{B}$ in $D(A^{\frac{3}{4}}) \times D(A^{\frac{1}{4}}) := Z$ satisfies $\|S_{\mathcal{B}}(t)\| \geq 1 \ \forall t \geq 0$. However, we do have strong stability. This is true, since the resolvent of $\mathcal{A} + \mathcal{B}$ is compact in $Z$ and there is no purely imaginary eigenvalue of $\mathcal{A} + \mathcal{B}$. With hindsight to the problem of stabilizability of a cantilever, where the control enters the first derivative at the clamped end of the beam, we might suggest that another control-law may indeed lead to uniform exponential decay of solutions, if we switch to a stronger topology, i.e. to smaller spaces, such as $D(A) \times D(A^{\frac{1}{2}})$ rather than $Z$. See Lasiecka[Las], for the corresponding plate problem. In particular her problem (1.4) with boundary condition (1.5) – in the linear case – can be converted by transposition to the problem addressed here. For the sake of selfconsistency, I provide the argument leading to the "right" Ansatz for the feedback control which is much easier to derive in the given one–dimensional case. The procedure is as follows. One defines a "the energy" as

$$E_1(t) := \frac{1}{2}\{\|A^{\frac{1}{2}}\dot{w}(t)\|^2 + \|Aw(t)\|^2\}$$

Upon taking the time derivative of $E_1(t)$ and using the equation (8) with $q_1 = q_2 = 0$ one derives

$$\frac{d}{dt}E_1(t) = -\alpha \int_0^\ell A\dot{w}x dx \cdot h(t)$$

But, upon integration by parts

$$A^{\frac{1}{4}}x = \sum_j \lambda_j^{-\frac{3}{4}} \phi_j''(0)\phi_j.$$

This shows that

$$< A\dot{w}, x >=< A^{\frac{3}{4}}\dot{w}, A^{\frac{1}{4}}x >= \sum_j < \dot{w}_j, \phi_j > \phi_j''(0) = \dot{w}''(0).$$

Therefore, the "right" feed–back, i.e. the one giving dissipation of the energy, is $h(t) = \dot{w}''(0)$. This leads to

$$\frac{d}{dt}E_1(t) = -(\dot{w}''(0))^2.$$

This feedback is not of the type given above, in the sense that it is not bounded. It does not seem to be known though whether uniform exponential decay obtains. What is known is that the solution decays strongly to zero. We do not dwell on this further here, and leave the question of uniform energy decay to the future.

**Remark:**

*The analoguous control–problem for thin plates is subject of current research. In addition, the control problem for a nonlinear version of (CP) is currently under investigation as well. However, the decomposition method is essentially limited to linear processes, so that it contributes information on the linearization, i.e. (CP), only. The corresponding results are then obtained on the local level by the use of some version of the implicit function theorem.*

The motivation for this additional work is the fact that light robot arms in recent robot structures are often more adequately modelled by thin plates rather than beams. In addition, the usually high angular velocity during a rotation maneuver constitutively introduces some nonlinearity to the system – the centrifugal– and Coriolis forces.

## References:

BL Baillieul,J. and Levi,M. (1987) Rotational elastic dynamics. Physica **27D**, 43–62.

DK Delfour,M.C., Kern,M.,Passeron,L. and Sevenne,B. (1986) Modelling of a rotating flexible beam. *Control of distributed parameter systems 1986*, 4th IFAC Symposium (ed. H.Rauch; Pergamon Press, Los Angeles ) 383–387.

DM Desch,W. and Miller,R.K. (1987) Exponential stabilization of Volterra–integrodifferential equations in Hilbert spaces. *Journal of Differential Equations* **70**, 366–389.

Kra Krabs, W. (1982) Optimal control of processes governed by partial differential equations, part II. Vibrations. *Zeitschrift für Operations Research* **26**, 63–86.

Las Lasiecka, I.(1989) Stabilization of wave and plate–like equations with nonlinear dissipation on the boundary. *Journal of Differential Equations* **79**, 340–381.

LT Lasiecka, I. and Triggiani, R.(1983) Dirichlet boundary stabilization of the wave equation with damping feedback of finite range. *Journal of Math. Analysis and Appl.*, **97**, 112–130.

Le1 Leugering, G. (1990) A decomposition method for integro- differential equations and applications. *Journal Mathematique Pures et Appliquées* to appear.

Le2 Leugering, G. (1990) Control and stabilization of a flexible robot arm. *Dynamics and Stability of Systems* **5**, Nr.1, 37–46.

# SOME REGULARITY PROPERTIES FOR THE WAVE EQUATION RELATED TO AN EXACT CONTROLLABILITY PROBLEM

Jean-Pierre PUEL *

## 1.INTRODUCTION.

If we consider the wave equation with Dirichlet boundary conditions, we know from a result of Zuazua using J.L.Lions' Hilbert Uniqueness Method (cf.[5]) that one can solve the problem of exact controllability when the control is distributed and acts on a neighborhood $\omega$ of a part $\Gamma_0$ of the boundary, if $\Gamma_0$ satisfies some suitable geometrical conditions and if the time of integration T is sufficiently large ($T > T_0$ where $T_0$ depends only on the diameter of $\Omega$). We want to study here what happens when the set $\omega$ shrinks to $\Gamma_0$. Do we obtain, at the limit, an exact controllability problem where the control acts on the boundary and which problem? We will give a precise answer to these questions in section 2 without getting into the details of the proofs. In fact, these proofs require some new regularity results for solutions of the wave equation, which may be of more general interest, and which will be presented independently in section 3, with the details of the proof for one of them. These results have been announced in [3]

## 2.THE EXACT CONTROLLABILITY PROBLEM

We first consider a problem of exact controllability associated with the wave equation where the control is distributed and acts on a neighborhood of a part $\Gamma_0$ of the boundary satisfying suitable geometrical conditions.

More precisely, let $\Omega$ be a bounded regular open set in $R^N$ with boundary $\Gamma$, $\nu(x)$ the unit exterior normal vector at a point $x \in \Gamma$ and let $\Gamma_0$ be a subset of $\Gamma$. We will sometimes assume that $\Gamma_0$ satisfies the following condition

$$\exists x_0 \in R^N, \Gamma_0 = \{x \in \Gamma, (x - x_0).\nu(x) > 0\}. \tag{2.1}$$

For $\epsilon > 0$, we consider the subset $\omega_\epsilon$ of $\Omega$ defined by

$$\omega_\epsilon = (\bigcup_{x \in \Gamma_0} B(x, \epsilon)) \cap \Omega$$

and the wave equation (we denote by ' the time derivative and by $\chi_{\omega_\epsilon}$ the indicator function of $\omega_\epsilon$)

* Département de Mathématiques et d'Informatique, Université d'Orléans et Centre de Mathématiques Appliquées, Ecole Polytechnique, 91128 PALAISEAU CEDEX, France

$$\begin{cases} y'' - \Delta y = v.\chi_{\omega_\epsilon} \text{ in } Q_T = \Omega \times (0,T), \\ y = 0 \text{ on } \Sigma = \Gamma \times (0,T), \\ y(0) = y_0; y'(0) = y_1, \end{cases} \tag{2.2}$$

where $y_0 \in H_0^1(\Omega)$ , $y_1 \in L^2(\Omega)$, and $v \in L^2(Q_T)$ is the control which is distributed and acts only on $\omega_\epsilon$. Equation (2.2) has a unique solution $y \in C([0,T]; H_0^1(\Omega)) \cap C^1([0,T]; L^2(\Omega))$.

For fixed $\epsilon$ , the problem of exact controllability consists in finding a control $v \in L^2(Q_T)$ such that

$$y(T) = 0 \text{ and } y'(T) = 0,$$

and if possible such a control which, in addition minimizes $||v||_{L^2(Q_T)}$ among admissible controls.

Zuazua has shown, using J.L.Lions' H.U.M. (cf. [5]) that if $T > T_0$ , where $T_0$ depends only on the diameter of $\Omega$ and not on $\epsilon$, and if $\Gamma_0$ satisfies (2.1), there exists such a control. In fact, he finds a control $v_\epsilon$ given by the following optimality system.

**Optimality system.**

There exist $\tilde{\varphi}_0^\epsilon \in L^2(\Omega)$ and $\tilde{\varphi}_1^\epsilon \in H^{-1}(\Omega)$ such that, if $\tilde{\varphi}^\epsilon$ and $\psi^\epsilon$ are solutions of

$$\begin{cases} \tilde{\varphi}^{\epsilon''} - \Delta\tilde{\varphi}^\epsilon = 0 \text{ in } Q_T, \\ \tilde{\varphi}^\epsilon = 0 \text{ on } \Sigma, \\ \tilde{\varphi}^\epsilon(0) = \tilde{\varphi}_0^\epsilon; \tilde{\varphi}^{\epsilon'}(0) = \tilde{\varphi}_1^\epsilon, \end{cases} \tag{2.3}$$

$$\begin{cases} \psi^{\epsilon''} - \Delta\psi^\epsilon = \tilde{\varphi}^\epsilon.\chi_{\omega_\epsilon} \text{ in } Q_T, \\ \psi^\epsilon = 0 \text{ on } \Sigma, \\ \psi^\epsilon(T) = 0; \psi^{\epsilon'}(T) = 0, \end{cases} \tag{2.4}$$

then

$$\psi^\epsilon(0) = y_0; \psi^{\epsilon'}(0) = y_1. \tag{2.5}$$

One can immediately see that by setting $v_\epsilon = \tilde{\varphi}^\epsilon$ , the corresponding solution of (2.2) noted $y_\epsilon$ is equal to $\psi^\epsilon$ and satisfies, because of (2.4) , $y_\epsilon(T) = 0$ and $y'_\epsilon(T) = 0$. This gives a solution to the exact controllability problem. In the present work, we are interested in studying what happens when $\epsilon$ tends to 0.

**Estimates.**

First of all, one has of course to find estimates on the functions $\tilde{\varphi}_0^\epsilon$ and $\tilde{\varphi}_1^\epsilon$. They are given by the following lemmas.

**Lemma 2.1.**(C.Fabre [1])

*There exists a constant C independent of $\epsilon$ such that*

$$[||\tilde{\varphi}_0^\epsilon||_{L^2}^2 + ||\tilde{\varphi}_1^\epsilon||_{H^{-1}}^2]^{\frac{1}{2}} = \frac{C}{\epsilon^3}. \tag{2.6}$$

Proof of Lemma 2.1 is very difficult and interesting. It requires the results on the wave equation which will be given in section 3.

**Lemma 2.2.**
*There exists a constant $C$ independent of $\epsilon$ such that*

$$\int_0^T \int_{\omega_\epsilon} |\tilde{\varphi}^\epsilon|^2 dxdt \leq \frac{C}{\epsilon^3}. \tag{2.7}$$

This result is simple to obtain by multiplying (2.4) by $\tilde{\varphi}^\epsilon$, integrating by parts and using Lemma 2.1.

These estimates suggest a change of notations. Let us write

$$\varphi_0^\epsilon = \epsilon^3 \tilde{\varphi}_0^\epsilon \quad ; \quad \varphi_1^\epsilon = \epsilon^3 \tilde{\varphi}_1^\epsilon \quad ; \quad \varphi^\epsilon = \epsilon^3 \tilde{\varphi}^\epsilon.$$

The optimality system becomes

$$\begin{cases} \varphi^{\epsilon\prime\prime} - \Delta\varphi^\epsilon = 0 \text{ in } Q_T, \\ \varphi^\epsilon = 0 \text{ on } \Sigma, \\ \varphi^\epsilon(0) = \varphi_0^\epsilon; \varphi^{\epsilon\prime}(0) = \varphi_1^\epsilon, \end{cases} \tag{2.8}$$

$$\begin{cases} \psi^{\epsilon\prime\prime} - \Delta\psi^\epsilon = \dfrac{1}{\epsilon^3}\varphi^\epsilon.\chi_{\omega_\epsilon} \text{ in } Q_T, \\ \psi^\epsilon = 0 \text{ on } \Sigma, \\ \psi^\epsilon(T) = 0; \psi^{\epsilon\prime}(T) = 0, \end{cases} \tag{2.9}$$

with

$$\psi^\epsilon(0) = y_0; \psi^{\epsilon\prime}(0) = y_1,$$

and we have the estimates

$$||\varphi_0^\epsilon||_{L^2}^2 + ||\varphi_1^\epsilon||_{H^{-1}}^2 \leq C; \tag{2.10}$$

$$\frac{1}{\epsilon^3}\int_0^T \int_{\omega_\epsilon} |\varphi^\epsilon|^2 dxdt \leq C. \tag{2.11}$$

We now want to study the limits of (2.8), (2.9) with the estimates (2.10), (2.11). To study the limit of (2.8) is easy. One can extract subsequences such that if $\epsilon$ tends to 0,

$$\varphi_0^\epsilon \to \varphi_0 \text{ in } L^2(\Omega) \text{ weak }, \varphi_1^\epsilon \to \varphi_1 \text{ in } H^{-1}(\Omega) \text{ weak },$$

and consequently

$$\varphi^\epsilon \to \varphi \text{ in } L^\infty(0,T;L^2(\Omega)) \text{ weak* },$$

with

$$\begin{cases} \varphi'' - \Delta\varphi = 0 \text{ in } Q_T, \\ \varphi = 0 \text{ on } \Sigma, \\ \varphi(0) = \varphi_0; \varphi'(0) = \varphi_1. \end{cases} \tag{2.12}$$

The main difficulty consists in studying the limit of (2.9) which is a wave equation with a singular right hand side. We can prove the following result.

**Theorem 2.1.**

*If $\varphi^\epsilon$ is solution of the wave equation (2.8) and $\varphi^\epsilon \rightarrow \varphi$ in $L^\infty(0,T;L^2(\Omega))$ weak* and if furthermore $1/\epsilon^3 \int_0^T \int_{\omega_\epsilon} |\varphi^\epsilon|^2 dxdt \leq C$, then $\psi^\epsilon$ converges to $\psi$ in $L^\infty(0,T;L^2(\Omega))$ weak*, $\psi^\epsilon(0)$ converges to $\psi(0)$ in $L^2(\Omega)$ weak, $\psi^{\epsilon\prime}(0)$ converges to $\psi'(0)$ in $H^{-1}(\Omega)$ weak, where $\psi$ satisfies*

$$\begin{cases} \psi'' - \Delta\psi = 0 \text{ in } Q_T, \\ \psi_{/\Sigma-\Sigma_0} = 0; \psi_{/\Sigma_0} = -\dfrac{1}{3}\dfrac{\partial\varphi}{\partial\nu} \text{ where } \Sigma_0 = \Gamma_0 \times (0,T), \\ \psi(T) = 0; \psi'(T) = 0. \end{cases} \tag{2.13}$$

*and we have*

$$\psi(0) = y_0; \psi'(0) = y_1. \tag{2.14}$$

Proof of Theorem 2.1 is long. The outline is given in [6] and the details are exposed in [1] and [2]. It requires regularity results on the wave equation which are interesting and sharp and which will be given in section 3. Notice that the sense of $\partial\varphi/\partial\nu$ is not clear yet but it will be proved that $\partial\varphi/\partial\nu \in L^2(\Sigma_0)$. The result of Theorem 2.1 together with (2.12) says in particular that when $\epsilon$ tends to 0, the solution given by H.U.M. of the exact controllability problem where the control acts on $\omega_\epsilon$ converges to the solution, given by H.U.M. again, of a problem of exact controllability with boundary control acting on the Dirichlet data on $\Sigma_0$. Notice that the functions $\psi^\epsilon$ vanish on the whole boundary $\Sigma$ and they converge to a function $\psi$ which, a priori, does not vanish on $\Sigma_0$. Then the convergence has to occur in weak spaces.

**Remark 2.1.**

One can study directly other equations with singular right hand sides, corresponding for example to the heat operator, the Schrödinger operator, the vibrating plates operator,... This leads to interesting questions which may have no relation with exact controllability. Some partial answers are given in [2].

## 3.REGULARITY RESULTS FOR THE WAVE EQUATION.

We are going to consider a situation which is analogous to the one encountered in the previous section, even if one could have taken into account more general situations.

We begin with a result giving the regularity "at the limit", for a sequence of very weak solutions of the wave equation satisfying estimates like (2.10), (2.11).

Let us consider a sequence of solutions $\varphi^\epsilon$ of the wave equation

$$\begin{cases} \varphi^{\epsilon\prime\prime} - \Delta\varphi^\epsilon = 0 \text{ in } Q_T, \\ \varphi^\epsilon = 0 \text{ on } \Sigma, \\ \varphi^\epsilon(0) = \varphi_0^\epsilon; \varphi^{\epsilon\prime}(0) = \varphi_1^\epsilon, \end{cases} \tag{3.1}$$

where

$$\varphi_0^\epsilon \rightarrow \varphi_0 \text{ in } L^2(\Omega) \text{ weak }, \varphi_1^\epsilon \rightarrow \varphi_1 \text{ in } H^{-1}(\Omega) \text{ weak },$$

$$\varphi^\epsilon \to \varphi \text{ in } L^\infty(0,T;L^2(\Omega)) \text{ weak* }.$$

**Theorem 3.1.**

*Under the above hypotheses (without assuming (2.1)), if in addition*

$$\frac{1}{\epsilon^3}\int_0^T\int_{\omega_\epsilon}|\varphi^\epsilon|^2 dxdt \le C,$$

*then* $\partial\varphi/\partial\nu \in L^2(\Sigma_0)$ .

*Therefore, if* $\Gamma_0$ *satisfies (2.1) and* $T \ge T_0$, *this implies that* $\varphi_0 \in H_0^1(\Omega)$ , $\varphi_1 \in L^2(\Omega)$ *and* $\varphi$ *is a solution of finite energy of the wave equation.*

**Proof.**

We know that $\varphi$ is a very weak solution of the wave equation

$$\begin{cases} \varphi'' - \Delta\varphi = 0 \text{ in } Q_T, \\ \varphi = 0 \text{ on } \Sigma, \\ \varphi(0) = \varphi_0; \varphi'(0) = \varphi_1. \end{cases}$$

and $\varphi \in C([0,T];L^2(\Omega)) \cap C^1([0,T];H^{-1}(\Omega))$.

Let us define as in [5] the solutions $\Phi_0^\epsilon$ and $\Phi_0$ of

$$\Delta\Phi_0^\epsilon = \varphi_1^\epsilon, \Phi_0^\epsilon \in H_0^1(\Omega); \Delta\Phi_0 = \varphi_1, \Phi_0 \in H_0^1(\Omega),$$

and

$$\Phi^\epsilon(x,t) = \int_0^t \varphi^\epsilon(x,s)ds + \Phi_0^\epsilon(x); \Phi(x,t) = \int_0^t \varphi(x,s)ds + \Phi_0(x).$$

Then

$$\begin{cases} \Phi^{\epsilon\prime\prime} - \Delta\Phi^\epsilon = 0 \text{ in } Q_T, \\ \Phi^\epsilon = 0 \text{ on } \Sigma, \\ \Phi^\epsilon(0) = \Phi_0^\epsilon; \Phi^{\epsilon\prime}(0) = \varphi_0^\epsilon, \end{cases}$$

and

$$\begin{cases} \Phi'' - \Delta\Phi = 0 \text{ in } Q_T, \\ \Phi = 0 \text{ on } \Sigma, \\ \Phi(0) = \Phi_0; \Phi'(0) = \varphi_0, \end{cases}$$

Therefore, $\Phi^\epsilon$ and $\Phi$ are solutions of finite energy and using a regularity result of J.L.Lions [5], we know that

$$\frac{\partial\Phi^\epsilon}{\partial\nu} \in L^2(\Sigma) \ , \ \frac{\partial\Phi}{\partial\nu} \in L^2(\Sigma) \text{ and } \frac{\partial\Phi^\epsilon}{\partial\nu} \to \frac{\partial\Phi}{\partial\nu} \text{ in } L^2(\Sigma) \text{ weak }.$$

This shows that $\partial\varphi^\epsilon/\partial\nu$ converges to $\partial\varphi/\partial\nu$ in $H^{-1}(0,T;L^2(\Gamma))$ weak. If $u \in \mathcal{D}(\Sigma_0)$ and if $<,>$ denotes the duality $\mathcal{D}'(\Sigma_0), \mathcal{D}(\Sigma_0)$ , we have

$$\lim_{\epsilon \to 0} < \frac{\partial \varphi^\epsilon}{\partial \nu}, u > = < \frac{\partial \varphi}{\partial \nu}, u >,$$

and in order to prove the theorem, we want to show that

$$\exists C > 0, \forall u \in \mathcal{D}(\Sigma_0), | < \frac{\partial \varphi}{\partial \nu}, u > | \leq C ||u||_{L^2(\Sigma_0)}.$$

Using a covering of a neighborhood of $\Gamma_0$ and a partition of unity, we can localize the problem and assume that we work in a neighborhood $U$ of a point of $\Gamma_0$. Then we can define a (local) change of coordinates

$$x = y - z\nu(y), y \in \Gamma_0, z \in R^+.$$

( $y$ is the tangential coordinate and $z$ the normal coordinate). The mapping

$$J^{-1} : x \to (y, z)$$

is a $C^2$ diffeomorphism, and if we write for a function v(x,t)

$$\hat{v}(z, y, t) = v(x, t),$$

then

$$\frac{\partial \hat{v}}{\partial z}(0, y, t) = -\frac{\partial v}{\partial \nu}(y, t).$$

If $u \in \mathcal{D}(\Sigma_0 \cap (U \times (0, T))$, let us define a regular function $w$ with compact support in time, such that

$$w = 0 \text{ on } \Sigma_0 \; ; \; \frac{\partial w}{\partial \nu} = u \text{ on } \Sigma_0.$$

This is always possible by taking, for example

$$\hat{w}(z, y, t) = -z.u(y, t).$$

We then have

$$(\frac{1}{\epsilon^3} \int_0^T \int_{\omega_\epsilon} |w|^2 dx dt)^{\frac{1}{2}} \leq M ||u||_{L^2(\Sigma_0)},$$

and of course

$$w' = 0 \text{ on } \Sigma_0 \; ; \; \frac{\partial w'}{\partial \nu} = u' \text{ on } \Sigma_0.$$

We can now write

$$
\begin{aligned}
< \frac{\partial \varphi^\epsilon}{\partial \nu}, u > &= < (\frac{\partial \Phi^\epsilon}{\partial \nu})', u > = - < \frac{\partial \Phi^\epsilon}{\partial \nu}, u' > = - \int_0^T \int_\Gamma \frac{\partial \Phi^\epsilon}{\partial \nu} u' dy dt \\
&= -\frac{3}{\epsilon^3} \int_0^T \int_{\omega_\epsilon} \Phi^\epsilon w' dx dt + (\frac{3}{\epsilon^3} \int_0^T \int_{\omega_\epsilon} \Phi^\epsilon w' dx dt - \int_0^T \int_\Gamma \frac{\partial \Phi^\epsilon}{\partial \nu} u' dy dt) \\
&= \frac{3}{\epsilon^3} \int_0^T \int_{\omega_\epsilon} \varphi^\epsilon w dx dt + (\frac{3}{\epsilon^3} \int_0^T \int_{\omega_\epsilon} \Phi^\epsilon w' dx dt - \int_0^T \int_\Gamma \frac{\partial \Phi^\epsilon}{\partial \nu} u' dy dt) \\
&= A_\epsilon + B_\epsilon,
\end{aligned}
$$

where

$$A_\epsilon = \frac{3}{\epsilon^3} \int_0^T \int_{\omega_\epsilon} \varphi^\epsilon w dx dt$$

$$B_\epsilon = (\frac{3}{\epsilon^3} \int_0^T \int_{\omega_\epsilon} \Phi^\epsilon w' dx dt - \int_0^T \int_\Gamma \frac{\partial \Phi^\epsilon}{\partial \nu} u' dy dt).$$

Using Cauchy-Schwarz inequality, we get

$$|A_\epsilon| \leq 3(\frac{1}{\epsilon^3} \int_0^T \int_{\omega_\epsilon} |\varphi^\epsilon|^2 dx dt)^{\frac{1}{2}} (\frac{1}{\epsilon^3} \int_0^T \int_{\omega_\epsilon} |w|^2 dx dt)^{\frac{1}{2}} \leq 3CM ||u||_{L^2(\Sigma_0)}.$$

Then

$$\limsup_{\epsilon \to 0} |A_\epsilon| \leq 3CM ||u||_{L^2(\Sigma_0)}.$$

We will now show that for a fixed $u$ , $B_\epsilon \to 0$ if $\epsilon \to 0$. This proof is very technical.

$$
\begin{aligned}
B_\epsilon =& (\frac{3}{\epsilon^3} \int_0^T \int_0^\epsilon \int_{\Gamma_0} \hat{\Phi}^\epsilon(z, y, t) z u'(y, t) |JacJ(o, y)| dz dy dt \\
& - \int_0^T \int_{\Gamma_0} \frac{\partial \hat{\Phi}^\epsilon}{\partial z}(0, y, t) u'(y, t) dy dt) \\
& + \frac{3}{\epsilon^3} \int_0^\epsilon z (\int_0^z < \frac{\partial \hat{\Phi}^\epsilon}{\partial z}(\xi, y, t), u'(y, t) \int_0^z \frac{\partial}{\partial z} |JacJ(s, y)| ds > d\xi) dz.
\end{aligned}
$$

One can show that $|JacJ(0, y)| = 1$ and if

$$
\begin{aligned}
C_\epsilon =& (\frac{3}{\epsilon^3} \int_0^T \int_0^\epsilon \int_{\Gamma_0} \hat{\Phi}^\epsilon(z, y, t) z u'(y, t) |JacJ(o, y)| dz dy dt \\
& - \int_0^T \int_{\Gamma_0} \frac{\partial \hat{\Phi}^\epsilon}{\partial z}(0, y, t) u'(y, t) dy dt),
\end{aligned}
$$

then

$$C_\epsilon = \frac{3}{\epsilon^3}\int_0^\epsilon z^2(\frac{1}{z}\int_0^z < \frac{\partial\hat{\Phi}^\epsilon}{\partial z}(\xi,y,t) - \frac{\partial\hat{\Phi}^\epsilon}{\partial z}(0,y,t), u'(y,t) > d\xi)dz.$$

We can integrate in time again and define

$$\theta^\epsilon(x,t) = \int_0^t \Phi^\epsilon(x,s)ds + \theta_0^\epsilon,$$

where

$$\Delta\theta_0^\epsilon = \varphi_0^\epsilon, \theta_0^\epsilon \in H_0^1(\Omega).$$

Then $\theta^\epsilon$ is bounded in $L^2(0,T;H^2(\Omega)\cap H_0^1(\Omega))$ , which shows that $\partial\hat{\theta}^\epsilon/\partial z$ is bounded in $H^1(0,\epsilon_0;L^2(\Sigma_0))$, and $\xi \to \partial\hat{\theta}^\epsilon/\partial z\ (\xi,y,t)$ is continuous at $\xi = 0$ , uniformly in $\epsilon$ , with values in $L^2(\Sigma_0)$. As $\partial\hat{\Phi}^\epsilon/\partial z = (\partial\hat{\theta}^\epsilon/\partial z\ )'$ , we see that $\xi \to \partial\hat{\Phi}^\epsilon/\partial z\ (\xi,y,t)$ is continuous at $\xi = 0$ , uniformly in $\epsilon$ , with values in $H^{-1}(\Sigma_0)$ , and this implies that $C_\epsilon \to 0$ if $\epsilon \to 0$.

The same argument also shows that $\xi \to \partial\hat{\Phi}^\epsilon/\partial z\ (\xi,y,t)$ is bounded on $[0,\epsilon_0]$ , uniformly in $\epsilon$ , with values in $H^{-1}(0,T;L^2(\Gamma_0))$ and on the other hand, one can show, because of the regularity of $\Omega$ that $\partial/\partial z|JacJ(s,y)|$ is bounded on $[0,\epsilon_0]$ with values in $L^\infty(\Gamma_0)$. Therefore

$$|< \frac{\partial\hat{\Phi}^\epsilon}{\partial z}(\xi,y,t), u'(y,t)\int_0^z \frac{\partial}{\partial z}|JacJ(s,y)|ds > | \leq Mz$$

and if

$$D_\epsilon = \frac{3}{\epsilon^3}\int_0^\epsilon z(\int_0^z < \frac{\partial\hat{\Phi}^\epsilon}{\partial z}(\xi,y,t), u'(y,t)\int_0^z \frac{\partial}{\partial z}|JacJ(s,y)|ds > d\xi)dz$$

we have

$$|D_\epsilon| \leq \frac{3}{\epsilon^3}\int_0^\epsilon \frac{M}{2}z^3 dz \leq \frac{3M\epsilon}{8} \to 0 \text{ if } \epsilon \to 0.$$

As $B_\epsilon = C_\epsilon - D_\epsilon$ , this shows that for fixed $u$ in $\mathcal{D}(\Sigma_0)$ , $B_\epsilon \to 0$ if $\epsilon \to 0$, and the proof of Theorem 3.1 is complete.

**Remark 3.1.**

The regularity result given in Theorem 3.1 is only valid for the limit. Each function $\varphi^\epsilon$ cannot, in general, be as regular as the limit $\varphi$. The precise behavior in $\epsilon$ of the $L^2$-norm of $\varphi^\epsilon$ in $\omega_\epsilon \times (0,T)$ gives an additional information about the regularity of the limit $\varphi$.

We next state a result giving the precise behavior, near the boundary, of the solutions of finite energy of the wave equation, and which appears as a reciproqual of Theorem 3.1, even if it is not really the case.

For $(f,u_0,u_1) \in L^1(0,T;L^2(\Omega)) \times H_0^1(\Omega) \times L^2(\Omega)$ let $u$ be the solution of

$$\begin{cases} u'' - \Delta u = f \text{ in } Q_T, \\ u = 0 \text{ on } \Sigma, \\ u(0) = u_0; u'(0) = u_1; \end{cases} \tag{3.2}$$

Then $u \in C(0,T;H_0^1(\Omega)) \cap C^1(0,T;L^2(\Omega))$ ( $u$ has finite energy) and the mapping $(f,u_0,u_1) \to u$ is linear and continuous.

**Theorem 3.2.**

*Let $u$ be the solution (of finite energy) of the wave equation (3.2). Then ( without assuming (2.1)) there exists a constant $C > 0$ independent of $\epsilon$ such that*

$$\frac{1}{\epsilon^3}\int_0^T\int_{\omega_\epsilon} |u|^2 dxdt \leq C[||f||^2_{L^1(0,T;L^2(\Omega))} + ||u_0||^2_{H_0^1(\Omega)} + ||u_1||^2_{L^2(\Omega)}] = CE, \tag{3.3}$$

*where $E$ is the "energy".*

**Idea of the proof.**

We just give here a sketch of the proof. The complete details will be given in [4]. We use an extension of the Rellich multiplier method. As $\Omega$ is regular, we know that there exist functions $h^\epsilon \in W^{2,\infty}(\overline{\Omega}; R^N)$ such that $h^\epsilon_{/\Gamma_0} = \nu$. Multiplying (3.2) by $2h^\epsilon.\nabla u - (div h^\epsilon).u$ and integrating by parts gives

$$2\int_0^T\int_\Omega \frac{\partial h_k^\epsilon}{\partial x_j}\frac{\partial u}{\partial x_j}\frac{\partial u}{\partial x_k} dxdt = -\int_0^T\int_\Omega u\frac{\partial u}{\partial x_j}\frac{\partial^2 h_k^\epsilon}{\partial x_j \partial x_k} dxdt + R_1, \tag{3.4}$$

where the terms in $R_1$ will be easily bounded in terms of the energy, thanks to the regularity result of J.L.Lions [5], saying that $\partial u/\partial \nu \in L^2(\Sigma)$ and that its norm is bounded by the energy. As in the proof of Theorem 3.1, we first localize the problem, then use the change of variables

$$x = y - z\nu(y), y \in \Gamma, z \in R^+, y = p(x).$$

Then, we take

$$h^\epsilon(x) = \begin{cases} \rho^\epsilon(z).\nu(y) \text{ if } x \in \omega_\epsilon, \\ 0 \text{ if } x \in \Omega - \omega_\epsilon, \end{cases}$$

where $\rho^\epsilon \in W^{2,\infty}(0,\epsilon)$ is a decreasing function such that

$$\begin{cases} \rho^\epsilon(\epsilon) = 0; \rho^{\epsilon\prime}(\epsilon) = 0, \\ ||\rho^\epsilon||_{L^\infty} = O(1); ||\rho^{\epsilon\prime}||_{L^\infty} = O(\frac{1}{\epsilon}). \end{cases}$$

From (3.4) we obtain

$$2\int_0^T\int_\Gamma\int_0^\epsilon |\frac{\partial \rho^\epsilon}{\partial z}|(\frac{\partial \hat{u}}{\partial z})^2 |Jac(J)| dzdydt = \int_0^T\int_\Gamma\int_0^\epsilon \frac{\partial^2 \rho^\epsilon}{\partial z^2}\hat{u}\frac{\partial \hat{u}}{\partial z}|Jac(J)| dzdydt + R_2,$$

where $R_2$ plays the same role as $R_1$.

Setting

$$G(\epsilon) = \frac{1}{\epsilon}\int_0^T \int_\Gamma \int_0^\epsilon (\frac{\partial \hat{u}}{\partial z})^2 |Jac(J)| dz dy dt = \frac{1}{\epsilon}\int_0^T \int_{\omega_\epsilon} |\nabla u(x,t).\nu(p(x))|^2 dx dt \text{ if } \epsilon > 0,$$

$$G(0) = \int_0^T \int_\Gamma |\frac{\partial u}{\partial \nu}|^2 dy dt,$$

$G$ is continuous at 0 and we can prove, by choosing suitable functions $\rho^\epsilon$ that there exists $\delta > 0$ such that

$$\sup_{\epsilon \in [0,\delta]} G(\epsilon) \leq C.E,$$

Where $C$ is independent of $\epsilon$ and $u$. Then, by a simple argument of integration with respect to the normal variable $z$, this shows the result of Theorem 3.2.

**Remark 3.2.**

Similar results have recently been obtained by C.Fabre [2] for the Schrödinger equation and the equation of vibrating plates or beams associated with the operator $\partial^2/\partial t^2 + \Delta^2$.

## REFERENCES


[1] C.FABRE: Equation des ondes avec second membre singulier et application à la contrôlabilité exacte. Note aux C. R. Acad. Sc., Paris, t.310, Série 1, p.813-818, 1990, and article, to appear.

[2] C.FABRE: Thèse de Doctorat, Université Paris 6, 1990.

[3] C.FABRE and J.P.PUEL: Comportement au voisinage du bord des solutions de l'équation des ondes. Notes aux C. R. Acad. Sc., Paris, t.310, Série 1, p.621-625, 1990.

[4] C.FABRE and J.P.PUEL: Behavior near the boundary for solutions of the wave equation. To appear.

[5] J.L.LIONS *Contrôlabilité exacte. Perturbations et Stabilisation de Systèmes Distribués.* Tome 1. Masson, 1988.

[6] J.P.PUEL and C.FABRE: Wave equations with singular right hand sides. Application to exact controllability. To appear in the Proceedings of the second Franco-Chilean and Latin American Conference, Santiago de Chile, 1989.

# Dynamical Shape Control of Nonlinear Thin Rods

Jan Sokolowski
Systems Research Institute
Polish Academy of Sciences
ul. Newelska 6
01-447 Warszawa
Poland

Jürgen Sprekels
Fachbereich 10 -Bauwesen
Universität-GH Essen
Postfach 103764
D-4300 Essen 1
Germany

## 1 Introduction

Dynamical shape control problems for linear partial differential equations have recently drawn much attention. The linear heat equation was studied in Cannarsa–Da Prato–Zolesio [1], where the feedback was constructed via the Hamilton–Jacobi–Bellman equation. Truchi–Zolesio [4] considered the linear wave equation. Closely related to the problem of dynamical shape control is the paper by Cannarsa–Da Prato–Zolesio [2], in which the damped linear wave equation was studied on a moving domain. In a recent paper, Sokolowski–Sprekels [3] considered the following problem:
Suppose a thin rod performs transversal oscillations which shall be damped out, and suppose the left part of the rod is fixed in such a way that it can be moved back and forth within the fixation, so that the length of the free part of the rod can be controlled dynamically. The objective is to move the rod in such a way that its tip is brought to rest at a prescribed point at a given time instant $T$. We sketch the situation in Figure 1. In contrast to the other works mentioned above, in [3] nonlinear constitutive laws were admitted. In particular, materials were considered that not only react to changes of the shear strain $\epsilon = u_x$ by a (possibly non-monotone) shear stress, but also to changes of the curvature of their crystal lattices by a couple stress. Consequently, the elastic potential $\Phi$ is assumed in the form

$$\Phi = \Phi(\epsilon, \epsilon_x) = F(\epsilon) + \frac{\gamma}{2}\epsilon_x^2, \tag{1.1}$$

where $\gamma > 0$, and where $F$ is smooth and possibly non-convex.
Coming back to our problem, we observe that pulling the rod inside the fixation stabilizes the rod, and the oscillations cease completely if the rod has been pulled into the fixation. Since this is not feasible in practical applications, such as the stabilization of flexible structures (for instance in space), it is desirable to allow controls $v$ (which denotes the deviation of the length of the rod from its initial length) where $v_t(t) > 0$, which means that the rod is pushed outside the fixation, i.e., the length of its free part is increased. Since increasing the length means to destabilize the structure, another mechanism is needed

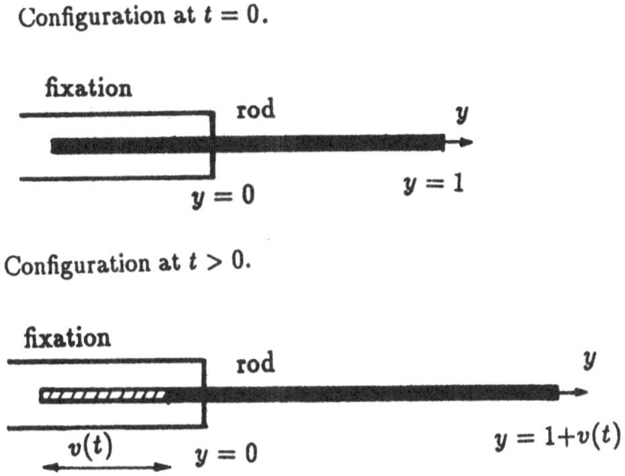


Figure 1: Dynamical shape control of a thin rod.

to damp the oscillations in this situation. In this paper we assume that a damping mechanism becomes active at the tip of the rod if $v_t(t) > 0$; to be precise, we assume that the total stress $\sigma$ at the tip is counteracted by means of the following boundary condition : $\sigma = G(v_t)u_t$ , where $u$ denotes the transversal displacement and $G$ is a real-valued function satisfying

$$G(v) - v \geq \delta, \quad \forall v \in \mathbb{R}, \text{ with some } \delta > 0. \tag{1.2}$$

Typically,

$$G(v) = v_+ + \delta, \quad \forall v \in \mathbb{R}, \text{ with some } \delta > 0. \tag{1.3}$$

In the sequel, we give a report on the results obtained in [3] concerning the optimal control problem for the nonlinear thin rod.

# 2 Well–Posedness of the State Equations

Let $T > 0$ be fixed. We consider the initial–boundary value problem :

$$u_{tt} - (F'(u_y))_y + u_{yyyy} = g(y,t), \quad \text{in } Q_T(v), \tag{2.1}$$

$$u(0,t) = u_y(0,t) = 0 = u_{yy}(1+v(t),t), \quad 0 < t < T, \tag{2.2}$$

$$u_{yyy}(1+v(t),t) - F'(u_y(1+v(t),t)) = G(v_t(t))u_t(1+v(t),t), \quad 0 < t < T, \tag{2.3}$$

$$u(y,0) = u_0(y), \quad u_t(y,0) = u_1(y), \quad 0 < y < 1. \tag{2.4}$$

In terms of our problem, (2.1) is the balance law of linear momentum after the introduction of dimensionless variables and upon normalizing all physical constants to unity; $g$ is a distributed (known) load, and $u_0$, $u_1$ stand for the initial displacement and velocity, respectively. Moreover, we have set

$$Q_T(v) = \{(y,t) \in \mathbb{R}^2 | 0 < t < T, 0 < y < 1 + v(t)\} \quad . \tag{2.5}$$

A typical situation is depicted in the following drawing:

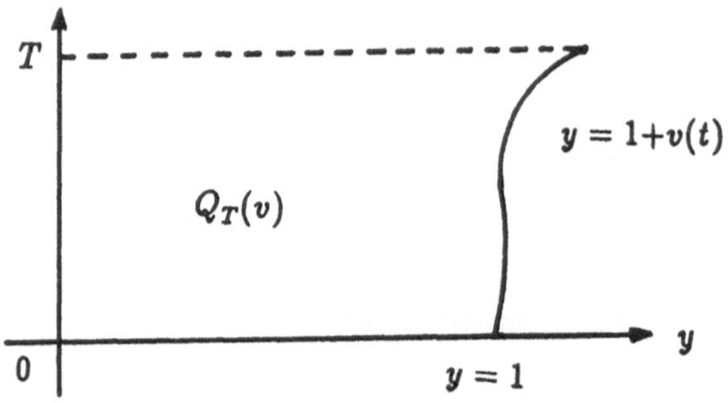


We always tacitly assume that the control $v$ is small enough so that the rod is not pulled out of the fixation completely. For the data of the problem we generally assume :

**(A1)** $g \in H^1(0,T;L^2_{loc}(\mathbb{R}))$.

**(A2)** $u_0 \in H^3(0,1)$, $u_1 \in H^2(0,1)$, $u_0(0) = u_0'(0) = 0$;
the compatibility conditions of sufficiently high order are satisfied.

**(A3)** $G \in C(\mathbb{R})$, and with some fixed $\epsilon > 0$, $G(x) - x \geq \epsilon$, $\forall x \in \mathbb{R}$.

**(A4)** $F \in C^4(\mathbb{R})$, and there exist a nonnegative function $F_1$ and positive constants $\beta_i$, $i = 1,\ldots,6$, such that

(i) $\beta_1 F_1(x) - \beta_2 \leq F(x) \leq \beta_3 F_1(x) + \beta_4 \quad ,\forall x \in \mathbb{R}$,

(ii) $xF'(x) \leq \beta_5 F(x) + \beta_6 \quad , \quad \forall x \in \mathbb{R}$ .

**(A5)** $v \in H^3(0,T)$, $v(0) = v_t(0) = 0$, and, with some $M \in (0,1)$, $v(t) \geq -M$, $\forall t \in [0,T]$ .

Note that (A4) holds if $F$ is an even polynomial with positive coefficient at the highest order; the condition $v(t) \geq -M$ means that the rod is never pulled into its fixation completely.
Next we transform the problem onto the fixed domain $\Omega_T$, where $\Omega = (0,1)$ and, for $t > 0$, $\Omega_t = \Omega \times (0,t)$. To this end, introduce the new coordinate $x = \frac{y}{1+v(t)}$. The new unknown function $z(x,t) = u(y,t) = u(x(1+v(t)),t)$ satisfies the system

$$z_{tt} - \frac{2xv_t}{1+v} z_{xt} + z_x \left( x \left( \frac{v_t}{1+v} \right)^2 - \left( \frac{xv_t}{1+v} \right)_t \right) + \left( \frac{xv_t}{1+v} \right) z_{xx}$$

$$- \frac{1}{1+v} \left( F' \left( \frac{z_x}{1+v} \right) \right)_x + \frac{1}{(1+v)^4} z_{xxxx} = f(x,t), \quad \text{in} \quad \Omega_T \quad , \tag{2.6}$$

$$z(0,t) = z_x(0,t) = 0 = z_{xx}(1,t), \quad 0 < t < T, \tag{2.7}$$

$$\frac{1}{(1+v(t))^3} z_{xxx}(1,t) - F'\left(\frac{z_x(1,t)}{1+v(t)}\right)$$
$$= G(v_t(t))\left(z_t(1,t) - \frac{v_t(t)}{1+v(t)} z_x(1,t)\right), \quad 0 < t < T, \tag{2.8}$$

$$z(x,0) = u_0(x), \quad z_t(x,0) = u_1(x), \quad 0 < x < 1. \tag{2.9}$$

Here we have set : $f(x,t) \equiv g(x(1+v(t)),t)$.
We now derive a weak formulation of (2.6)–(2.9). To this end, we test (2.6) by any function $\varphi \in V$, where

$$V = \{\varphi \in H^2(0,1) | \varphi(0) = \varphi'(0) = 0\} \quad . \tag{2.10}$$

Then

$$[z_{tt}(t) + a_2(t) z_x(t) + a_3(t) z_{xx}(t), \varphi] - [a_1(t) z_t(t), \varphi_x] + [a_4(t) F'\left(\frac{z_x(t)}{1+v(t)}\right), \varphi_x]$$
$$+[a_5(t) z_{xx}(t), \varphi_{xx}] + a_1(1,t) z_t(1,t)\varphi(1) + a_6(t) z_t(1,t)\varphi(1) + a_7(t) z_x(1,t)\varphi(1)$$
$$-[f(t),\varphi] = 0 \quad , \quad \forall \varphi \in V \quad . \tag{2.11}$$

Here, $[\cdot,\cdot]$ denotes the inner product in $L^2(0,1)$, and we have used the abbreviations

$$a_1(x,t) = \frac{-2x v_t(t)}{1+v(t)} \quad , \quad a_2(x,t) = x\left(\frac{v_t(t)}{1+v(t)}\right)^2 + \left(\frac{x v_t(t)}{1+v(t)}\right)_t,$$
$$a_3(x,t) = \frac{x v_t(t)}{1+v(t)} \quad , \quad a_4(t) = \frac{1}{1+v(t)} \quad , \quad a_5(t) = \frac{1}{(1+v(t))^4} \quad ,$$
$$a_6(t) = \frac{G(v_t(t))}{1+v(t)} \quad , \quad a_7(t) = -\frac{v_t(t) G(v_t(t))}{(1+v(t))^2} \quad . \tag{2.12}$$

Next we introduce our notion of a weak solution to the system (2.1)–(2.4). To this end, we integrate (2.11) with respect to $t$. We define:

<u>Definition :</u>

A function $u$ is called a *weak solution* of (2.1)–(2.4), if $u(y,t) = u(x(1+v(t)),t) = z(x,t)$, where $z$ has the following properties :

(i) $z \in L^\infty(0,T;H^2(0,1))$, $\quad z_t \in C(0,T;L^2(0,1))$, $\quad z_x \in L^\infty(\Omega_T)$, $\quad z_t(1,\cdot) \in L^2(0,T)$.

(ii) For all $\varphi \in V$ and for all $\tau \in (0,T]$ there holds

$$[z_t(\tau),\varphi] - [u_1,\varphi] + \int_0^\tau [a_5(t) z_{xx}(t), \varphi_{xx}]dt - \int_0^\tau [a_1(t) z_t(t), \varphi_x]dt$$

$$+\int_0^T [a_2(t)z_x(t) + a_3(t)z_{xx}(t), \varphi]dt + \int_0^T \left[a_4(t)F'\left(\frac{z_x(t)}{1+v(t)}\right), \varphi_x\right] dt$$
$$+\int_0^T [(a_6(t) + a_1(1,t))z_t(1,t)\varphi(1) + a_7(t)z_x(1,t)\varphi(1)]dt$$
$$-\int_0^T [f(t), \varphi]dt = 0. \tag{2.13}$$

(iii) $z(x,0) = u_0(x)$, for all $x \in [0,1]$.

We have the following existence result (cf. [3], Theorems 2.1 and 2.3):

**Theorem 2.1** *Under the assumptions (A1)–(A5), the system (2.1)–(2.4) has a weak solution. If, in addition, $G \in C^1(\mathbb{R})$, then the weak solution satisfies*

$$z_{tt} \in L^\infty(0,T;L^2(0,1)), z_{xxt} \in L^\infty(0,T;L^2(0,1)),$$
$$z_{xt} \in L^\infty(\Omega_T), z_{tt}(1,\cdot) \in L^2(0,T). \tag{2.14}$$

The following result states that the weak solution depends Lipschitz continously on the domain parameter $v \in H^3(0,T)$. In particular, the weak solution is unique. We have (cf. [3], Theorem 2.4):

**Theorem 2.2** *Let $G \in C^1(\mathbb{R})$, and let (A1)–(A4) be satisfied. Suppose $z^{(i)}$ is any weak solution associated with the domain parameter $v^{(i)} \in H^3(0,T)$, where $v^{(i)}$ satisfies (A5), $i = 1,2$. Then for $z = z^{(1)} - z^{(2)}, v = v^{(1)} - v^{(2)}$, there holds the inequality*

$$\sup_{t\in(0,T)} (\|z_t(t)\|^2 + \|z_{xx}(t)\|^2) + \|z_x\|_{L^\infty(\Omega_T)} + \int_0^t z_t^2(1,s)\,ds$$
$$\le C\,\|v\|^2_{H^2(0,T)} \quad , \tag{2.15}$$

*where $C$ depends only on the data and $\|v^{(i)}\|_{H^3(0,T)}$, $i = 1,2$.*

# 3 The Optimal Control Problem

Let $K \subset H^3(0,T)$ denote some convex set. We consider a dynamical shape control problem for the nonlinear thin rod :

Minimize the cost functional

$$I(v) = \frac{1}{2}\int_0^T \int_0^{1+v(t)} (u_t^2 + u_{xx}^2)\,dx\,dt + \frac{\alpha}{2}\,\Pi(v,v) \quad , \tag{3.1}$$

where $\alpha > 0$, $\Pi(v,v)^{1/2}$ is a norm or a seminorm on $H^3(0,T)$, and $u$ the weak solution of (2.1)–(2.4) corresponding to $v$ (we assume that the assumptions of Theorem 2.2 are satisfied).

To treat this control problem, we transform the state equations and the cost functional onto the fixed domain $\Omega_T = (0,1) \times (0,T)$. To simplify the exposition somewhat, we consider the problem :

Problem $(P)$ :

Minimize the cost functional

$$J(v) = \frac{1}{2}\int_0^T \int_0^1 (z_t^2 + z_{xx}^2)\,dx\,dt + \frac{\beta}{2}\int_0^T |G(v_t(t))z_t(1,t)|^2\,dt + \frac{\alpha}{2}\Pi(v,v) \tag{3.2}$$

over the closed and convex set

$$K = \{v \in H^3(0,T) \,|\, v(0) = v_t(0) = v_{tt}(0) = 0\,,\ v(T) = L\,,\ v_t(T) = v_{tt}(T) = 0\,,\ v(t) \geq -M\,,\ 0 \leq t \leq T\}\ . \tag{3.3}$$

We thus want to bring the tip of the rod at time $T$ to a rest at the prescribed position $1 + L$ (where, of course, $L \geq -M$) while keeping the total energy of the rod, penalized by two terms representing the cost of the control action, at a minimal value. We have (cf. [3], Theorem 3.1) the result:

**Theorem 3.1** *Suppose that $G \in C^1(\mathbb{R})$, and suppose (A1)–(A4) are satisfied. Furthermore, let for $C > 0$ the sets $\{v \in K \,|\, \Pi(v,v) \leq C\}$ be bounded in $H^3(0,T)$. Then the problem $(P)$ has an optimal solution $v^* \in K$, and the necessary conditions of optimality are satisfied in the sense that there exist $(z^*, p^*)$ which satisfy the system :*

1. State Equation :

$$\int_0^1 [z_{tt}^*(t) + a_1^*(t)z_{xt}^*(t) + a_2^*(t)z_x^*(t) + a_3^*(t)z_{xx}^*(t)]\varphi\,dx + \int_0^1 a_4^*(t)F'\left(\frac{z_x^*(t)}{1+v^*(t)}\right)\varphi_x\,dx + \int_0^1 a_5^*(t)z_{xx}^*(t)\varphi_{xx}\,dx - \int_0^1 f(t)\varphi\,dx$$
$$+ (a_6^*(t)z_t^*(1,t) + a_7^*(t)z_x^*(1,t))\varphi(1) = 0\ , \quad \textit{for all}\ \ \varphi \in V\ \ \textit{and a.e.}\ \ t \in (0,T)\ , \tag{3.4}$$

*with the initial conditions*

$$z^*(x,0) = u_0(x)\ ,\quad z_t^*(x,0) = u_1(x)\ ,\quad 0 \leq x \leq 1\ . \tag{3.5}$$

2. Adjoint-State Equation :

$$\int_0^1 [p_{tt}^*(t)\eta - (p^*(t)a_1^*(t))_t\eta_x + p^*(t)(a_2^*(t)\eta_x + a_3^*(t)\eta_{xx})$$

$$+a_4^*(t)F''\left(\frac{z_x^*(t)}{1+v^*(t)}\right)\frac{p_x^*(t)}{1+v^*(t)}\eta_x + a_5^*(t)p_{xx}^*(t)\eta_{xx}]\,dx$$

$$-(a_6^*(t)p^*(1,t))_t\eta(1) + a_7^*(t)p^*(1,t)\eta_x(1)$$

$$= \beta(\,(G(v_t^*(t)))^2 z_t^*(1,t))_t\,\eta(1) + \int_0^1 (z_{tt}^*(t)\eta - z_{xx}^*(t)\eta_{xx})\,dx \quad ,$$

$$\textit{for all} \quad \eta \in V \quad \textit{and a.e.} \quad t \in (0,T) \quad , \tag{3.6}$$

*with final conditions*

$$p^*(x,T) = 0 \quad , \quad p_t^*(x,T) = z_t^*(T) \quad . \tag{3.7}$$

3. Optimality Conditions :

$$\int_0^1 \{ \quad p^*(t) \quad [a_1'(v^*; v - v^*(t))z_{xt}^*(t) + a_2'(v^*; v - v^*(t))z_x^*(t)$$

$$+a_3'(v^*; v - v^*(t))z_{xx}^*(t)] + a_4'(v^*; v - v^*(t))F'\left(\frac{z_x^*(t)}{1+v^*(t)}\right)p_x^*(t)$$

$$-a_4(v^*)F''\left(\frac{z_x^*(t)}{1+v^*(t)}\right)\frac{p_x^*(t)}{(1+v^*(t))^2}(v - v^*(t))$$

$$+a_5'(v^*; v - v^*(t))z_{xx}^*(t)p_{xx}^*(t) \quad \}\,dx$$

$$+a_6'(v^*; v - v^*(t))z_t^*(1,t)p^*(1,t) + a_7'(v^*; v - v^*(t))z_x^*(1,t)p^*(1,t)$$

$$+\alpha\,\Pi(v^*(t), v - v^*(t)) \geq 0 \quad ,$$

$$\textit{for all} \quad v \in K \quad \textit{and a.e.} \quad t \in (0,T) \quad . \tag{3.8}$$

*(Here the expressions $a_j'(v^*; v - v^*(t))$ , $1 \leq j \leq 7$ , have their obvious meanings.)*

## References


1. Cannarsa, P., Da Prato, G., Zolesio, J.-P.: Dynamical shape control of the heat equation. Systems and Control Letters 12 (1989), 103–109.

2. Cannarsa, P., Da Prato, G., Zolesio, J.-P.: The damped wave equation in a moving domain. J. Diff. Equ. 85 (1990), 1–16.

3. Sokolowski, J., Sprekels, J.: Dynamical shape control and the stabilization of thin rods. To appear in Math. Meth. in the Appl. Sci.

4. Truchi, C., Zolesio, J.-P.: to appear.

# SHAPE DERIVATIVE OF DISCRETIZED PROBLEMS

M.SOULI*
Universite de NICE
Departement de Mathematiques
Parc VALROSE NICE 06034

J.P.ZOLESIO*
CNRS
USTL Place E.Bataillon
34060 MONTPELLIER CEDEX

**SUMMARY**

This paper is concerned by numerical sensitivity analysis in problems governed by boundary-value problems described by partial differential equations (P.D.E.).

The P.D.E. solution y is approximated by the solution of finite element method (F.E.M) $y_h$.

In shape optimization, we are concerned by a cost functional $J(\Omega_h) = h(\Omega_h, M_h, y_h)$, where $\Omega_h$ is the discretized geometrical domain and $M_h$ the nodes of the mesh.

This work is devoted to the numerical calculation of the derivative with respect to the coordinates of the nodes of the cost J associated to $M_h$ we have a $Q_\ell$ Lagrange F.E.M and y is solution of a variational problem. For simplicity, we shall restrict the presentation to a second order problem.

The domain $\Omega_h$ is only described by the boundary nodes, we also discuss the use of the derivatives with respect to the internal nodes. The case $P_1$ Lagrange finite element is treated by Zolesio [6]. In this work, we treat the general case of $Q_\ell$ Lagrange finite element.

**INTRODUCTION**

The main objective of this paper, is to show how to obtain in shape optimization the derivatives with respect to the nodes as a particular case of the general theory of the gradient calculation developped in [1], [4] and [5], by a simple choice of the velocity field V well adapted to the situation.

The domain is only defined by the boundary nodes, and numerically the internal nodes move in an injustified way, we shall precise in the last part, the use of the derivatives with respect to the internal nodes.

The calculation of the derivative with respect to the domain $\Omega$ for boundary-value problems is realized by introducing the deformations $T_t$ constructed from a velocity field V (particular velocity), this field is described by the functional space

$$E = C^0\left([0,\epsilon[,\ C_k\ (\mathbb{R}^n,\mathbb{R}^n)\right)\ k \geq 1$$

with this method, we envisage the general deformations $\Omega_t = T_t(\Omega)$ of the domain $\Omega$, for V given in E.
The properties of $T_t$ and $\Omega_t$ are widly studied. If $U_t$ is solution in the configuration $\Omega_t$, $\Gamma_t$ of a boundary value problem, so $J(\Omega_t) = h(\Omega_t, u_t)$ is the functional to minimize with respect to the domain $\Omega$.
The derivative of J at $\Omega$ in the V direction is given by $dJ(\Omega,V) = \frac{d}{dt} J(\Omega_t)|_{t=0}$.
We establish in [4],[5] the properties of this derivative. In particular if $\Gamma$ is regular, then $dJ(\Omega,V)$ is only dependent on the germ of V(0) on the neighbourhood of $\Gamma$, if in addition the application $V \to dJ(\Omega,V)$ is linear, then from the structure theorem (Hadamard formula theorem, see [4], [5]), it exists a distribution $g_n$ on the surface $\Gamma$ ($g_n$ is unique) such that the gradient G, vectorial distribution on $\mathbb{R}^n$, is given by $G = {}^t\gamma_\Gamma(g_n.n)$ ($\gamma_\Gamma$ is the trace application or restriction on $\Gamma$) and the derivative with respect to the domain verifies :

$$(0.1) \qquad dJ(\Omega,V) = \langle G,\ V(0) \rangle_{D^k(\mathbb{R}^n,\mathbb{R}^n)' \times D^k(\mathbb{R}^n,\mathbb{R}^n)}$$

The calculation of the gradient G is generally done via the calculation of the eulerian derivative of the state equation.

$$\dot{u} = \frac{d}{dt}(u_t \circ T_t)\Big|_{t=0} \quad \text{(derivative in } H^m(\Omega))$$

For instance, for the simplest functional

$$J(\Omega) = \frac{1}{2}\int_\Omega (u - y_d)^2\, dx$$

$\left(y_d \text{ is a given function on } \mathbb{R}^n\right)$, we have :

$$dJ(\Omega,V) = \int_\Omega (u-y_d)\left(\dot{u} - \nabla y_d.V(0)\right) dx + \frac{1}{2}\int_\Omega (u - y_d)^2 \operatorname{div} V(0)\, dx$$

However, introducing the derivative u' of the state u with respect to the domain on the V direction defined by

$$(0.2) \qquad u' = \dot{u} - \nabla u.V(0)$$

we obtain ($\Gamma$ is supposed regular)

$$(0.3) \qquad dJ(\Omega,V) = \int_{\Omega}(u - y_d)\, u'\, dx + \frac{1}{2}\int_{\Gamma} (u - y_d)^2\, V(0).n\, d\sigma$$

this second expression (0,3) of dJ is more interesting since, closer to the structure noticed by the theorem, seeing that the second term on the right hand side already depends explicitly on the restriction V(0).n on $\Gamma$.

In fact, for elliptic and parabolic boundary value problems, we characterize u using the implicit function theorem, and we obtain the linearity and continuity of $V \longrightarrow \dot{u}$ and then of $V \longrightarrow u'$. We then prove the following results :

The derivative u' with respect to $\Omega$ on the V direction depends only on the restriction of V(0).n on $\Gamma$, (this result does not hold for $\dot{u}$). It means that u' is solution of boundary value problem with homogeneous conditions on $\Omega$, and introducing in a classical way an adjoint state p in (0,3), we immediatly obtain the expression of the gradient G defined by (0,1).

# 1. SHAPE OPTIMIZATION

## 1.1 General case

It would be well to distinguish two types of deformation : The first one adapted to small displacements, and the second to large displacements and to deformations governed by parameters. Let $\Omega$ included in $\mathbb{R}^n$ (to simplify, we will take n = 2). V is a field defined on the neighbourhood of U of $\bar{\Omega}$ (resp. on $[0,\epsilon[ \times U$). The transformation $T_t$ is defined by :

$$(1.1) \qquad T_t : x \longrightarrow x + t\, V(x) \qquad \left(\text{resp. } x(t) = x + \int_0^t V(s,\, x(q))\, ds\right)$$

we will find the properties of $T_t$ in F. Murat and J. Simon [6] (resp. in Zolesio [4],[5]).

We will be more interested to the second deformations which, for small t, have the same properties as the first one, see $T_t$ and $T^{-1}$ are $C^k$, and preserve them when $t \longrightarrow +\infty$ while the field V is in the space $R_{\infty}$ (see [4]). In addition, for the shape optimization defined by parameters, it's easy to find the field V(t,x) which construct the same variation of the geometry as the variation of the parameters. So the derivation of the cost and states with respect to the parameters is simply a matter of a particular choice of the field V. For such situation see for instance Delfour, Payre, Zolesio

[2]. It's the analogous situation that we consider for the displacements of nodes in a mesh, however the coordinates of nodes are the parameters.

### 2.2 Preliminarities and notations

The first aspect, is that the triangulation $\mathcal{T}_h$ is established over the set $\bar{\Omega}$, i.e, the set $\bar{\Omega}$ is subdivised into a finite number of subsets K (see Ciarlet [9]) in such a way that the following properties are satisfied.

$$\Omega = \bigcup_{1 \leq k \leq M} K_k \quad .$$

M is the number of elements, and each element $K_i$ is isoparametrically equivalent to a reference element $\hat{K}$ trough a mapping $F_i$, we shall use the usual correspondance

$$K_i : \hat{K} \longrightarrow K_i$$

$$\xi \longrightarrow \sum_{j=1}^{\ell} \hat{q}_j(\xi)\; M_j(K_i)$$

$F_i \in \left(Q_\ell(\hat{K})\right)^2$, where $Q_\ell$ is polynomial space of degree $\leq \ell$ and $M_j(K_i)$ the freedom degrees on the element $K_i$. So we are considering the general case of $Q_\ell$ Lagrange finite element. Let $N_1$ the number of nodes in the triangulation, and $N_2$ the number of freedom degrees which are not nodes of $\Omega$. In what follows, we will consider only the gradient with respect to the nodes. Once the nodes are on position, the other freedom degrees are automatically known.

Let move a node $M_j$ in the k-direction (k = 1,2)

$$\begin{aligned} M_j^t &= M_j + t\, \ell_k \\ &= \left(X_1^j,\, X_2^j\right) + t\, \ell_k \end{aligned}$$

$\ell_k$ is the euclidean basis in $\mathbb{R}^2$ ((0,1), (1,0)). This assumption will define a new triangulation $\mathcal{T}_h^t$, the elements are called $K_i^t$.

As for the element $K_i$, we define the mapping

$$K_i^t \quad : \quad \hat{K} \longrightarrow K_i^t$$

$$\xi \longrightarrow \Sigma\, \hat{q}_j(\xi)\ (M_j(K_i) + t\ \ell_k)$$

Let $X_h$ a finite-dimensional subspace on the Hilert space (for example $H^1(\Omega)$) associated to the triangulation $\mathcal{T}_h$, and $X_h^t$ to the triangulation $\mathcal{T}_h^t$. Since we are considering Lagrange finite element, the space $X_h$ will be spanned by the functions $v_i$, $i=1,\dots,N$ ($N = N_1+N_2$) such that.

$$(2.1)\qquad \begin{cases} v_i \in \mathcal{C}^0\left(\bar{\Omega}\right) & \\ v_i/K_j \circ F_j \in Q_\ell(\hat{K}) & i \leqslant j \leqslant M \\ v_i(M_j) = \delta_{i,j} & 1 \leqslant j \leqslant N \end{cases}$$

In the same way, $X_h^t$ is spanned by $v_i^t$

$$(2.1)'\qquad \begin{cases} v_i^t \in \mathcal{C}^0\left\{\bar{\Omega}^t\right\} & \\ v_i^t\big/k_j^t \circ F_j^t \in Q_p(\hat{K}) & 1 \leqslant j \leqslant M \\ v_i^t\left(M_j^t\right) = \delta_{i,j} & 1 \leqslant j \leqslant N \end{cases}$$

### 2.3 The deformation field

As in the continuous case ([4], [5]), the goal is to determine the transformation $T_t$ which maps $\Omega$ into $\Omega^t$, via an adapted choice of velocity $V(t,.)$.
The transformation $T_t$ defined by (0,1) is continuous, so we will determine it on each element $K_i$ $1 \leqslant i \leqslant M$.
Let consider the situation where only the node i is noved in the k-direction.

**Definition 1.1**
For $1 \leqslant j \leqslant N_1$ where $N_1$ is the number of nodes, we define

$$V_j(t,x) = v_j^t(x).\ell_k$$

and

$$T_t(x) = x + \int_0^t V_j(s)\ x(s)\ ds$$

<u>Lemma 1.1</u>. The restriction of the transformation $T_t$ to the element $K_i$ is

$$T_t/K_i = F_i^t \circ F_i^{-1}$$

<u>Proof.</u> We will show that $\frac{d}{dt}(S_t) \circ (S^t)^{-1} = V_j(t,x)$ where $S^t = F_i^t \circ F_i^{-1}$.

From the definition of $F_i$ and $F_i^t$e have

$$F_i^t(\xi) = F_i(\xi) + t\,\hat{q}_j(\xi)\,\ell_k$$

$$\frac{d}{dt}\left(F_i^t \circ F_i^{-1}\right) = \left(\frac{d}{dt}F_i^t\right) \circ (F_i)^{-1} = \hat{q}_j \circ (F_i)^{-1}.\ell_k$$

$$\frac{d}{dt}(S^t) \circ (S^t)^{-1} = \hat{q}_j \circ (F_i)^{-1} \circ \left(F_i^{-t} \circ F_i^{-1}\right)^{-1}.\ell_k$$

$$= \hat{q}_j \circ \left(F_i^t\right)^{-1}.\ell_k = v_i^t/K_i^t \, . \, \ell_k \tag{2.3}$$

We deduce that $S^t$ transforms $K_i$ into $K_i^t$ and its velocity $\left(\text{ie } \frac{d}{dt}(S^t) \circ (S^t)^{-1}\right)$

is given from (2.3) so $T_t/S^t$.

<u>Lemma 1.2</u>

Let $v_i$, $v_i^t$, $1 \leqslant i \leqslant N_i$ defined by (2.1) and (2.1)'. The basis function related to the nodes, then

$$v_i^t \circ T_t = v_i \tag{2.4}$$

<u>Proof.</u> We will verify (2.4) on each element $K_j$ from the definition of $v_i$ and $v_i^t$ we have

$$(v_i/K_j) \circ F_j \in \hat{Q}(\hat{K}) \iff v_i/K_j \quad \hat{q} \circ F_j^{-1}$$

$$\left(v_i^t/K_j^t\right) \circ F_j^t \in \hat{Q}\,(\hat{K}) \iff v_i^t/\,K_j^t \quad \hat{q} \circ \left(F_j^t\right)^{-1}$$

however

$$v_i^t/K_j^t \circ T_t/K_j = v_i^t/K_j^t \circ \left(F_j^t \circ F_j^{-1}\right)$$

$$= \hat{q} \circ \left(F_j^t\right)^{-1} \circ \left(F_j^t \circ F_j^{-1}\right) = \hat{q} \circ F_j^{-1} = v_i/K_j$$

Corollary

For each node i $1 \leq i \leq N_1$ we have

$$\frac{d}{dt}\left(v_i^t \circ T_t\right) = 0 \iff \dot{v}_i^t = 0$$

then the basis function related to the nodes are convected.

Remark.

The lemma 1.2 still holds for $N_1+1 \leq i \leq N_1+N_2$, the freedom degrees which are not nodes of $\Omega$, since

so $\dot{v}_i = 0$, for $1 \leq i \leq N1+N2$, and, the basis functions are convected.

The velocity field V discrebing the transformation $T_t$ which applies $\Omega$ into $\Omega_t$ is spanned by the basis functions $v_i$ related to the nodes $1 \leq i \leq N1$, so

$$(2.5) \qquad v_t \in V \iff v^t(x,y) = \left(\sum_{i=1}^{N1} r_i\ v_i^t,\ \sum_{i=1}^{N1} s_i\ v_i^t\right)$$

## 3. DERIVATION WITH RESPECT TO THE NODES

In the general case, we are looking for partial derivative with respect to the coordinates of the nodes, for a functional $J(u,\Omega)$, where u is a discretized (generally noted $u_h$) solution of a variational problem defined on the domain $\Omega$, called state problem.

We will see in this part how to derive bilinear forms, which are generally involved in such variational problems. To simplify the notation, we only consider the two-dimensional case n = 2.

### 3.1 Derivation of the state u

Let $u_t$ an element of $X_h^t$, we set

$$(3.1) \qquad u^t = u_t \circ T_t$$

which is an element of $X_h$ defined by

$$u^t = \sum_{i=1}^{N} u_i(t)\ N_i \qquad\qquad N = N1 + N2$$

we note the derivative (when it exists)

$$\dot{u} = \frac{d}{dt} u^t \Big|_{t=o} = \sum_{i=1}^{N} u_i^1 (0\dot{a}\ v_i$$

$\dot{u}$ is the eulerion derivative, with respect to the field velocity V defined on (2.5).

3.2 Transport and derivative of bilinear forms

As in the situation of shape optimization, the field V is defined on (2.5) and $u_t$, $v_t$ in $X_h^t$, so on each element $K_j^t$ we have

$$\begin{cases} \int_{K_j^t} \nabla u_t\ \nabla v_t\ dx = \int_{K_j} \left\langle A_j(t)\ \nabla u^t,\ \nabla(v_t \circ T_t)\right\rangle dx \\ \int_{K_j^t} F\ v_t\ dx = \int_{K_j} F \circ T_t . J_j^t\ v^t\ dx \end{cases}$$

where $A_j(t)$ is a (2x2) matrice defined by

$$\begin{cases} A_j(t) = J_j^t\ (DT_t)^{-1}\ {}^t(DT_t)^{-1} & A_j(0) = Id \\ J_j^t = \det(DT_t) \end{cases}$$

Defining A(t) and $J_t$ on the domain $\Omega$ by

$$(3.2) \qquad \begin{aligned} A(t)/K_j &= A_j(t) \\ J_t/K_j &= J_j^t \end{aligned}$$

we obtain :

$$(3.3)\qquad \begin{aligned} \int_{\Omega_t} \nabla u_t \ \nabla v_t \ dx &= \int_{\Omega} \langle A(t)\ \nabla u^t ,\ \nabla v^t \rangle \, dx \\ \int_{\Omega_t} F\ v_t \ dx &= \int_{\Omega} F \circ T_t \circ J_t \ v^t \, dx \end{aligned}$$

The matrix function $x \longrightarrow A(t,x)$ and the function $x \longrightarrow J(t,x)$ are continuous on $\Omega^t$, and differentiable with respect to t at t = 0 on each element $K_j$ so

$$A' = \frac{d}{dt} A(t,.)\Big|_{t=o} = \operatorname{div} V(0) - (DV(0) + {}^tDV(0))$$

So the derivative of bilinear form (3.3) is, at t=o :

$$(3.4)\qquad \begin{aligned} &\frac{d}{dt} \sum_j \int_{K_j^t} \langle A(t)\ \nabla u^t ,\ \nabla v^t \rangle dx = \sum_j \int_{K_j} \Big( \langle A' \nabla u,\ \nabla v \rangle + \Big\langle \nabla \dot{u},\ \nabla v \Big\rangle \Big) dx \\ &\frac{d}{dt} \sum_j \int_{K_j^t} F \circ T_t \ .\ J_t \ v^t \, dx = \sum_k \int_{K_j} \operatorname{div}\,(F V(0))\ v \, dx \end{aligned}$$

In two-dimensional problem, where $v(t) = \left(v_i^t , 0\right) \left(\text{resp. } \left(0,\ v_i^t\right)\right)$ $1 \leqslant i \leqslant N1$ the matrix A' has the following form.

$$(3.5)\qquad A' = \begin{pmatrix} \partial_1 v_1 & \partial_2 v_1 \\ \partial_2 v_1 & -\partial_1 v_1 \end{pmatrix} \quad \left(\text{resp. } A' = - \begin{pmatrix} -\partial_2 v_1 & \partial_1 v_1 \\ \partial_1 v_1 & \partial_2 v_1 \end{pmatrix}\right)$$

### 3.3. The Dirichlet Problem

Consider $u_t \in V_h^t$ (subspace of $H_o^1\left(\Omega_h^t\right)$) solution of the system

$$(3.6)\qquad \int_{\Omega_h^t} \nabla u_t \ \nabla v_t \ dx = \int_{\Omega_t} F\ V_t \ dx \qquad \forall\ V_t \in V_h^t$$

F is given in $L^2(\mathbb{R}^2)$.

From (3.3) we have :

$$\begin{cases} u^t = u_t \circ T_t \in X_h \quad \text{subspace of } H_o^1(\Omega_h) \\ \int_{\Omega_h} \langle A(t)\, \nabla u^t, \nabla W \rangle\, dx = \int_{\Omega_h} F \circ T_t . J_t \; W\, dx \quad \forall W \in X_h \end{cases}$$

In the canonical basis of the vectorial space $X_h$, $u^t = \sum u_i(t)\, v_i$ where $u(t) = (u_1(t), \ldots, u_N(t))$ is the solution of the linear system $a(t).u(t)=f$, $a(t)$ is the matrix

$$a(t)_{ij} = \int_{\Omega_h} \langle A(t)\; \nabla \ell_j, \nabla \ell_i \rangle\, dx$$

$$f_i = \int_{\Omega_h} F \circ T_t . J_t \; \ell_i \; dx$$

Let $\emptyset(t,u) = a(t).u-f$, defined from $\mathbb{R}\times\mathbb{R}^N$ into $\mathbb{R}^N$, we check that the implicit function theorem holds, the only verification is the calculation of the derivative : $\frac{\partial}{\partial t}\emptyset(t,u)$ given by

$$\frac{\partial}{\partial t}\emptyset(t,u) = a'(0).u-f$$

which is obvious using the expression of A' in (3.5) for a velocity field V.

Proposition 3.1

The application $t \longrightarrow u^t$ belongs to $C'([0,\epsilon[, V_h)$. Its derivative at t=o is characterized by the linear system.

$$(3.7) \quad \begin{cases} \dot{u} \in V_h \quad \forall\, v \in V_h \\ \int_{\Omega_h} \nabla \dot{u}\; \nabla v\; dx = - \int_{\Omega_h} \langle A'\; \nabla u, \nabla v \rangle + \int_{\Omega_h} -\text{div}\; (FV(0))\; dx \end{cases}$$

To obtain the eulerian partial derivative $\dot{u}_{x_i}$, we select the velocity mentioned in (2.5). So A' has the form indicated in (3.5).

Proposition 3.2

u is solution of the Dirichlet problem (3.6), $M_i(x_i, y_i)$ $1 \leqslant i \leqslant N$, the nodes of the triangulation of $\Omega_h$, then the Eulerian partial derivative

$\dot{u}_{x_1}$ is solution of

$$\dot{u}_{x_1} \in V_h \qquad \forall\, v \in V_h$$

$$\int_{\Omega_h} \nabla\dot{u}_{x_1}\ \nabla V\ dx = - \int_{\Omega_h} (\partial_1 \ell i\ (\partial_1 u \partial_1 v - \partial_2 u \partial_2 v) + \partial_2 \ell i\ (\partial_1 u\ \partial_2 v + \partial_2 u\ \partial_1 v))dx$$

$$+ \int_{\Omega_h} \partial_1 (F\ell i) V\ dx$$

the Eulerian partial derivative $\dot{u}_{y_1}$ is solution of

$$\dot{u}_{y_1} \in V_h \qquad \forall\, v \in V_h$$

$$\int_{\Omega_h} \nabla\dot{u}_{y_1}\ \nabla v\ dx = - \int_{\Omega_h} (\partial_1 \ell i(\partial_1 u \partial_2 v + \partial_2 u \partial_1 v) + \partial_2 \ell i(\partial_2 u \partial_2 v - \partial_1 u \partial_1 v))dx$$

$$+ \int_{\Omega_h} \partial_2 (F\ell i)\ v\ dx$$

<u>Remark</u>

This technique to obtain $\dot{u}$ can be extended to Newman problems. In the numerical example, we will consider the mixt problem.

## 4. <u>DERIVATIVE OF FUNCTIONAL COST WITH RESPECT TO THE NODES</u>

Let consider a functional defined on the domain $\Omega$ for instance

$$\text{(4.1)} \qquad J(\Omega_h) = \frac{1}{2} \int_{\Omega_h} (u-z)^2\ dx \qquad u \in X_h$$

where u is solution of the Dirichlet problem (3.6). Let V the velocity fiel defined in (2.5) and $T_t$ defined in the lemma 1.2

$$J\left(\Omega_h^t\right) = \frac{1}{2} \int_{\Omega_h} \left(u^t - z \circ T_t\right) J_t\ dx$$

The derivative with respect to t at t=o is

$$(4.2) \qquad dJ(\Omega_h, V) = \int_{\Omega_h} \left[ \left( \dot{u} - \nabla z . V(0) \right) (u-z) + \frac{1}{2} (u-z)^2 \operatorname{div} V(0) \right] dx$$

we introduce an adjoint state defined by the linear system, $p \in X_h$

$$(4.3) \qquad \int_{\Omega_h} \nabla p \; \nabla v \; dx = \int_{\Omega_h} (u-z) v \; dx \qquad \forall \; v \in X_h$$

using (4.3) into (4.2), we have

$$(4.4) \qquad dJ(\Omega_h, V) = \int_{\Omega_h} \nabla \dot{u} \; \nabla p \; dx + \int_{\Omega_h} \left[ -\nabla z . V(o)(u-z) + \frac{1}{2}(u-z)^2 \operatorname{div} V(o) \right] dx$$

this expression is valid for a general velocity field V, to calculate the partial derivative with respect to the node coordinates $x_i$ and $y_i$, we select the velocity V mentioned in (2.5) the expression is then simplified, and we have

<u>Theorem 4.1</u>

Let u and p the solutions of the linear systems (3.6) and (4.3), and $J(\Omega_h)$ the functional cost defined by (4.1), $x_i$ and $y_i$ are the coordinates of the node $M_i$ in the triangulation of $\Omega_h$. Then the partial derivatives of H with respect to $x_i$ and $y_i$ are :

$$\frac{\partial}{\partial x_i} J(\Omega_h) = + \int_{\Omega_h} [\partial_1 \ell i \; (\partial_1 u \partial_1 p - \partial_2 u \partial_2 p) + \partial_2 \ell i \; (\partial_1 u \partial_1 p + \partial_2 u \partial_2 p)] dx$$

$$(4.5) \qquad + \int_{\Omega_h} \partial_1 \; (F \ell i) p \; dx - \int_{\Omega_h} \ell i \; \partial_1 z (u-z) \; dx + \frac{1}{2} \int_{\Omega_h} (u-z)^2 \partial_1 \ell i \; dx$$

$$\frac{\partial}{\partial y_i} J(\Omega_h) = - \int_{\Omega_h} [\partial_2 \ell i \; (\partial_1 u \partial_1 p - \partial_2 u \partial_2 p) - \partial_1 \ell i (\partial_1 u \partial_2 p + \partial_2 u \partial_1 p)] dx$$

$$+ \int_{\Omega_h} \partial_2 (F \ell i) p \; dx - \int_{\ell_h} \ell i \; \partial_2 z \; (u-z) dx + \frac{1}{2} \int_{\Omega_h} (u-z)^2 \partial_2 \; \ell i \; dx$$

these derivatives have been obtained by O. Pironneau [3]. however, the technique used here is systematic and respects the general theory of deformation as in [1], [4], [5], [6]. The partial derivatives appear as a particular case of the general situation.

## 5. MOTION OF INTERNAL NODES

Given the position of the nodes on the boundary, in numerical computations, the position of the internal nodes id deduced from a regular meshing, this gives a good finite element approximation for the state and adjoint problems (3.5) and (4.3).

This procedure defines explicitly and some times implicitly an application $\Theta$ such that

$$\Theta : \mathbb{R}^{2\mu} \longrightarrow \mathbb{R}^{2q}$$

$\mu$ is the number of boundary noces $i = 1,\ldots,\mu$
$q$ is the number of internal nodes $i = \mu+1,\ldots,\mu+q$.
The application $\Theta$ express the position of internal nodes, in terms of those on the boundary.

$$\Theta(x_1, y_1, \ldots, x_\mu, y_\mu) = (x_{\mu+1}, y_{\mu+1}, \ldots, x_{\mu+q}, y_{\mu+q})$$

Assume, the application $\Theta$ derivable, and $D\Theta$ defined by

$$\frac{\partial x_j}{\partial x_i}, \frac{\partial x_j}{\partial y_i}, \frac{\partial y_j}{\partial x_i}, \frac{\partial y_j}{\partial y_i}$$

$1 \leqslant i \leqslant \mu \quad \mu+1 \leqslant j \leqslant \mu+q$
the elements of the jacobian matrix $D\Theta$.
Let the case, where the matric $D\Theta$ is known explicitly, so the functional to minimize is

$$j_\theta = j_\Theta(x_1, y_1, \ldots, x_\mu, y_\mu) = J(\Omega(x_1, y_1, \ldots, x_\mu, y_\mu), \Omega(x_1, y_1, \ldots, x_\mu, y_\mu))$$

this functional depends on the application $\Theta$, so its derivative is also depending on $\Theta$, in fact it depends only on the matrix $D\Theta$ for $1 \leqslant i \leqslant \mu$.

$$(5.1)\qquad \begin{aligned} \frac{\partial}{\partial x_i} j_\theta &= \frac{\partial}{\partial x_i} J(\Omega) + \sum_{\ell=\mu+1}^{\mu+q} \frac{\partial}{\partial x_\ell} J(\Omega) \frac{\partial x_\ell}{\partial x_i} + \frac{\partial}{\partial y_\ell} J(\Omega) \frac{\partial y_\ell}{\partial x_i} \\ \frac{\partial}{\partial y_i} j\theta &= \frac{\partial}{\partial y_i} J(\Omega) + \sum_{\ell=\mu+1}^{\mu+q} \frac{\partial}{\partial x_\ell} J(\Omega) \frac{\partial x_\ell}{\partial y_i} + \frac{\partial}{\partial y_\ell} J(\Omega) \frac{\partial y_\ell}{\partial y_i} \end{aligned}$$

the derivatives $\frac{\partial}{\partial x_\ell} J$, $\frac{\partial}{\partial y_\ell} J$, $\frac{\partial}{\partial x_i} J$, $\frac{\partial}{\partial y_i} J$, are calculated from (4.5) with only one state adjoint P.

<u>Remark</u>.

The gradient of $j_\theta$ is only dependent on the matrix $D\Theta$. So it is not necessary to know explicitly the application $\theta$ to calculate the gradient of $j_\theta$ in (5.2).

Let give an example for the calculation of the gradient when the application $\theta$ and $D\Theta$ are known explicitly.

Consider the domain $\Omega$ defined by :

$$\Omega = \{(x,y) \;/\; a \leqslant x \leqslant b \;\; ; \;\; 0 \leqslant y \leqslant g(x)\}$$

where g is a continuous positive function, piecewise linear on [a,b].

For the triangulation, each vertical segment is divised into (m+1) equal segments (this is the choice of the application $\Theta$). So, the nodes are the points $(x_{ij}, y_{ij}) = \left(x_i, j \cdot \frac{1}{m+1} y_i\right)$, $1 \leqslant i \leqslant n$ ; $0 \leqslant j \leqslant m+1$, and the parameters defining the geometry are the coordinates of the nodes belonging to the graph of g, $M_i = (x_i, y_i)$ $1 \leqslant i \leqslant n$. The nodes on the triangulation are then defined as a function of those points, and satisfy

$$\frac{\partial}{\partial x_k} y_{ij} = 0 \qquad \frac{\partial}{\partial y_k} y_{ij} = \frac{1}{n+1} \delta_{kj}$$

$$\frac{\partial}{\partial x_k} x_{ij} = \delta_{ki} \qquad \frac{\partial}{\partial y_k} x_{ij} = 0$$

which give the jacobian matrix D$\Theta$, and the partial derivative (5.1) are for $1 \leqslant i \leqslant n$.

$$\frac{\partial}{\partial x_j} j_\theta(x_1, y_1, \ldots, x_n, y_n) = \frac{\partial}{\partial x_i} J(\Omega) + \sum_{j=o}^{m+1} \frac{\partial}{\partial x_{ij}} J(\Omega)$$

$$\frac{\partial}{\partial y_i} j_\theta(x_1, y_1, \ldots, x_n, y_n) = \frac{\partial}{\partial y_i} J(\Omega) + \sum_{j=o}^{m+1} \frac{\partial}{\partial y_{ij}} J(\Omega).$$

The derivatives $\frac{\partial}{\partial x_{ij}} J$, $\frac{\partial}{\partial y_{ij}} J, \ldots,$ are calculated from (4.5).

# 6 Application to a naval hydrodynamical problem

## 6.1 The continuous case.

To test the gradient (5.4) with respect to the nodes, we will give two numerical examples related to a naval hydrodynamical problem, and concerning the search of a free surface, which is the water wave. In the first example the water wave is generated by a flow arround an obstacle resting at the bottom Fig 6.1. For the second, the obstacle is piercing the free surface Fig 6.4.

Let explain the physical phenomena, and deduce the equations and boundary conditions governing the flow. In what follows, we will consider a stationary irrotational and incompressible flow in a domain $\Omega \subset \mathbb{R}^2$. So we define a stream funtion $\Psi$, such that the velocity $V = (\partial\Psi/\partial y, -\partial\Psi/\partial x)$.

Problem with obstacle at the bottom.

To simplify, we consider the flow uniform at upstream and downstream infinity (see Cahouet [12], Cahouet-Lenoir [11]), and write the equations concerning a flow arround an obstacle resting at the bottom, in adimensional stream function formulation.

$$
\begin{aligned}
\Delta\Psi &= 0 && \Omega \\
\Psi &= 0 && B \\
\partial\Psi/\partial n &= 0 && S_1, S_2 \qquad (6.1) \\
\partial\Psi/\partial n &= a(y) && S
\end{aligned}
$$

where $a(y) = (1 - (2/F*F)*(y-1))^{1/2}$

Let explain in a physical way the meaning of the equations (6.1) as done in [12].

Equation in $\Omega$

Since the flow is irrotational, the vorticity vanishes, so the stream function is harmonic.

Condition at the bottom B.

The bottom B is a streamline, this implies that $\Psi$ is constant on B, we will fix arbitrary its value to zero.

Conditions on lateral boundaries.

The flow is uniform at upstream, and dowstream infinity. If the domain is large enough compared to the obstacle, So the homogenous Newman conditions seem reasonable. This is only valid when the Froude number F in (6.1) is greather than 1. In this case the problem (6.1) is variational and symetrical. For F<1 (6.1) is not any more variational, and non symetrical. So we will get more complicated boundary conditions on the lateral sides. This is described in detail in (12).

Conditions on the free surface.

Along the free surface S, we dispose on two boundary conditions . The first one is deduced from the Bernouilli equation, and is related to the continuity of the pressure through the free surface, this is given in (6.1), where F is the Froude number and $y(x)$ the adimensional equation of S. The second condition explains the fact that S is a streamline, so $\Psi$ is constant on S. The constant is fixed by the corresponds to Fig 6.3 with F=0.4 and the amplitude of the sinusoidal obstacle is 0.4. The gradient method converges after 40 iterations, and the cost decreases from $J_0 = 0.270$ to $J_{40} = 0.12.10^{-4}$. For the search of the water wave in naval hydrodynamical problem, or in a general free surface problem, we should first test the fixed point method, because it's faster and less expensive. The convergence is geometrical, for this method, at each iteration we only need to solve a linear system corresponding to the state problem. In case this method fails, the gradient method may converge, at least for some cases. The gradient method is more expensive, at each iteration, we solve two linear systems, the state and the adjoint state, and the convergence as it is known, is not geometrical.

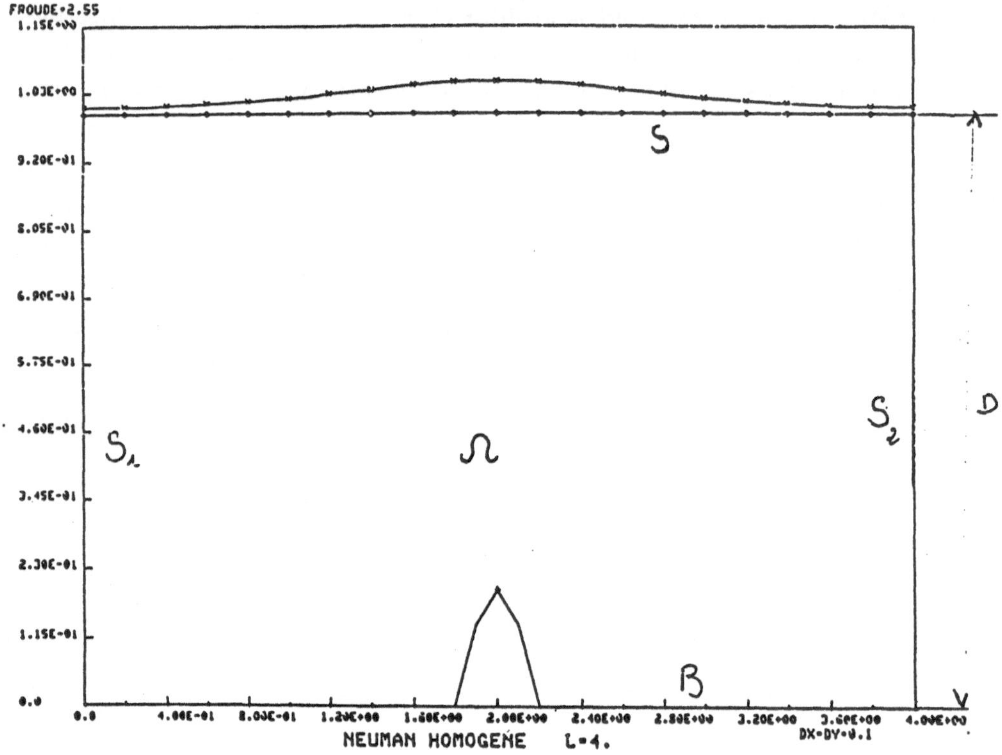
FROUDE=2.55
S
S1
Ω
S2
D
B
NEUMAN HOMOGENE L=4.
DX=DY=0.1

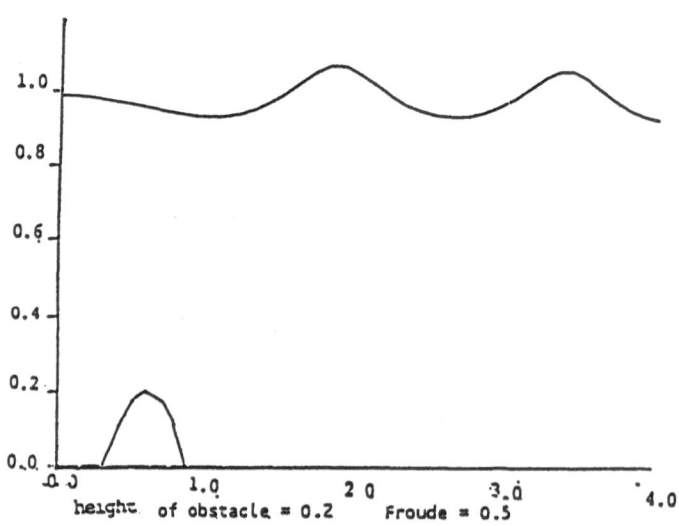

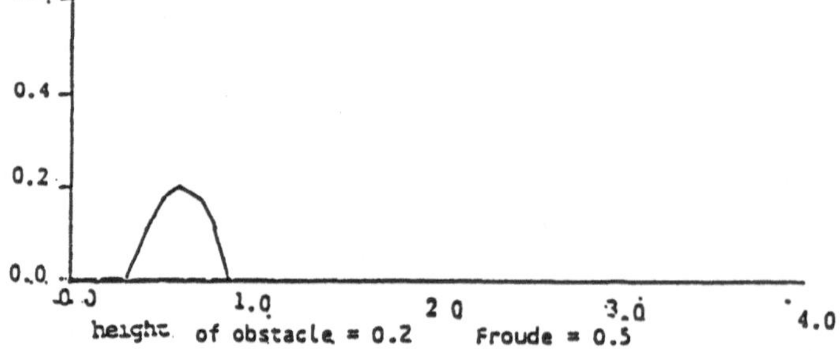
height of obstacle = 0.2
Froude = 0.5

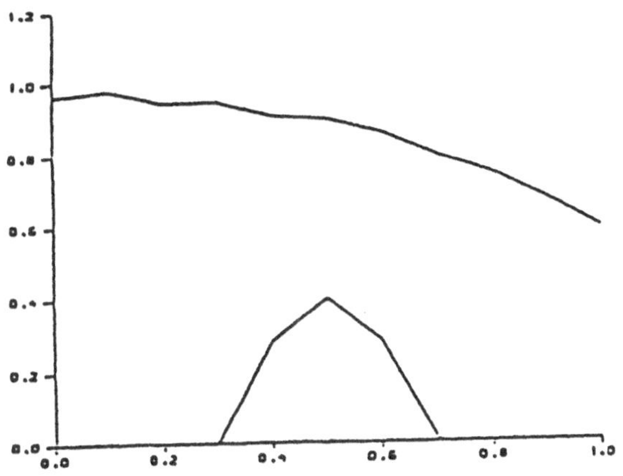


height obst = 0.4 $F = 0.5$

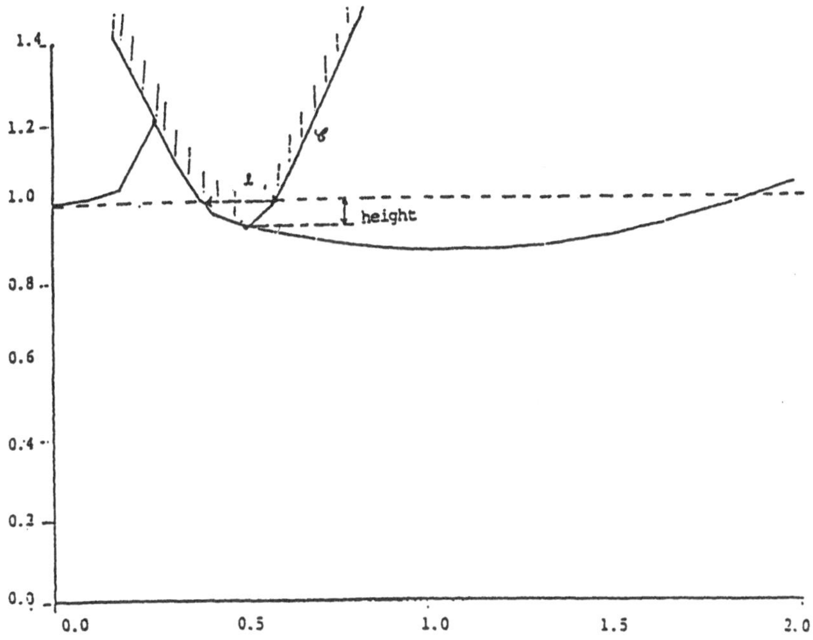


Froude = 0.64 height of obstacle = 0.05 ℓ=0.2

**REFERENCES**


[1] Cea, J., Problems of shape optimal design, dans "Optimization of distributed parameter structures". Vol.2, J. Cea et Ed. Haug, eds.,Sijthoff and Noordhoff, Alphen aan der Rijn, The Netherlands, 1981, 1005-1088.

[2] Delfour, M., Payre, G., Zolesio, J.P., Optimal and suboptimal design of thermal diffusers for communication satelittes, Comptes-rendus du 3ème symposium "Control of distributed parameter systems" Juin 182.

[3] Pi onneau, O., Optimal shape design for elliptic systems, dans "System modeling and optimization" R.F. Drenick et F. Kozin eds., Springer-Verlag 42-66.

[4] Zolesio, J.P., Identification de domaines par déformations, Thèse de doctorat d'état, Nice 1979.

[5] Zolesio, J.P., The matérial derivatives, dans "Optimization of distributed parameter structures", Vol.2, J. Cea et E. Haug, eds.

[6] Zolesio, J.P., Les dérivées par rapport aux noeuds des triangularisations et leurs utilisation en identification de domaines. Ann. Sc. Math. Quebec, 1984, Vol.8 n°1, pp 97-120.

[7] Souli, M., Zolesio, J.P., Semi-discrete and discrete gradient in wave problems. Proceedings of the IFTP WG 7.2 working conference on boundary control and boundary variations. Nice 1986.

[8] NEITTANMAKI: Finite element approximation for optimal shape design. Theory and application WILLEY 1988

[9] Ciarlet Ph. G., The finite element method for elliptic problems, North-Holland 1978.

[10] Buckeley, A., An alternate implementation of Goldfard's minimization algorithm. Math. Progr. 8, 1975 p.207-231.

[11] Cahouet, J., Lenoir, M., Résolution numérique du problème non linéaire de la résistance de vague bidimensionelle. C.R.A.S. 297, 1985.

[12] Cahouet, J., Etude numérique et expérimentale du problème bidimensionel de la résistance de vague non linéaire. Thèse. Paris, 1981.

# On Necessary Optimality Conditions for Optimal Control Problems

*Tilo Staib*
*Institut für Angewandte Mathematik*
*Universität Erlangen–Nürnberg*
*Martensstr. 3 , 8520 Erlangen*
*West Germany*

## 1 Introduction and Definitions

Optimal control problems, where optimality is understood as optimality of a given cost functional, can be regarded as constrained optimization problems, where the equality constraint is given by the state equation of the control problem.

Consequently it is an old idea to derive necessary optimality conditions for control problems as e.g. maximum principles, and adjoint equations from necessary conditions of constrained (nonlinear) optimization, i.e. Kuhn-Tucker Theorems.

That this approach can be useful has been demonstrated successfully for optimal control problems with ODE-state equation by Dubovitsky-Miljutin and others since about 30 years.

Control Problems with PDE state equation too can be treated by methods from nonlinear optimization, in particular Kuhn-Tucker theorems as the works of *Mackenroth*, *Barbu* and *Troeltzsch* ([27], [19], [2] ,among others) indicate. Remark 3.5 gives a short survey.

The generality of this method of approach has the advantage that it is applicable to many nonlinear control problems, with or without state constraints. But this generality is also responsible for the drawback of the method: One always has to verify the central hypotheses of the Kuhn-Tucker Theorem, i.e. the constraint qualifications. This generally, for control problems, amounts to proving existence of a solution of the *linearized* state equation. The purpose of this paper is to put this in an more exact way.

In section 2, a general Kuhn Tucker Theorem is derived and a discussion and comparison of the constraint qualifications used to establish this theorem is given. In section 3 it will be shown, how it can be used to derive necessary optimality conditions for nonlinear optimal control problems. The example considered there is a control problem where the cost functional may be nonsmooth and the equation of the state of the system is a nonlinear parabolic PDE.

**Definition 1.1** Let $f : X \to \mathbb{R}$ be locally Lipschitz continuous at $x \in X$ with a constant $\lambda$, i.e. let assume there is $|f(u)-f(v)| \le \lambda\|u-v\|$ for all $u, v$ in a neighborhood of $x$. Then the Clarke-derivative is given by

$$f^\circ(x,h) = \lim_{\epsilon\to 0_+} \sup\{\,\theta^{-1}(f(v+\theta h)-f(v)) \mid 0<\theta\le\epsilon, \|v-x\|\le\epsilon\,\}.$$

This limit exists for every $h \in X$ and is always sublinear in $h$ (see e.g. [6]). The *directional derivative* of $f$ (if it exists) is defined as

$$df(x,h) \quad := \quad \lim_{\theta\to 0_+} \theta^{-1}(f(x+\theta h)-f(x)).$$

This derivative is in general nonconvex, so we define the gereralized derivative of $f$ as

$$Df(x,h) \quad = \quad \begin{cases} df(x,h), & \text{if } df(x,\cdot) \text{ is sublinear} \\ f^\circ(x,h) & \text{else.} \end{cases} \tag{1}$$

The subdifferential of $f$ then is

$$\partial f(x) \quad := \quad \{\,x^* \in X^* \mid x^*(h) \le Df(x,h)\ \forall\, h \in X\,\}.$$

If $Z$ is a topological vector space and $g : X \to Z$ a given mapping then the *directional derivative* of $g$ (if it exists) is defined as

$$Dg(x,h) \quad := \quad \lim_{\theta\to 0_+} \theta^{-1}(g(x+\theta h)-g(x)).$$

If ist is *linear* in $h$ we will call $Dg(x,\cdot)$ the Gâteaux-derivative of $g$. In this paper we will only use mappings $g$ where $Dg$ is at least sublinear, so in this case (if additionally $Z$ is an order complete vector lattice with a semi ordering $\le$ see e.g. [4] or [25]) we can define the subdifferential of $g$ as

$$\partial g(x) \quad := \quad \{\,L \in \mathcal{L}(X,Y) \mid L(h) \le Dg(x,h)\ \ \forall\, h \in X\,\}. \tag{2}$$

Here $\mathcal{L}(X,Y)$ denotes the set of linear continuous operators from $X$ to $Y$. It is also possible to define a kind of Clarke's derivative of $g$ which is always sublinear, by using a different notion of Lipschitz continuity (see e.g. [15] or [25]).

To approximate sets, we will employ the concept of Bouligand's tangent cone, usually referred to as contingent cone.

**Definition 1.2** Let $X$ be a normed space, $S \subseteq X$ be a nonempty subset and $x \in X$ be given. Then the *contingent cone* of $S$ at $x$ is defined as

$$T(S,x) := \left\{ h \in X \;\middle|\; \begin{matrix} \exists\,\{h_n\}_{n\in\mathbb{N}} \subseteq X, & h = \lim_{n\to\infty} h_n, \\ \exists\,\{t_n\}_{n\in\mathbb{N}} \subseteq \mathbb{R}_+ & t_n \to 0, \end{matrix} \quad x + t_n h_n \in S\ \ \forall n \in \mathbb{N} \right\}.$$

If $S$ is starshaped at $x$ or even convex, then we will use the cone

$$K(S,x) \quad = \quad \bigcup\{\,\lambda(s-x) \mid \lambda \in \mathbb{R}_+,\ s \in S\,\},$$

(it is well known, that the closure of this cone coincides with $T(S,x)$ if $S$ is starshaped at $x$.)

# 2 Necessary Optimality Conditions

**Theorem 2.1** *Let $X$ be a normed vector space, and $\emptyset \neq S \subseteq X$. Assume that $f : X \to \mathbb{R}$ is locally Lipschitz at $x \in X$ with constant $\lambda$ and suppose that $f(x) = \min\{ f(s) \mid s \in S \}$. Then the following assertions hold:*

*(a)* $Df(x,h) \geq 0 \quad \forall\, h \in T(S,x)$,
*and, if the directional derivative of $f$ exists then*
$df(x,h) \geq 0 \quad \forall\, h \in T(S,x)$.

*(b) For every convex cone $K \subseteq T(S,x)$ we have*

$$0 \quad \in \quad \partial f(x) \quad - \quad K^*. \tag{3}$$

**Proof.**

(a) Suppose that there is some $h \in T(S,x)$ with the property $Df(x,h) < 0$, hence there is $\delta > 0$ with $Df(x,h) + 4\delta < 0$. Set

$$S(\epsilon) := \sup\{ \theta^{-1}(f(v + \theta h) - f(v)) \mid 0 < \theta \leq \epsilon, \|v - x\| \leq \epsilon \}$$

By the definition of $Df$ there is $\epsilon > 0$ with $|S(\epsilon) - Df(x,h)| < \delta$. Because of $h \in T(S,x)$ there are sequences $\{h_n\}_{n \in \mathbb{N}} \subseteq X$ and $\{\theta_n\}_{n \in \mathbb{N}} \subseteq \mathbb{R}_+$ for which

$$0 = \lim_{n \to \infty} \theta_n, \quad h = \lim_{n \to \infty} h_n \text{ and } x + \theta_n h_n \in S \quad \forall\, n \in \mathbb{N}.$$

Hence there is a number $n_0 \in \mathbb{N}$ with $\|h_n - h\| \leq \delta/\lambda$ and $\theta_n < \epsilon$ for $n > n_0$. Furthermore because $f$ is Lipschitz continuous ,

$$\|\theta_n^{-1}(f(x + \theta_n h) - f(x)) - \theta_n^{-1}(f(x + \theta_n h_n) - f(x))\| \leq \lambda\|h - h_n\|,$$

and hence

$$\theta_n^{-1}(f(x + \theta_n h_n) - f(x)) \quad \leq \quad \theta_n^{-1}(f(x + \theta_n h) - f(x)) + \delta \tag{4}$$

$$\leq S(\epsilon) + \delta \quad \leq \quad Df(x,h) + 2\delta < 0 \tag{5}$$

for all $n > n_0$ in contradiction to $f(x)$ being minimal. The same argument applies to the directional derivative $df(x,h)$ if it exists, but we can omitt $S(e)$ from the proof.

(b) To prove (3) set $p(h) := Df(x,h)$. Observe that $p : X \to \mathbb{R}$ is sublinear and continuous. Hence the set

$$epi(p) := \{ (h,r) \in X \times \mathbb{R} \mid p(h) \leq r \}$$

is convex and has nonempty interior in $X \times \mathbb{R}$ . Because statement (a) implies $p(k) \geq 0 \;\; \forall\, k \in K$ and hence

$$int\, epi(p) \cap \{ (k,\alpha) \in X \times \mathbb{R} \mid k \in K, \alpha \leq 0 \} \quad = \quad \emptyset. \tag{6}$$

By application of a separation theorem to this intersection we obtain the existence of a linear continuous functional $(x_1^*, \beta) \in (X \times \mathbb{R})^* = X^* \times \mathbb{R}$, $(x_1^*, \beta) \neq 0$ satisfying

$$x_1^*(h) + \beta r \quad \leq \quad 0 \quad \leq \quad x_1^*(k) + \beta\alpha \quad \forall\, h \in X, k \in K, r \geq p(h), \alpha < 0. \quad (7)$$

These inequalities imply $\beta \leq 0$. If $\beta = 0$ then (7) yields $x_1^* = 0$ in contradiction to $(x_1^*, \beta) \neq 0$. Hence we have $\beta < 0$. Set $x^* := -\frac{1}{\beta}x_1^*$ and conclude from (7)

$$x^* \quad \in \quad -K^* \quad \text{und} \quad x^*(h) \quad \leq \quad p(h) \quad \forall\, h \in X.$$

From the latter inequality follows $x^* \in \partial f(x)$ hence (3).

We now consider the situation that the set $S$ is given by explicit constraints. Let $Z_1$,$Z_2$ be partially ordered, $g_1: X \to Z_1$, $g_2$: $X \to Z_2$ be given mappings and suppose that $S_0 \subseteq X$ $D_1 \subseteq Z_1$ $D_2 \subseteq Z_2$ are nonempty sets. For $i = 1, 2$ we define

$$\begin{aligned}
S_i &:= \{\, x \in X \mid g_i(x) \in D_i \,\}, && (8)\\
S &:= S_0 \cap S_1 \cap S_2, && (9)\\
P_i &:= \overline{co\,T}(D_i, g_i(x)), && (10)\\
P_i^* &:= \{\, z^* \in Z_i^* \mid z^*(p) \geq 0,\ \forall\, p \in P_i \,\}, && (11)\\
J_i &:= \{\, h \in X \mid L(h) \in P_i, \forall\, L \in \partial g_i(x) \,\}, && (12)\\
H_i &:= \{\, x^* \in X^* \mid \exists\, p^* \in P_i^*, L \in \partial g_i(x) \ mit\ x^* = p^* \circ L \,\}, && (13)\\
J_0 &:= \overline{co\,T}(S_0, x). && (14)
\end{aligned}$$

The sets $P_i^*$ are multiplier sets and the sets $H_i$ are sets of compositions of multipliers and subdifferentials. If in particular inequality and equality constraints are given, take $D_1 := -C$ and $D_2 = 0$, where $C \subseteq Z_1$ is a convex cone representing the inequality constraints.

**Lemma 2.2** *Let $K \subseteq X$ be a convex cone satisfying*

$$K \subseteq T(S, x) \quad \textit{and} \quad K^* = J_0^* + J_1^* + J_2^* \quad (15)$$

*and assume that the sets $H_1, H_2$ are $w^*$-closed. Then*

$$K^* \quad \subseteq \quad co\,H_1 + co\,H_2 + J_0^*. \quad (16)$$

**Proof.** In order to prove (16) we show $J_i^* = \overline{co\,H_i}^{w^*}$. Suppose that $x^* \in J_i^*$ with $x^* \notin \overline{co\,H_i}^{w^*} =: B$. By applying a separation theorem to the $w^*$–compact set $\{x^*\}$ and the $w^*$–closed set $B$ we conclude the existence of $h \in X$

$$x^*(h) \quad < \quad 0 \quad \leq \quad v^*(h) \quad \forall\, v^* \in co\,H_i.$$

This implies $L(h) \in P_i^{**} = P_i \ \ \forall\, L \in \partial g_i(x)$ since $P$ is a closed and convex cone and hence $h \in J_i$ for i=1,2. But this implies $x^*(h) \geq 0$ in contradiction to $x^*(h) < 0$ from the inequality above. Let now be $x^* \in co\,H_i$ and $x^* \notin J_i^*$. Linear separation then yields the existence of some $h \in X$ with the property

$$x^*(h) \quad < \quad 0 \quad \leq \quad v^*(h) \quad \forall v^* \in J_i^*.$$

Because of $J_i^{**} = J_i$ this implies $h \in J_i$ which means $L(h) \geq 0 \;\; \forall L \in \partial g_i(x)$ and hence $z^* \circ L(h) \geq 0 \; \forall z^* \circ L \in H_i$. But this inequality is a contradiction to the separation inequality, since $x^* \in co\,H_i$ The last statement of the theorem follows from $coH_i$ being $w^*$–closed. ◇

The assumptions of this central Lemma are called *constraint qualifications* and they will, together with the following simple facts imply necessary optimality conditions for inequality and equality constrained problems.

**Lemma 2.3** *For* $i = 1,2$ *:*

*(a)* $H_i \quad = \quad \bigcup\{\, L^*(P_i^*) \mid L^*$ *adjoint of* $L \in \partial g_i(x)\,\}$

*(b) If* $\partial g_i(x) = \{L\}$ *i.e. if the subdifferential consists of a single element, then the set* $H_i$ *is always convex.*

**Proof.** Immediate.

**Theorem 2.4** *Let* $C \subseteq Z_1$ *be a convex, closed cone and assume that* $f : X \to \mathbb{R}$ *is Lipschitz-continuous, that* $g_1 : X \to Z_1$ *and* $g_2 : X \to Z_2$ *have a sublinear directional derivative and a subdifferential. Let the assumptions of Lemma 2.2 be satisfied.*

*(a) Then there is* $f^* \in \partial f(x)$ *and for* $j = 1,\ldots,n$ *and* $k = 1,\ldots,n$ *there are linear operators* $L_j^1 \in \partial g_1(x)$, $L_k^2 \in \partial g_2(x)$, *and real numbers* $\alpha_j \geq 0$ , $\beta_k \geq 0$ *with* $\sum_j \alpha_j = \sum_k \beta_k = 1$,
*and linear functionals (i.e multipliers)* $z_{1,j}^* \in P_1^*$, $z_{2,k}^* \in P_2^*$, *with*

$$f^*(h) - \sum_j \alpha_j z_{1,j}^* \circ L_j^1(h) - \sum_k \beta_k z_{2,k}^* \circ L_k^2(h) \quad \geq \quad 0 \quad \forall\, h \in J_0 \tag{17}$$

*(b) (Smooth constraints) If we consider especially* $D_1 = -C, D_2 = 0$ *i.e.* $f(x) = \min\{x \in S_0 \mid g_1(x) \in -C, g_2(x) = 0\}$ *and asssume that* $g_1, g_2$ *are Gâteaux-differentiable , then there is* $f^* \in \partial f(x)$ *and there exist multipliers* $z_1^* \in -C^*, z_2^* \in Z_2^*$ *such that* $z_1^*(g_1(x)) = 0$ *and*

$$f^*(h) - z_1^* \circ Dg_1(x,h) - z_2^* \circ Dg_2(x,h) \quad \geq \quad 0 \quad \forall\, h \in J_0 \tag{18}$$

*hold.*

**Proof.** Theorem 2.1 together with Lemma (2.2) implies

$$0_{X^*} \quad \in \quad \partial f(x) \quad - \quad K^* \quad \subseteq \quad \partial f(x) \quad - (co\,H_1 + co\,H_2 + J_0^*).$$

Considering the definition of the sets $H_1, H_2$ we obtain (a) (note that $x^* \in J_0^*$ is equivalent to $x^*(h) \geq 0 \; \forall\, h \in J_0$.) To prove statement (b), we note that by Lemma 2.3

(b) we obtain that the sets $H_1, H_2$ are convex since $g_1, g_2$ are Gâteaux-differentiable ($\partial g_i(x) = Dg_i(x,\cdot)$) Moreover we have to consider here $D_1 = C, D_2 = 0$ and hence it follows with standard arguments $P_1^* = (K(-C, -g_1(x))^* = -C^*$, $z_1^*(g_1(x)) = 0$ and $P_2^* = Z_2^*$. So, by the definition of $H_1, H_2$ we obtain the result (18)

Remark: The convex combinations of multipliers and derivatives seems to be typical for non-Gâteaux-differentiable problems and have their counterpart in the "Hamiltonian multipliers" Clarke's calculus of variations [6]. The assumptions of Lemma 2.2 used in the previous Theorem are abstract constraint qualifications (CQ) of a type introduced by *Guignard* in [10] . It will turn out in the sequel, that they are implied by the classical Slater conditions but that they are more general than those. In fact, as it has been demonstrated by *Bazaraa/Shetty* [3] even in finite dimensional spaces, Guignard's CQ is the weakest among several known CQ's, moreover Guignard's CQ does not assume the existence of interior points of $C$ and $S_0$ required by Slater's CQ (see also *Penot* [22]).

We will use the following Lemma.

**Lemma 2.5** *(a) For every convex cone $I$ with $\text{int}\, I \neq \emptyset$ the dual cone $I^*$ is $w^*$-locally compact (or, equivalently, has a $w^*$-compact base).*

*(b) If $I^*$ and $J^*$ are convex closed cones and if $I^* \cap -J^* = \{0\}$ then the set $I^* + J^*$ is $w^*$-closed.*

**Proof.** The statement (a) is due to *Ky Fan* see e.g. [29]. The statement (b) is due to *Dieudonnée* (see e.g [13] or [12] Lemma 15 d.) ◇

**Theorem 2.6** *Assume that $\text{int}\, S_0 \neq \emptyset$ and $\text{int}\, C \neq \emptyset$. Define $I_0 := \text{int}\, J_0$ and $I_1 := \{ h \in X \mid Dg_1(x,h) \in \text{int}\, P_1 \}$ and $K := I_0 \cap I_1 \cap J_2$. If the cone $K$ satisfies the Slater condition $K \neq \emptyset$ and if additionally the tangential inclusion*

$$J_2 \subseteq T(S_2, x) \tag{19}$$

*holds, then $K$ satisfies Guignard's CQ, i.e. $K$ is a convex cone with the properties*

$$K \subseteq T(S,x), \quad K^* = (J_0 \cap J_1 \cap J_2)^*, \quad K^* = J_0^* + J_1^* + J_2^* \tag{20}$$

**Proof.** It is standard (convexity arguments) to prove the following statements:

$$\overline{I_0} = J_0, \quad \overline{I_1} = J_1, \tag{21}$$

$$K^* = (J_0 \cap J_1 \cap J_2)^*, \quad \overline{K} = J_0 \cap J_1 \cap J_2, \tag{22}$$

$$I_0 \cap I_1 \cap J_2 \subseteq T(S,x). \tag{23}$$

We will now deduce the equality $K^* = J_0^* + J_1^* + J_2^*$. Since

$$J_0^* + J_1^* + J_2^* \subseteq K = (J_0 \cap J_1 \cap J_2)^* \subseteq \overline{J_0^* + J_1^* + J_2^*}^{w^*}, \tag{24}$$

we only have to show that the set $J_0^* + J_1^* + J_2^*$ is $w^*$-closed, which will be done by Lemma 2.5. We now observe that $K \neq \emptyset$ implies $I_k^* \cap -J_2^*$ $(k = 0,1)$ and that $I_0^*, I_1^*$ are $w^*$-locally compact. Hence repeated application of (a),(b) of Lemma 2.5 gives $w^*$-closedness of the set $(J_0 \cap J_1 \cap J_2)^*$. ◇

**Remark 2.7** (a) The Slater CQ is sometimes stated in the equivalent form

$$\exists\ h \in int\, S_0\ :\ Dg_1(x,h) \in int\, K(-C, g_1(x)), \qquad Dg_2(x,h) = 0. \tag{25}$$

(b) If no equality constraints are present and we just consider affine-linear inequality constraints i.e. the feasible set is $S := \{\, x \in X \mid g(x) = Ax - b \in -C \,\}$, $A$ linear, then for

$$K \;:=\; J \;=\; \{\, h \in X \mid Dg(x,h) = Ah \in K(S,x) \,\}$$

one does not need Slater CQ because the inclusion $K \subseteq T(S,x)$ is always true, hence this hypothesis of Theorem 2.4 holds. ( Note that $h \in J$ implies $\lambda Ah \in -C - g(x)\ \forall \lambda \geq 0$ hence $g(x + \lambda h) = Ax + \lambda Ah - b \in -C$ if $g(x) \in -C$ which is true since $x$ is feasible. Hence $h \in T(S,x)$.)

(c) The hypothesis $J_2 \subseteq T(S_2, x)$ (19 ) for the equality constraints, called here tangential inclusion is probably the hardest one to prove, it is the essence of the famous Theorem of *Ljusternik.* In the next theorem we cite two modern references which extend the original one. But apart from these general conditions for (19 ) to hold, this condition can sometimes be verified *directly* if $g_2$ is an integral equation or a differential equation (see [25]).

**Theorem 2.8** *Let $X, Z$ be Banach spaces, $Z_0 \subseteq Z$ a given set and $g : X \to Z$ a mapping, $x \in S := \{\, x \in X \mid g(x) \in Z_0 \,\}$ and assume that $g$ is continuous in $x$. Let $U(x)$ denote some neighborhood of $x$.*

*(a) (Penot [23],Theorem 3.1 ) If the strict directional derivative $Dg(u,\cdot)$ of $g$ exists for all $u \in U(x)$, and is regularly surjective and continuous (i.e. it's inverse is Lipschitz continuous) then*

$$J := \{\, h \in X \mid Dg(x,h) \in T(Z_0, g(x)) \,\} \quad \subseteq \quad T(S,x). \tag{26}$$

*$T$ as always denotes Bouligands tangent cone. (If $g$ is Gâteaux-differentiable in $U(x)$ and $Dg(x,\cdot)$ is continous and surjective, then it is regularly surjective).*

*(b) Kirsch/Warth/Werner,[14] Theorem 1.13 We consider especially $Z_0 = 0$. If $g$ is Gâteaux-differentiable on some $U(x)$ and $Dg(x,\cdot)$ is surjective then*

$$J := \{\, h \in X \mid Dg(x,h) = 0 \,\} \quad \subseteq \quad T(S,x). \tag{27}$$

Part (a) of this theorem obviously can be used for inequality and for equality constraints. Another Theorem of this kind is given by *Frankowska* in [8].
We now discuss various sufficient conditions which imply that $H_1, H_2$ is $w^*$-closed.

**Theorem 2.9** *Let $X, Z_1, Z_2$ be Banach spaces.*

*(a) If $H_2 \;=\; \bigcup \{\, L^*(P_2^*) \mid L^* \text{ adjoint of } L \in \partial g_2(x) \,\} = X^*$, then $H_2$ is convex and $w^*$-closed. If one $L \in \partial g_2(x)$ is a closed operator (i.e. $L(X)$ is closed) and injective, then $H_2 = X^*$, and this set is convex and $w^*$-closed.*

(b) *Let now* $\partial g_2(x) = \{L\}$. *Then the set* $H_2$ *is always convex and the following holds.*

  (i) *If* $Z_2$ *is finite dimensional then the set* $H_2$ *is always* $w^*$*-closed.*

  (ii) *If* $L$ *is closed then* $H_2$ *is* $w^*$*-closed and norm-closed.*

(c) *If* $\partial g_1(x) = \{L\}$ *and* $\operatorname{int} C \neq \emptyset$ *then the set* $H_1 = L^*(-C^*)$ *is a convex cone with* $w^*$*-compact base.*

**Proof.**

(a) Because $L$ is injective it follows that $\overline{L^*(Z_2^*)}^{w^*} = X^*$, (see e.g. [21] Theorem 4.12 and Corollaries) and closedness of $L$ is equivalent to closedness of $L^*$ in the norm-topology and in the $w^*$-topology (see e.g. [21] Theorem 4.14.) hence $L^*(Z_2^*) = X^* = H_2$ and this last set is obviously convex and $w^*$-closed.

(b) (i) The set $H_2 = L^*(Z_2^*)$ is a finite dimensional subspace and hence $w^*$-closed.

  (ii) Same argument as in (a).

(c) See [29]. ◇

# 3 Optimal Control Problem

In this section we discuss with the aid of an (abstract) example of an optimal control problem having a nonlinar parabolic equation of state how the necessary optimality conditions of Theorem 2.4 can be used to derive necessary conditions for this control problem.

Let $\Omega \subseteq \mathbb{R}^n$ be bounded with boundary $\partial\Omega$, let $I = [0, b] \subseteq \mathbb{R}_+$ be an interval and $W$ and $U$ be spaces of functions on $\Omega \times I$. We consider the following general problem of optimal (distributed) control on $\Omega \times I$

(**Problem (P)**): Minimize the functional $f : W \times U \to \mathbb{R}$, $f = f_1 + f_2$ given by

$$f_1(y, u) := f_1(y(b)), \qquad f_2(y, u) := \int_I F(y(t), u(t), t)\, dt$$

over all functions $y(x,t), u(x,t)$, $(x,t) \in \Omega \times I$, $(y,u) \in W \times U$ satisfying the following nonlinear parabolic equation with initial conditions and homogeneous Dichlet boundary conditions.

$$
\begin{aligned}
y_t(x,t) - \mathbf{div}[G_2(y(x,t), \nabla y(x,t))] - B(u(x,t)) &= 0 \qquad (x,t) \in \Omega \times I \quad (28)\\
y(x,0) - y_0(x) &= 0 \qquad x \in \Omega, \quad (29)\\
y(x,t) &= 0, \qquad (x,t) \in \partial\Omega \times I \quad (30)\\
u \in U_0 \subseteq U = L^s(\Omega \times I),\ 1 \leq s \leq \infty &\quad \text{given.} \quad (31)
\end{aligned}
$$

The operators $\mathbf{div}[..]$ and $\nabla$ are with respect to $x$, and $G_2$ and $B$ are given expressions which may be nonlinear. (We will not distinguish here between the mapping $G_2 : \mathbb{R} \times \mathbb{R}^n \to \mathbb{R}^n$ and it's associated Nemytski operator.) The solutions of (28) -(30)

should be understood as weak solutions, i.e. we use reflexive function spaces $H$,$V$ of functions over $\Omega$ with the property

$$V \subseteq H \subseteq V^*, \tag{32}$$

( for example let $V = W_0^{1,r}(\Omega)$ for some $r > 1$ and $H$ some $L^p(\Omega)$ space such that (32) holds, which space one finally has to choose depends on the existence theory of (28 - 30) for an explicitly given $G_2$). Moreover we use

$$W \subseteq W^{1,p,q}(I;V,V^*) = \{\, y \mid \|y\|_V \in L^p(I), \|y_t\|_{V^*} \in L^q(I), p^{-1} + q^{-1} = 1 \,\}. \tag{33}$$

Then with the abbreviation $( z \mid v ) = \int_\Omega z(x)v(x)\,dx$, we see (by using Green's formula) that the equations

$$\Gamma(y,u,v)(t) \quad := \quad ( y_t \mid v ) + \int_\Omega G_2(y,\nabla y)\cdot\nabla v \; + B(u)v\,dx \quad = \quad 0 \quad \forall v \in V. \tag{34}$$

represent (28) and (30). So we have the equality constraint

$g_2(y,u) = (g_2^1(y,u), g_2^2(y,u)) = 0$ where $g_2^1$ given by (34) and $g_2^2$ is defined by (29). Hence we set

$$g_2(y,u) = \begin{pmatrix} g_2^1(y,u) \\ g_2^2(y,u) \end{pmatrix} = \begin{pmatrix} y_t - \mathbf{div}[G_2(y,\nabla_x y)] - B(u) \\ y(x,t) - y_0(x) \end{pmatrix} = \begin{pmatrix} 0 \\ 0 \end{pmatrix}$$

and apply Theorem 2.4 to the problem

$$\text{minimize } f(y,u) \text{ subject to } g_2(y,u) = 0,\ u \in U_0.$$

This is possible if we assume that

(A1) Problem (P) has at least one solution, i.e. there exists one optimal state-control pair $(y,u)$.

(A2) The constraint qualifications of Theorem 2.4 are satisfied. This assumption can be reduced to the assumption that the linearized operator $D_y g$ is surjective, which amounts to the solvability of a *linear parabolic* equation (see below).

Both assumptions have to be verified if one wants to apply the necessary condition to a concrete problem. In e.g [24] the existence of solutions of quasilinear problems is investigated. With "Linearization " we mean the following

(A0) (Differentiability) Assume that $G_2 : \mathbb{R}\times\mathbb{R}^n \to \mathbb{R}^n$ and $B : \mathbb{R} \to \mathbb{R}$ is directionally differentiable in these spaces and with bounded difference quotients (this holds e.g. if $G_2, B$ are lipschitz continuous) with *linear* directional derivatives. From $f$ we assume Lipschitz-continuity (in $H$-norm.)

Then from (A0) by Lebesgues' Theorem on dominated convergence we can deduce the following : For directions $(h,k) \in W \times U$ : the Gâteaux-derivative $Dg_2((y,u),(h,k))$ exists and is given by

$$\begin{aligned} \langle\, D_y g_2^1(y,u)(h) \mid v \,\rangle_V &= \int_\Omega h_t v + D_y G_2((y,\nabla y);(h,\nabla h)) \cdot \nabla v \, dx & (35) \\ &= \int_\Omega h_t v + \{D_1 G_2(w)h + D_2 G_2(w) \cdot \nabla h\} \nabla v \, dx & (36) \\ &=: (\, h_t \mid v \,)_H + \langle\, A(t)h \mid v \,\rangle_V \;\; \forall v \in V & (37) \end{aligned}$$

where we set $w := w(x,t) := (y(x,t), \nabla y(x,t))$, and

$$D_1 G_2(y,\nabla y) := \frac{\partial}{\partial y} G_2(y,\nabla y) \quad D_2 G_2(y,\nabla y) := \frac{\partial}{\partial(\nabla y)} G_2(y,\nabla y). \tag{38}$$

Equation (37) should be understood as the Definition of the linear Operator $A(t) : V \to V^*$. So the operator $D_y g_2^1(y,h) = h_t + A(t)h$, is linear in $h$. The other derivatives of $g_2$ are obvious, we obtain :

$$Dg_2(h,k) = \begin{pmatrix} D_y g_2^1(h) & +D_u g_2^1(k) \\ D_y g_2^2(h) & +D_u g_2^2(k) \end{pmatrix} = \begin{pmatrix} h_t + A(t)h & +D_u B(k) \\ h(0,x) & +0 \end{pmatrix}. \tag{39}$$

According to [6] the derivatives of $f$ read as follows: There are linear functionals ( derivatives)

$$\begin{aligned} \phi_y \in \partial_y f_1(y(b)), \;\; \phi_y(h) &= \langle\, \phi_y \mid h \,\rangle_H & (40) \\ \psi_y \in \partial_y F(y(t),u(t)), \;\; \psi_y(h) &= \int_I \langle\, \psi_y \mid h \,\rangle_V dt & (41) \\ \psi_u \in \partial_u F(y(t),u(t)), \;\; \psi_y(k) &= \int_I \langle\, \psi_u \mid k \,\rangle_U dt & (42) \end{aligned}$$

for all $h \in W, k \in U$. If $f$ is additionally Gâteaux-differentiable then we can take even $\phi_y = D_y f_1(y,\cdot)$ and so on.

We can now replace (A2) by a more specific assumption :

(A2') Assume that $int U_0 \neq \emptyset, D_u B(U_0) \subseteq Z_2$ holds and that the operator $A(t) : V \to V^*$ (i.e. the linearization of $G_2$ ) has such properties that $\forall z = (z_1,z_2) \in Z_2 = L^q(I;V^*) \times H$, there is $h \in W$ with $h = 0$ on $\partial\Omega \times I$ and

$$\begin{aligned} D_y g_2^1(h) &= h_t + A(t)h = z_1 & (43) \\ D_y g_2^2(h) &= h(0,x) = z_2, & (44) \end{aligned}$$

i.e. a weak solution of the linear parabolic problem exists for all appropriate functions $(z_1,z_2)$, i.e. we assume that $D_y g(y,\cdot) : W \to Z_2$ is surjective.

From this assumptions we can now deduce that the constraint qualifications are satisfied.

**Lemma 3.1** *Assume (A2') . Then the constraint qualifications of theorem 2.4 are fulfilled.*

**Proof.** In the following read $Dg := D_{y,u}g$. Set $K = intJ_0 \cap J_2$, where $intJ_0 = int(K(W,w) \times K(U_0,u)) = W \times intK(U_0,u)$, (since $W$ is a linear space we have $K(W,w) = W$) and $J_2 = \{ (h,k) \mid Dg_2((y,u);(h,k)) = 0 \}$. According to Theorems 2.6 , 2.8, 2.9 we have to show that $K \neq \emptyset$ holds and that the mapping $Dg((y,u),\cdot)$ is surjective, since then we can deduce $J_2 \subseteq T(S_2,(y,u))$ (Theorem 2.8(a)) where $S_2 := \{ (y,u) \in W \times U \mid g_2(y,u) = 0 \}$. By (A2') the operator $D_y g_2 : W \to Z_2$ is surjective hence it follows that $Dg_2 : W \times U \to Z_2$ is surjective and in particular a closed operator. Hence by Theorem 2.8 we have $J_2 \subseteq T(S_2,(y,u))$ and by Theorem 2.9 the set $H_2$ is $w^*$-closed. Now we show that $K \neq \emptyset$ holds. By (A2') we have for even for every $k \in int\,K(U_0,u)$ that $z_2 := D_u B(k) \in Z_2$ and hence that there exists a solution $h \in W$ of the linear equation

$$Dg_2(h,k) = \begin{pmatrix} h_t + A(t)h + D_u B(u,k) \\ h(0,x) + 0 \end{pmatrix} = \begin{pmatrix} 0 \\ 0 \end{pmatrix}.$$

This implies that there exists

$$(h,k) \in intJ_0 = W \times int\,K(U_0,u) \text{ such that } Dg_2((y,u);(h,k)) = 0,$$

which is exactly $K \neq \emptyset$.

This Lemma now permits the application of Theorem 2.4 in order to to derive the necessary optimality conditions for Problem (P) under (A1), (A2').

**Theorem 3.2** (Necessary Conditions) *Assume that hypotheses (A0), (A1) and (A2') hold, such that $(y,u)$ is a solution of (P). Then there exist derivatives (see (37) to (42)) and Multipliers $z_2^* = (z_{2,1}^*, z_{2,2}^*) \in Z_2^* = (L^q(I;V^*) \times H)^* = L^p(I;V) \times H^*$ i.e. $z_{2,1}^* \in L^p(I;V)$, $z_{2,2}^* \in H^*$ such that the following adjoint equation and variational inequality hold.*

$$\psi_y(t) = -\frac{\partial}{\partial t} z_{2,1}^*(t) + [A(t)]^* z_{2,1}^*(t) \quad t \in I \tag{45}$$

$$\phi_y = z_{2,1}^*(b) \tag{46}$$

$$z_{2,2}^* = z_{2,1}^*(0) \tag{47}$$

*where*

$$\langle A(t)h \mid v \rangle_V = \int_\Omega [D_1 G_2(y,\nabla y)h(x) + D_2 G_2(y,\nabla y)\cdot \nabla h(x)]\nabla v(x)\,dx \tag{48}$$

$$\int_I \left\langle \psi_u - [D_u B]^* z_{2,1}^* \;\middle|\; k(t) \right\rangle_H dt \geq 0 \quad \forall k \in K(U_0,u) = \cup_{\lambda \geq 0}\lambda(U_0 - u) \tag{49}$$

*If especially $U_0 = \{ u \in L^\infty(\Omega \times I) \mid |u(x,t)| \leq 1 \text{ a.e} \}$ then there holds a pointwise maximum principle (a.e. in $\Omega \times I$.*

$$\left(\psi_u - [D_u B]^* z_{2,1}^*(x,t)\right) u(x,t) = \min_{|\alpha|\leq 1} \left(\psi_u - [D_u B]^* z_{2,1}^*(x,t)\right) \alpha \tag{50}$$

*and a weak Bang-Bang principle*

$$u(x,t) = \begin{cases} 1 & (x,t) : \psi_u(x,t) - [D_u B]^* z_{2,1}^*(x,t) < 0 \\ -1 & (x,t) : \psi_u(x,t) - [D_u B]^* z_{2,1}^*(x,t) > 0. \end{cases} \tag{51}$$

**Proof.** We set $X := W \times U$, $Z_2 = L^q(I; V*) \times H$. Then by the assumptions for the mapping $f : X \to I\!R$, $g_2 : X := W \times U \to Z_2$ and $D_{y,u}g_2 : X := W \times U \to Z$ the constraint qualifications are satisfied. The application of Theorem 2.4 then produces multipliers $z_2^* = (z_{2,1}^*, z_{2,2}^*) \in Z_2^*$ such that for all $h \in W$, $k \in K(U_0, u)$ we have

$$\psi_y(h) + \phi_y(h) - z_{2,1}^* \circ D_y g_2^1(h) - z_{2,2}^* \circ D_y g_2^2(h) \quad = \quad 0 \tag{52}$$

$$\psi_u(k) - z_{2,2}^* \circ D_u g_2^1(k) \quad \geq \quad 0 \tag{53}$$

Inserting the computed derivatives of (37 ) to (42) and using the notation $v^* \in V^*, v \in V, v^*(v) = \langle\, v^* \mid v \,\rangle_V$ and noting moreover, that
$\langle\, z_{2,1}^* \mid v^* \,\rangle_{V^*} = \langle\, v^* \mid z_{2,1}^* \,\rangle_V$ since $V$ is reflexive ($V^{**} = V$) we then obtain

$$\langle\, \phi_y \mid h(b) \,\rangle_H + \int_I \langle\, \psi_y \mid h \,\rangle_V - \langle\, z_{2,1}^* \mid h_t + A(t)h \,\rangle_{V^*} dt - \langle\, z_{2,2}^* \mid h(0) \,\rangle_H \quad = \quad 0, \tag{54}$$

$$\int_I \Big\langle\, \psi_u(t)k(t) \;\Big|\; k(t) \,\Big\rangle_H - \Big\langle\, z_{2,1}^* \;\Big|\; D_u Bk(t) \,\Big\rangle_H dt \quad \geq \quad 0 \quad \forall k \in K(U_0, u). \tag{55}$$

By partial integration of the product $z_{2,1}^* h_t$ in (54) with respect to $t$ (see e.g. [9] p. 147) and taking the adjoint operator $A^*$ of $A(t)$ we obtain for all $h \in W$

$$\langle\, \phi_y \mid h(b) \,\rangle_H + \langle\, z_{2,1}^*(b) \mid h(b) \,\rangle_H - \langle\, z_{2,1}^*(0) \mid h(0) \,\rangle_H + \tag{56}$$

$$+ \quad \int_I \langle\, \psi_y \mid h \,\rangle_V - \langle\, -\tfrac{\partial}{\partial t} z_{2,1}^* + A^*(t) \mid h \,\rangle_V \, dt - \langle\, z_{2,2}^* \mid h(0) \,\rangle_H \quad = \quad 0, \tag{57}$$

which is equivalent to

$$\langle\, \phi_y - z_{2,1}^*(b) \mid h(b) \,\rangle_H - \langle\, z_{2,1}^*(0) - z_{2,2}^* \mid h(0) \,\rangle_H + \tag{58}$$

$$+ \quad \int_I \langle\, \psi_y \mid h \,\rangle_V - \langle\, -\tfrac{\partial}{\partial t} z_{2,1}^* + A^*(t) \mid h \,\rangle_{V^*} \, dt \quad = \quad 0. \tag{59}$$

From this we deduce by standard arguments (variation over appropriate $h$) the adjoint equation (45) with the endpoint condition. The multiplier $z_{2,2}^*$ obviously can be eliminated by expressing it by $z_{2,1}^*(0)$. By taking the adjoint of $D_u B$ we obtain directly the variational inequality (49). If we consider now the special control set $U_0$ we we obtain (50) and (51) from (49) by noting that $k \in K(U_0, u)$ means $k = \lambda(v - u)$, $\lambda \geq 0, v \in U_0$ and varying over all functions $|v(x,t)| \leq 1$.

Some typical examples for the functional $f_1$ and it's derivative: Let $y_d : \Omega \to I\!R$ be the desired final state of the system, which is given, then

$$f_1(y) = \|y(b) - y_d\|_{L^2(\Omega)}^2 \quad \Rightarrow \quad \langle\, \phi_Y \mid h \,\rangle_H = \int_\Omega 2(y(x,b) - y_d(x))h(x,b)\, dx$$

$$f_1(y) = \|y(b) - y_d\|_{L^1(\Omega)} \quad \Rightarrow \quad Df(y,h) = \int_\Omega I_N(x)|h| + (1 - I_N(x))\mathrm{sgn}(y)h\, dx$$

(read always $h = h(x,b)$, and , for the second, nonsmooth functional,

$$\langle\, \phi_Y \mid h \,\rangle_H \quad = \quad \int_\Omega I_N(x)\alpha(x)h(x,b) + (1 - I_N(x))\ \mathrm{sgn}(y)h(x,b)\, dx,$$

where $I_N$ is the indicator function of the set $N = \{\, (x,t) \in \Omega \times I \mid y(x,t) = 0 \,\}$ and $\alpha$ is a function with $|\alpha(x)| \leq 1$ *a.e.*.

**Remark 3.3'** (Extensions) In the example above we have not considered constraints on the state which can be formulated as inequality constraints. If we consider additionally $g_1(y(x,t)) \leq 0$ i.e. $g_1(y(x,t)) \in -C$ with $C := \{ y \mid y(x,t) \geq 0 \ (x,t) \in \Omega \times I \}$ we encounter two problems: One is to prove that the constraint qualifications of Theorem 2.4 hold, the second one is to characterize the dual space of $Z_1$ where $g_1 : Y \to Z_1$. These two problems are of course connected. If we want to use the Slater Constraint qualification, we have to use a function cone $C$ which has nonempty interior. This is done by *Mackenroth, Troeltzsch* [19] and [27], by taking $Z_1 = C^0(I;E)$ = set of continuous functions from $I$ to $E$, where $E$ again has to be a function cone of functions over $\Omega$ who has a nonempty topological interior i.e. $E = C^0(\Omega)$, because in $L^p$-spaces the cone of (a.e.) positive functions does *not* have this property. Having made this choice one then has to prove (or assume) that there is $h,k$ with $Dg_2((y,u),(h,k)) = 0$ (i.e. $h,k$ is the solution of the linear PDE in (A2') and *additionally* satisfies the system of strict inequalities $g_1(y,u)(t) + Dg_1((y,u),(h,k))(t) \in -\mathit{int}\, C_E$. This implies that $K \neq \emptyset$ in Theorem 2.6 hence the inclusion $K \subseteq T(S,x)$ used in Theorem 2.4 holds. The dual $Z_1^*$ of $Z_1 = C^0(I;E)$ is ,according to [19] $NBV(I;E^*)$, i.e. consists of normalized functions with bounded variation that are borel measures with values in $E^*$, and then one apply the Kuhn Tucker Theorem and proceed as [19] to develop the adjoint equations. The problem there is always to evaluate the multiplier $z_1^*$. (If $g_1(y)$ depends only on $t$, e.g. $g_1(y) = -\|y(t)\|_{L^p(\Omega)} + \alpha$ then the situation is slightly easier,because one then has only $Z_1 = C(I)$, $Z_1^*$ = positive Borel-measures.) In particular, if one has only a weak existence theory for the PDE, the inequality constraint $g_1$ may not depend from $\nabla y$ since then its image $Z_1$ is some $L^p$ space where Slater CQ cannot be proved. The same problem arises if the state $y$ is given by a *variational inequality* as e.g.

$$\int_\Omega G_1(y,\nabla y)(w(x) - y(x)) + G_2(y,\nabla y)(\nabla w(x) - \nabla y(x))\, dx \geq 0 \quad w \in W_0 \subseteq W$$

instead of the PDE (28) e.g. if one considers obstacle problems or optimal shape design problems, see e. g. [20], [31], [11]. Here we have an inequality constrained problem where Slater CQ (if $G$ depends *nonlinearly* on $y$) seams difficult to prove and where the Guignard CQ might be used better.

In the following example we give a constructive procedure how this CQ can be verified directly in an inequality constrained problem without using Theorem 2.8.

**Example 3.4** Let $g : Y := L^2(\Omega \times I) \to L^1(\Omega \times I)$ be given by $g(y) := y^2(x,t) - c$ where $c > 0$ is a given, positive number. We want to consider the nonlinear mapping $g$ as inequality constraint, i.e. we consider the feasible set $S := \{ y \in Y \mid g(y) = |y^2(x,t)|^2 - c \leq 0 \ a.e. \}$. The Slater CQ cannot be verified since $g(y) \in -C$ where the cone $C = \{ y \in L^2(\Omega \times I) \mid y(x,t) \geq 0 \ a.e. \}$ has empty interior. Take a fixed $y \in S$ which is equivalent to $g(y) \in -C$. Note that $Dg(y,h) = 2yh$ and that, by linearity of $Dg(x,\cdot)$ ,we have
$J = \{ h \in Y \mid g(y) + Dg(y,h) \leq 0 a.e. \} = \{ h \in Y \mid Dg(y,h) \in K(-C, g(x)) \}$.
We want to prove directly that the inclusion

$$K \ := \ J \ = \ \{ h \in Y \mid g(y) + Dg(y,h) \leq 0 a.e. \} \ \subseteq \ T(S,y),$$

which is central in Theorem 2.4, holds. Note that $Dg(y,h) = 2yh$, Theorem 2.8 is not applicable since $Dg(y,\cdot)$ is in general not surjective ( it's surjective if the set $\{ (x,t) \mid y(x,t) = 0 \}$ has measure zero). Take $h \in J$. In order to prove $h \in T(S,y)$ we have to find sequences $h_n \in Y$, $\lambda_n \in \mathbb{R}_+$ with $h_n \to h$ and $y + \lambda_n h_n \in S$ i.e. $g(y + \lambda_n h_n) \leq 0$ *a.e.* Set $M = \{ (x,t) \mid y^2(x,t) = c \}$, $N = \{ (x,t) \mid y^2(x,t) < c \}$ and for every $n \in \mathbb{N}$ set $M_n = M \cap \{ (x,t) \mid h(x,t) \leq n \}$, $N_n = \{ (x,t) \mid y^2(x,t) \leq c - n^{-1} \} \cap \{ (x,t) \mid h(x,t) \leq n \}$. It's clear that (modulo sets of measure zero) $M \cup N = \Omega \times I$, $M_n \subseteq M, N_n \subseteq N$. Let $I_N, I_M, I_{M_n}, I_{N_n}$ be the indicator functions of those sets, i.e. $I_M(x,t) = 1$ if $(x,t) \in M$, $I_M(x,t) = 0$ if $(x,t) \notin M$ and so on. Then with $n \to \infty$ we have $I_{M_n} \to I_M$, $I_{N_n} \to I_N$ pointwise. Now we choose the sequence $\lambda_n \geq 0$ such that $\lambda_n \leq \min\{n^{-1}\sqrt{c}\,, n^{-2}(4\sqrt{c}+1)^{-1}\}$. and set $h_n := I_{M_n}h + I_{N_n}h$. Then $h_n \to h$ pointwise and, by Lebesgues' Theorem on dominated convergence, $h_n \to h$ in $L^2(\Omega \times I)$. Now it remains to show $g(y + \lambda_n h_n) \leq 0$. Read always $h = h(x,t), y = y(x,t)$. If $(x,t) \in M_n \subseteq M$ we deduce from $h \in J$ that $\operatorname{sgn}(y) = -\operatorname{sgn}(h)$ hence by $\lambda_n \leq n^{-1}\sqrt{c}$ , we obtain $(y + \lambda_n h_n)^2 \leq c$ which is the assertion. If $(x,t) \in N_n \subseteq N$ we have $|y| \leq \sqrt{c - n^{-1}}$, and $h \leq n$ and we obtain from $\lambda_n \leq n^{-2}(4\sqrt{c}+1)^{-1}$ that

$$(y + \lambda_n h_n)^2 = y^2 + 2\lambda_n h_n y + \lambda_n^2 h_n^2 \quad \leq \quad c - n^{-1} + 2\lambda_n n\sqrt{c - n^{-1}} + \lambda_n^2 n^2 \quad \leq \quad c$$

holds. Hence we have for all $(x,t) \in M_n \cup N_n$ that $g(y + \lambda_n h_n) \leq 0$. On $\Omega \times I \setminus M_n \cup N_n$ we have by construction $h_n = 0$ hence $g(y + \lambda_n h_n) = g(y) \leq 0$. Thus the Constraint Qualification is proved. It should be clear that such a construction can be carried through for more complicated operators $g$, even a nonsmooth one.

**Remark 3.5** Comparable necessary conditions for nonlinear elliptic or parabolic control problems (nonlinearity in the functional $f$ or in the state equation (PDE) or in both) have been obtained by several authors with approaches related to necessary conditions of optimization theory. The following is a short and incomplete survey indicating the problems and the employed techniques.

(a) For a linear PDE and the quadratic functional $f(y,u) = f(u) = \|y(u) - z_d\|_H^2 + (u|u)_U$ quadratic already *Lions* in [16] obtains necessary conditions by considering the ***linear*** mapping $u \to y(u)$ and directly introducing the adjoint state into the necessary optimality condition $D_u f(y(u), u) = 0$ (or $\geq 0$ if there is a constraint on the control. This general approach can also be used successfully for nonlinear problems as *Casas/Fernandez* in [5] show for a control problem with nonlinear elliptic PDE e.g. $-|\nabla y|^{p-2}\nabla y = B(u)$. But here the nonlinearity of the map $u \to y(u)$ requires additional investigations concerning differentiability.

(b) For convex functional and linear PDE with state constraints (= inequality contraint) *Mackenroth* [19] uses a convex duality theorem of Rockafellar to obtain the adjoint equation.

(c) For the nonlinear PDE

$$y_t - \Delta y - y^3 = u$$

(which might have explosive solutions) and the convex cost functional
$f(y,u) = f(u) = \|y(u) - z_d\|_H^6 + (u|u)_{L^2(\Omega\times I)}$
*Lions* in [18] proves (among other properties) necessary conditions using a *penalty method.*

(d) Instead of considering the PDE $y_t - A(t)y = B(u)$ as equality constraint one can, at least if this equation is linear or semilinear and $A$ generates a $C_0$- semigroup $S(t,s)$ use the integral equation

$$y(t) = S(t,0)y_0 + \int_0^t S(t,s)u(s)\, ds.$$

This is done by in the book of *Barbu/Precupanu* [2] if $f$ is convex and $A$ linear by using convex duality theory and by *Tröltzsch* [26], [27], [28] if the PDE is semilinear and $f$ is Fréchet-differentiable by using the *nonlinear* integral equation,

$$y(t) = S(t,0)y_0 + \int_0^t S(t,s)B(y(s),u(s))\, ds$$

and applying a Kuhn-Tucker Theorem of [32].

(d) An other approach , which more based on variational calculus is used by *Fattorini* in [7].

# References


[1] *Barbu, V. [82] , Boundary Control Problems with Nonlinear State Equations,* SIAM Journal on Control and Optimization Vol. 20, No.1, 1982.

[2] *Barbu, V., Precupanu, Th. [86] , Convexity and Optimization in Banach Spaces* , D. Reidel Publishing Comp. (1986) .

[3] *Bazaraa, M.S., Shetty, C.M. [76] , Foundations of Optimization* , Lecture Notes in Economics and Mathematical Systems 122, Springer Verlag, Berlin, Heidelberg, New York (1976) .

[4] *Borwein, J.M. [82] , Continuity and Differentiability Properties of Convex Operators,* Proc. London Math. Soc. (3) ,44 (1982), 420-444 .

[5] *Casas, E., Fernandez, L.A. , Distributed Control of Systems Governed by a General Class of Quasilinear Elliptic Equations,* Preprint Universidad de Cantabria, Oct. (1989).

[6] *Clarke, F.H. [83] , Optimization and Nonsmooth Analysis* , Wiley Interscience (1983).

[7] *Fattorini, H.O. [87], A Unified Theory of Necessary Conditions of Nonlinear Nonconvex Control Systems* , Applied Mathematics and Optimization 15, pp. 141-185, (1987) .

[8] *Frankowska, H. [87] , On the Linearization of Nonlinear Control Systems and Exact Reachability,* in: A.Bermudez (ed.) : Control of Partial Differential Equations, Lecture Notes in Control and Information Sciences (Springer),Vol 114, (1987).

[9] *Gajewski, H., Gröger, K., Zacharias, K. [74] , Nichtlineare Operatorgleichungen und Operatordifferentialgleichungen* , Akademie–Verlag, Berlin, (1974) .

[10] *Guignard, M. [69]*, *Generalized Kuhn–Tucker Conditions for Mathematical Programming Problems in a Banach Space,* SIAM Journal on Control and Optimization Vol. 7, No. 2, pp. 232–241 (1969).

[11] *Haslinger,Hlavacek,Necas,Lovisek,* *Solution of Variational Inequalities in Mechanics,* Appl. Math. Sciences Springer Verlag, Berlin, Heidelberg, New York (1988).

[12] *Holmes, R.B. [75],* *Geometric Functional Analysis and its Applications,* Springer Verlag, Berlin, Heidelberg, New York (1975).

[13] *Jameson, G. [70],* *Ordered Linear Spaces*, Lecture Notes in Mathematics No. 141, Springer Verlag, Berlin, Heidelberg, New York , (1970).

[14] *Kirsch,A., Warth,W., Werner, J. [78]* , *Notwendige Optimalitätsbedingungen und ihre Anwendung* , Lecture Notes in Economics and Mathematical Systems No. 152, Springer Verlag, Berlin, Heidelberg, New York (1978) .

[15] *Kusraev, A. G. [78]* , *On Necessary Conditions for an Extremum of Nonsmooth Vector-Valued Mappings* , Doklady Akad. Nauk SSSR Tom 242, (1978) .

[16] *Lions, P.L. [71]* , *Optimal Control of Systems Governed by Partial Differential Equations,* Springer Verlag, Berlin, Heidelberg, New York (1971).

[17] *Lions, P.L., Magenes, E. [72]* , *Non-Homogeneous Boundary Value Problems and Applications II* , Springer Verlag, Berlin, Heidelberg, New York (1972) .

[18] *Lions, P.L. [85]* , *Control of Distributed Singular Systems* , Gauthier-Villars Bordas Paris (1985).

[19] *Mackenroth, U. [81]* , *Optimalitätsbedingungen und Dualität bei zustandsrestringierten parabolischen Kontrollproblemen* , Math. Operationsforschung und Statistik, Serie Opt.Vol. 12, No.1,pp. 65-89, (1981).

[20] *Pironneau,O.,* *Optimal Design for Elliptic Systems,* Springer Verlag, Berlin, Heidelberg, New York (1984).

[21] *Rudin, W. [73],* *Functional Analysis,* Mc Graw Hill Verlag, New York, (1973).

[22] *Penot, P. [82]* , *On Regularity Conditions in Mathematical Programming,* Mathematical Programming Study 19, North Holland Amsterdam, Hrsg. M. Guignard, (1982) .

[23] *Penot, P. [85],* *Open Mappings Theorems and Linearization Stability,* Numerical Funct. Anal. and Optimiz. 8(1), 21-35, (1985) .

[24] *Seidman, T.I., Zhou, H-X. [82]* , *Existence and Uniqueness of Optimal Controls for a Quasilinear Parabolic Equation* , SIAM Journal on Control and Optimization Vol. 20, No. 6,pp. 747-762, (1982) .

[25] *Staib, T. [89],* *Notwendige Optimalitätsbedingungen in der mehrkriteriellen Optimierung mit Anwendung auf Steuerungsprobleme,* Doctoral Dissertation, University of Erlangen-Nürnberg, 1989.

[26] *Tröltzsch, F. [83]* , *A modification of the ZOWE and KURCYUSZ Regularity Condition with Application to the Optimal Control of NOETHER Operator Equations with Constraints on the Control and the State,* Math. Operationsforschung und Statistik, Serie Opt.14, (1983), 245–253.

[27] *Tröltzsch, F. [84]*, *Optimality Conditions for Parabolic Control Problems and Applications* , Teubner Verlag Leipzig, (1984) .

[28] *Tröltzsch, F. [85]*, *On Changing the Spaces in Lagrange Multiplier Rules for the Optimal Control of Non-linear Operator Equations* , Math. Operationsforschung und Statistik, Serie Opt.16, pp. 877-885, (1985).

[29] *Zalinescu, C. [78]* , *A Generalization of the Farkas Lemma and Applications to Convex Programming* , Journal of Mathematical Analysis amd Applications 66, 651-678, (1978) .

[30] *Zeidler, E.* , *Nonlinear Functional Analysis and its Applications II/B* , Springer Verlag, Berlin, Heidelberg, New York (1990).

[31] *Zeidler, E.* , *Nonlinear Functional Analysis and its Applications IV*, Springer Verlag, Berlin, Heidelberg, New York (1988).

[32] *Zowe, J., Kurcyusz, S., [79]* , *Regularity and Stability for the Mathematical Programming Problem in Banach Spaces,* , Applied Mathematics and Optimization 5, pp. 49–62, (1979)

# Lecture Notes in Control and Information Sciences

Edited by M. Thoma and A. Wyner

# Lecture Notes in Control and Information Sciences

Edited by M. Thoma and A. Wyner

# Lecture Notes in Control and Information Sciences

Edited by M. Thoma and A. Wyner

Vol. 137: S. L. Shah, G. Dumont (Eds.)
Adaptive Control Strategies for
Industrial Use
Proceedings of a Workshop
Kananaskis, Canada, 1988
VI, 360 pages. 1989

Vol. 138: D. C. McFarlane, K. Glover
Robust Controller Design Using
Normalized Coprime Factor
Plant Descriptions
X, 206 pages. 1990

Vol. 139: V. Hayward, O. Khatib (Eds.)
Experimental Robotics I
The First International Symposium
Montreal, June 19-21, 1989
XVII, 613 pages. 1990

Vol. 140: Z. Gajic, D. Petkovski,
X. Shen
Singularly Perturbed and Weakly
Coupled Linear Control Systems
A Recursive Approach
VII, 202 pages. 1990

Vol. 141: S. Gutman
Root Clustering in Parameter Space
VIII, 153 pages. 1990

Vol. 142: A. N. Gündeş, C. A. Desoer
Algebraic Theory of Linear
Feedback Systems with Full and
Decentralized Compensators
V, 176 pages. 1990

Vol. 143: H.-J. Sebastian, K. Tammer (Eds.)
System Modelling and Optimization
Proceedings of the 14th IFIP Conference
Leipzig, GDR, July 3-7, 1989
X, 960 pages. 1990

Vol. 144: A. Bensoussan, J. L. Lions (Eds.)
Analysis and Optimization of Systems
Proceedings of the 9th International
Conference, Antibes, June 12–15, 1990
XII, 992 pages. 1990

Vol. 145: M. B. Subrahmanyam
Optimal Control with a Worst-Case
Performance Criterion and Applications
X, 133 pages. 1990

**Vol. 146: D. Mustafa, K. Glover**
**Minimum Entropy $H_\infty$ Control**
**IX, 144 pages. 1990**

Vol. 147: J. P. Zolésio (Ed.)
Stabilization of Flexible Structures
Third Working Conference,
Montpellier, France, January 1989
V, 327 pages, 1990

Vol. 148: In preparation

Vol. 149: K.H. Hoffmann, W. Krabs (Eds.)
Optimal Control of Partial
Differential Equations
Proceedings of the IFIP WG 7.2 International
Conference, Irsee, April 9-12, 1990
VI, 245 pages. 1991